♥ love • biophilia • interdependence • possibility • transformation • imagin... • collective wisdom • vote • protest • ... footprint • our superpowers • cli... • photosynthesis • Paris Agreement • reforestation • mangroves • wetlands • seagrass • kelp • afforestation • biodiversity • carbon sequestration • soil • carbon dioxide removal • wind turbines • transmission lines • solar panels • public transit ♥ mitigation • ecological restoration • re-greening • nature-inspired solutions • permafrost • albedo • carbon cycle • water cycle • evapotranspiration • biomimicry • bioregionalism • ahupua'a • ecological forestry • perennial • agroforestry • land trust • rural repopulation • generational project • climate victory garden • regenerative agriculture • compost • polycultures • cover crops • mulching • no-till farming • collective bargaining • microbiome • farmers markets • community garden • food sovereignty • strikes • boycotts • ritual • food justice certification ♥ adaptation • green infrastructure • living breakwaters • oyster reefs • climate-driven relocation • design justice • mending • repairing • recycling • upcycling • rewilding • restorative design • bike lanes • fifteen-minute city • embodied energy • precautionary principle • artificial intelligence • grid management • carbon capture • batteries • progressive taxation • universal basic income • elections ♥ divestment • loss-and-damage fund • climate reparations • credit unions • heat pumps • induction cooktops • insulation • public utility commission • net zero • durability • circularity • scope 3 emissions • deep decarbonization • pre-competitive cooperation • enough • subsidies • carbon tax • philanthropy • investment • green capitalism • leadership • electric vehicles • green hydrogen • nuclear power • fusion energy • union labor ♥ culture • Hollywood • rom-com • Scully Effect • democracy • local news • solutions journalism • producer responsibility • education • organizing • youth climate movement • climate justice • intergenerational collaboration ♥ treaty • negotiations • nationally determined contributions • global stocktake • low-carbon development • leapfrogging • climate finance • political will • carbon markets • emissions trading • emissions budget • Green New Deal • energy efficiency • water efficiency • electrification • high-speed rail • just transition • universal healthcare • tax credits • abolishing the filibuster • voting rights • Blue New Deal • rights of nature • lose loudly • court reform • wildlife corridors ♥ disaster justice • building codes • emergency management • justly sourced renewable energy • Indigenous wisdom • traditional ecological knowledge • climate goodbyes • accountability • consume less • climate apocalypse skills • Land Back • Indigenous sovereignty • community development financial institution • devotions • aquifer • free, prior, and informed consent ♥ catalytic hope • determination • seaweeds • bivalves • regenerative ocean farming • blue-collar innovation • community-supported fishery • working waterfront • plant-based diets • revenue-based financing • crop insurance • worker cooperatives • working-class environmentalism • implementation • team • leaderful • your climate purpose • tenacious ♥

WHAT IF
WE
GET IT
RIGHT?

One World
New York

Ayana
Elizabeth
Johnson

WHAT IF WE GET IT RIGHT?

Visions of
Climate
Futures

Published in the United States by One World, an imprint of
Random House, a division of Penguin Random House LLC, New York.

ONE WORLD and colophon are registered trademarks of
Penguin Random House LLC.

Photographs are from the author's collection, and by Aaron Wojack.
Permission credits and a list of the author's publications that were
drawn upon for this work are located on pages 446–47.

Hardback ISBN 978-0-593-22936-1
Ebook ISBN 978-0-593-22937-8

Printed in the United States of America on Rolland EnviroTM Book Stock,
which is manufactured using FSC-certified 100% postconsumer fiber.
oneworldlit.com
randomhousebooks.com

1st Printing

FIRST EDITION

BOOK DESIGN
Art direction by Ayana Elizabeth Johnson
Cover design by Arsh Raziuddin, managed by Greg Mollica
Interior design by Simon Sullivan
Endpapers design and Venn diagram design by Holly Griffin
Advising by Carrie Seaver

For the love of nature, culture, and the future.

CONTENTS

Possibility

Replenish and Re-Green

If We Build It . . .

Follow the Money

Culture Is the Context

Proto-Farm Communities
Art by Olalekan Jeyifous   between pages 392 and 393

Transformation

Author's Note

To help you navigate this book, I have placed some markings in the margins. An asterisk (✻) indicates a key insight or possibility. Key points of concern are noted with an exclamation point (!). And poignant and heartfelt bits are accompanied by a heart (♥). Key terms that point the way forward are <u>underlined</u>. (They also appear inside the front cover—treasure hunt!)

Climate change and our responses to it are dynamic, so the statistics presented throughout the book may have shifted somewhat by the time you are reading them—hopefully in a good direction. If you want to dig in further, references and more are at www.getitright.earth.

What lies ahead? Reimagining the world. Only that.

—Arundhati Roy

Prelude: An Ocean Love Story

It was my bad fate to fall in <u>love</u> with coral reefs just as they were dying. At five years old, when I first saw this tropical ecosystem through the floor of a glass-bottomed boat, I bubbled with wide-eyed glee. One of my favorite things in the world at that age was a sixty-four-count box of crayons. I was enamored with all the color names. Now, it was as if my crayons had come to life, swimming beneath my feet in the forms of green and teal parrotfish, blue wrasses, and yellow tangs. While I knew flowers came in many colors, it had never occurred to me that there could be a full rainbow of ocean creatures, too. The aquamarine water and all it held were pure magic.

The kids on the boat were invited out to the aft deck to feed the fish. This was back in the day when it was somehow considered acceptable to give junk food to wild animals. Soon I was armpit-deep in a jumbo bag of cheese popcorn, tossing handfuls overboard, enthralled with the colorful gaggle coming up to the surface for snacks. I was dusted in orange "cheese" powder and blotched with pink hives before my mother caught on. I am allergic to milk, but was so captivated that I hadn't even noticed my swelling arms. Mom rinsed me off and guided me into the cabin to sit while my immune reaction ran its course. I didn't mind at all because with everyone out on the deck I had my own personal window to the enchanting underwater realm. She held one of my hands to keep me from scratching, and with my other hand I kept pointing out all the fish cruising below us.

This was in Key West, Florida, in the summer of 1986. For me and my working-class parents living in Brooklyn, this vacation was a big deal—it was the only trip we three went on that wasn't to visit family. I have wonderfilled memories of those two weeks, but more notable to me now are the memories I don't have. I didn't see any large fish. There should have been mottled, brown groupers bigger than me. Nor do I remember seeing massive corals—I wouldn't forget chartreuse boulders and pink bushes that were living animals. That would have been the stuff of fairy tales.

But by the '80s people had already loved Florida's marine life to bits. The nearshore reefs were the first to be pummeled by our flippers, polluted by our sewage, and emptied with our hooks and nets. Corals—the foundation, the structure, the namesake of coral reef ecosystems—were in trouble. And overfishing—not just commercial but recreational—was rampant. Though each sport fisherman may catch only a few fish, those few are multiplied by thousands of fishermen, totaling tens of thousands of fish hoisted onto decks every day, hundreds of millions a year, for decades. Fish simply can't make babies as fast as we can catch them. The biggest fish—sharks and mammoth groupers and snappers—are the most prized and therefore the first to disappear. And with them goes the top of the food chain.

Meanwhile, at the bottom of the food chain, algae was growing like mad. It can grow as much in a week as coral grows in a year, so to keep coral reefs coral-y, algae needs to be constantly reined in. Sea urchins and herbivorous fish had been effective algae mowers for millennia, scraping surfaces bare and readying them to welcome baby coral that would drift in, keen to settle down and make a home. But in 1983, long-spined urchins were wiped out by a mysterious disease. And there weren't enough parrotfish and surgeonfish left to pick up the slack and keep algae in check. The reefs became fuzzy and overgrown.

That first coral reef I saw, through the glass bottom of that boat, and every reef I have seen since, was a shadow of its former self. To me it was magnificent, but I had no idea how splendid and teeming that same reef had been a generation before. And twenty years later, when I entered graduate school for marine biology and began to try to understand how we might live in harmony with reefs, my studies were as much about learning history as they were about learning ecology. To understand how we could turn things around, I had to study how humans had made this terrible mess.

But before the ocean became my job, it was my love. How lucky I was as a child to see the wildness below the surface. Miss me with the metaverse. I want the boat rocking, the salty air, the unorchestrated parade of marine life.

The key inspiration for this Caribbean trip was my parents' decision that it was time for me to learn to swim. My mother found us a simple bed-and-breakfast with a great pool. There, she coached me through the basics: holding on to the edge and kicking, putting my face in the water and blowing bubbles. Giggling, I would wrap my arms around her shoulders to hitch a ride as she swam across the pool. I reveled in the shriveling of my over-soaked fingertips, a badge of honor. By the second week, when my father came down to join us, my skin had turned from tan to brown, the freckles on my mother's peach[1] skin had multiplied, and I was swimming across the pool by myself. This accomplishment earned the rare and coveted "Good job, kid" from my dad.

Then Dad taught me to dive. I was good at the straight-arms part and the jumping part. Not so good at the straight-legs part, or the head-first part. So the initial, amusing result was pointy-armed cannonballs. And once I learned what a cannonball was, and that the goal was to make the biggest splash, I wholeheartedly embraced the challenge. Screw the laws of physics, I was convinced that with my tiny body I could make a larger splash than my dad's—or at least elicit his raucous laugh. I would feel the smack of the surface against my shins and butt and imagine water rocketing into the air all around me. I would feel the water enveloping me, my braids getting pulled upward as I plunged down. My dad's hands would find me and shepherd me safely back to the surface.

Our biggest-splash competition continued all week, because we suddenly had the pool to ourselves. Most of the other guests stopped coming after my dad arrived with his dark brown skin. Fifteen years later, reminiscing at the dinner table, my parents helped me put the pieces together. I joked that sometimes racism has unexpected perks, in this case doing cannonballs with impunity in the hotel pool. We laughed, but I wondered how deeply that pool scene had stung him, if that made it harder for him to smile and cheer little me on, to give me patient pointers after each adorably botched dive.

I realize now how fortunate I was, as an inner-city kid, not just for that vacation, but to have grown up swimming. So many others—especially Black kids, most especially Black girls—never learn to swim, so they have a higher rate of childhood drowning.[2] My life could easily have been very different.

———

1 The "flesh" crayon was renamed "peach" in 1962, in response to the Civil Rights Movement. That was four years before my parents met, and five years before the Supreme Court deemed their future multi-colored marriage legal. Cultural change precedes policy change.

2 Drowning death rates of Black children are (depending on age group) 2.6 to 7.6 times higher than for White children, the highest disparity between any races.

What the heck does this childhood story have to do with climate futurism? Well, a few things. For one, as your guide through these chapters, as curator and narrator of the questions and possibilities we'll be exploring, I thought I should introduce myself and reveal my predilections toward wonder and joy and saltiness and loving the world.

My earliest aquatic experiences—and nature experiences more generally—were deeply formative. (Hello, backyard bugs! Greetings, autumn leaves and spring crocuses! Salutations, shooting stars!) I am far from alone in this. Do you recall being a little kid, mouth agape at the magic of forests or dinosaurs or goldfish or the moon or ants? Do you remember the encompassing amazement? The desire to understand how it all worked? Though few of us become academically trained scientists, we can all share a curiosity about and love of nature. And this biophilia[3] can be a powerful driving force for conservation, for maintaining a proper respect for the millions of non-human species on this planet who also deserve to thrive and do all the fantastic and weird things they do. We are but a minuscule part of an interdependent multitude. The web of life is vast. And really, really cool.

This innate sensibility tends to get muted as we "mature," making it all the more important to revisit and rekindle those early sparks. That feeling of awe that rushes through us when we get up close to the natural world is a signal. Like all strong feelings in the heart and gut, it is telling us something about what matters—about where we can find joy and connection, about what is valuable and sacred, about how to orient a moral compass in a decaying world. In other words, it is telling us how to survive. As humans, it is maladaptive to lose touch with these feelings. So perhaps sharing some of my formative moments may lead you to recollect your own, to tap into the simple wisdom and clarity of your first experiences with nature, and revive your own naive and delighted commitment to being a good citizen of this planet.

———

3 This term, describing our innate love of life, was introduced by E.O. Wilson, who wrote, in his 1984 book *Biophilia*, "our existence depends on this propensity, our spirit is woven from it, hope rises on its currents." He posited that "to the degree we come to understand other organisms, we will place a greater value on them, and on ourselves."

I have always felt a poetic pull, an obligation, to try to help restore the ecological abundance of the Caribbean Sea, not only for the sake of "environmentalism" but also, because of my ancestry, for the culture. As kids in the 1940s, my father and his friends would go down to the harbor in Kingston, Jamaica, when the behemoth cruise ships came in. The not-yet-sunburned tourists would line up on the decks to quaff the island views—Blue Mountains in the background, palm trees in the foreground, the scents of roasted breadfruit, ripe mangoes, and grilled fish on the breeze—and to consider the people. They would throw coins into the water for the Brown boys, who would jump and dive for them, agile as spinner dolphins. *Oh, how delightful; how well they fetch.* And the water in Kingston Harbour was clear enough and clean enough that you could spot not only coins, but also fish and coral and sharks. My father relayed this story as a lesson both in the decline of ecosystems and in the White expectation to be entertained by Black folks.

In the '60s, my dad moved to New York City for college, fortifying himself against winter (and fall, and spring) with wool, cable-knit fishermen's sweaters. He became an architect. Every day of my childhood, he would get up, Sisyphean, put on a sharp suit, buy a newspaper and a pack of Doublemint gum, take the subway into Manhattan, and try to make headway as a Black man in this notoriously insular and racist field. I'm ashamed to admit that I thought he was a failure—he didn't have fancy buildings to point to and claim. He never made any money to speak of. It wasn't until a few years ago, after he had passed away, that I was disabused of that perception when I happened to meet a Black architect a generation his junior at a cocktail party. This man knew of my dad's architecture firm, which was one of the first Black-owned firms in New York City. With glints of respect and gratitude, this man informed me that my father had indeed paved the way and opened doors, just as he had intended. My tears of relief welled up. My dad had achieved his ultimate goal. I tell you this in part because I am proud, but really I tell you this because I had missed the entire freaking point. It's not about the glory. It's about the ripples. This is what progress often looks like: success without rewards.

I wonder if in the end my dad thought it was worth it, leaving behind friends, family, culture, and sea breezes to try and try and try to make it in America. I never asked. He would have brushed me off, given a one-sentence reply at most—maybe in the form of a Jamaican idiom I would have fun trying to parse. He was not one to indulge such introspection or tell sentimental stories, or to tell stories at all.

Apart from the cruise-ship parable, the only other ocean story he ever told was about running on the beach. He was a track star, and on the weekends he would go to the beach with his friends and they'd race each other in the sand, charging against that extra resistance of their feet sinking into shifting ground. When they had worn themselves out training they would swim, do acrobatics in the surf, and catch and roast fish. But by the time he moved north to New York, the fish along Jamaica's coasts were already too rare to reliably supply a beach barbecue. And that was already changing culture, sending people to markets that more and more often sold imported seafood. Jamaica was probably the first Caribbean island to be severely overfished. The coral reefs could not sustain the growing population. And, because of pollution and development, the water in Kingston Harbour is no longer crystal clear.

To me, ocean conservation is in part about cultural preservation. We are losing something more fundamental than a meal: a way of life. However, as we career past eight billion people on the planet, it would be foolish nostalgia to hold on to a vision of everyone catching a fish for supper whenever they like. Some traditions don't scale sustainably. ✳
What worked with a few thousand people on an island doesn't necessarily work with a few hundred thousand.

———

On that same trip to Florida, we visited the Key West aquarium, which, with its dim lighting and profusion of water-breathing biodiversity, felt like an outing to meet friendly aliens. I stared into each tank, rapt, as my mom read the plaques aloud and helped me decode this new world. And I was shocked to find there was a touch tank, and that I would be allowed to hold sea creatures. The hundreds of tube feet of a sea urchin tickled my palm. In reverie, I leaned in close to observe its waving spines. I looked up at my mother's encouraging face, asking if there was a job where you got to know all about ocean creatures. I looked down at this marvelous invertebrate, mind blown—*Can this be real?*—wondering how I could hold on to this feeling forever. "Marine biologist," she responded. From then on, when adults posed their inevitable question—*What do you want to be when you grow up?*—I had my answer.

My parents held my hands as our plane landed back in New York. The ocean seemed so far away. And there I was, this little Black girl in Brooklyn thinking she could be a marine biologist. "Cute. Sure, kid," said the grownups, humoring me. I know now that my background made my dream seem far-fetched. And it hadn't occurred to me—even as the plane flew over the Atlantic Ocean, Rockaway Beach, and Jamaica Bay as we landed—that New York City (an archipelago, with over five hundred miles of coastline) was a perfectly reasonable place to live if you wanted to study the sea. The water off Coney Island wasn't a sparkling, welcoming turquoise.

My dreams shifted and arced and merged. By ten years old, I wanted to be the lawyer who got the next Dr. Martin Luther King Jr. out of jail. I wanted to fight for those who were fighting for justice. At fifteen, the goal was park ranger. What could possibly be better than getting paid to walk around in the forest? At twenty, back in the Caribbean studying abroad, it was ocean policy. I had discovered the wild puzzle of science, economics, government, and culture. At twenty-two, while working at the U.S. Environmental Protection Agency, the plan was to become an environmental lawyer, because I knew how badly the Earth needed defending. By twenty-five, I found myself back at marine biology and heading to graduate school—it seemed way more fun than law school and fewer people were building the bridge to policy from the science side. By thirty, PhD in hand, I had figured out I could essentially do and be all of these things combined, and that would be ocean conservation. All told, I spent nearly a decade studying and working in the Caribbean, trying to help figure out what sustainable ocean use could look like, working with communities to put careful fishing laws in place, establish ocean zoning plans, and create marine protected areas. Then at thirty-seven, watching from afar as hurricanes supercharged by warmed waters tore through the islands I love, it became clear that my work had to expand to focus on climate solutions. And now I'm middle-aged, and here we are—from awe and wonder, to science and policy, to heartache, to building <u>community</u> around solutions.

So, that's me. My perspective is that of someone brimming with juxtapositions. I am a scientist who always intended to have a career in policy. I am the daughter of a practical schoolteacher and a wistful artist. I am cold New York winters and Caribbean heat. I am working-class and Harvard. I am Black and White. I am urban and smitten with wilderness. I am proof of the American Dream, and proof that it is all too rare. These are not dichotomies but currents, and they all flow into this book. For twenty-five years, I have been trying to understand what went so wrong, what we can do about it, and what the future can look like, if we get it right.

POSSIBILITY

You have to act as if it were possible
to radically transform the world.
And you have to do it all the time.

—Angela Davis

Introduction

Welcome. This book is about possibility and transformation. It's about what the world could be if we charge ahead with the array of climate solutions we have at our fingertips, solutions at the intersection of science, policy, culture, and justice. These pages conjure a thriving (and quite different) world, and show us that it's worth the effort—the overhaul—to get there, together. This book is a quest to answer a question not asked often enough about the climate crisis: **What if we get it right?**

What if . . . ? A mind-expanding question, often asked with a twinkle in the eye. An invitation to imagine. And goodness do we need more imagination right now, to create clearer visions of desirable climate futures. I'm not talking about some frictionless techno-utopian future (although Silicon Valley is selling that pretty hard). Or an off-grid hippie commune fantasy (although there's something to that, for sure). I'm talking about a future we can see ourselves in, where there's a place for us and the communities we hold dear.

It will take momentous shifts to ensure a reasonably safe future. How can we find the wherewithal and endurance to transform our energy, food, and transportation systems; our buildings, infrastructure, and manufacturing; our economies, governments, and cultures, if we can't envision the outcome of our efforts? Before we fully commit our brains and brawn, before we go all-in, it's reasonable to want some indication of what success looks like, some sense of what all this change will mean for our lives. Vague depictions of solar panels and electric vehicles, and a future that's maybe less bad than it would otherwise be, just aren't cutting it. So most of us saunter, we lollygag, we mosey away from the brink instead of running full tilt toward solutions.

If we're honest, maybe we doubt that humanity actually *can* get it together, can rally the depth of motivation and creativity required to face this unprecedentedly gargantuan challenge. But one thing is certain: Half-assed action in the face of potential doom is an indisputably absurd choice, especially given that we already have most of the climate solutions we need—*heaps* of them. Moving forward requires that we propel each other—propel our species—out of a phenomenally entrenched procrastination. We don't need more data or a more rigorous cost-benefit analysis; we need to leap.

But I get it. For decades, what scientists, writers, filmmakers, and artists have projected for us is the apocalypse, in great detail. We can easily picture the climate-change-fueled fires, floods, droughts, and storms, and the immense suffering, all of which are now well underway. However, when it comes to better outcomes, we've largely been left hanging. That is a problem. Humans have evolved to *not* leap into a void—that's dangerous! So we need something firm to aim for. Something with love and joy in it. And we need the <u>gumption</u> that emerges from an effervescent sense of possibility.

What if we get it right? I ask this question because I too need to know what it all adds up to. In other words, I created this book because it's what I've needed to read. If your imagination fails you as mine often fails me, this book is for you. If you haven't yet been able to see yourself in climate work, or you haven't felt welcomed or enticed into applying what you have to offer, you're in the right place. If you, like me, tend to skip ahead to detailed action items as quickly as possible and lose sight of the big picture, I'm glad you're here. Or if you find yourself grasping for a futurism that doesn't feel naive, ditto, we'll muddle through together. I don't have all the answers—not by a mile. But I have listened to a lot of smart people, seen <u>art</u> and <u>poetry</u> that show the way, and am putting the puzzle pieces together.

If ever there were a moment for <u>collective wisdom</u>, this is it. All hands (and minds and hearts) on deck. So the core of this book is twenty interviews with people I have learned so much from, whose perspectives have deeply influenced mine. I asked each of these folks similar questions about their visions for the future, barriers to achieving it, what we need to do next, and what they wish we knew. This way, we can aggregate their insights across the spectrum of expertise and solutions we must harness. This way, we can see a holistic answer to the title question.

Exploring our climate challenges through deep conversations, unpacking and grappling together, is also what I hope you will do out in the world.[4] Because for something so fundamental, we really don't talk about

---

4 "In conversation, we can hold thoughts and reflect on problems sometimes for hours on end. . . . Human thought is inherently dialogic."—David Graeber and David Wengrow, *The Dawn of Everything*

climate change much. While some of us spend a lot of time reading, worrying, doomscrolling, and posting about the climate crisis, seldom do we gather to discuss what we might *do*. Polling data makes this clear.[5] In the U.S., despite 70% of Americans being concerned that climate change will harm plants, animals, and future generations, and 43% saying it has already affected them personally, a whopping 65% rarely or never talk about the topic with their friends and families. Furthermore, 11% of Americans—37 million people—are willing to become actively engaged but have not taken any actions . . . yet. How's that for potential?

————

Too often, the climate movement and the media tell everyone to do the same things: <u>Vote</u>, <u>protest</u>, <u>donate</u>, <u>spread the word</u>, and lower your <u>carbon footprint</u>.[6] And, yes, it's great to do those things. But all too rarely are we asked to contribute our specific talents, <u>our superpowers</u>. Here's one way to think about what you, specifically you, can do. It's a <u>climate action Venn diagram</u> with three overlapping circles.[7]

1. **What are you good at?** What are your areas of expertise? What can you bring to the table? Think about your skills, resources, and networks—you have a lot to offer.

2. **What work needs doing?** Are there particular climate and justice solutions you want to focus on? Think about systemic changes and efforts that can be replicated or scaled. There are heaps of options.[8]

3. **What brings you joy?** Or perhaps a better word is "satisfaction." What gets you out of bed in the morning? Choose climate actions that energize and enliven you.

———

5 Yale and George Mason universities have been conducting rigorous polling on Americans' climate perspectives since 2012. (See: climatecommunication.yale.edu)

6 The useful term "carbon footprint," by the way, was popularized by fossil fuel companies, in an attempt to put the weight of the blame on us as individuals. Get outta here with that nonsense.

7 If you're familiar with the Japanese concept of *ikigai* for finding your purpose, you can consider this a simplified, climate-focused version of that. For more, see: climatevenn.info

8 For a list of solutions, see: drawdown.org/solutions

For me, a marine biologist, policy nerd, and Brooklyn native worried about sea level rise and climate injustice, my search for this Venn sweet spot led me to co-found Urban Ocean Lab, a policy think tank for the future of coastal cities. It's also how I ended up co-creating the Blue New Deal plan when the Green New Deal all but left out the ocean.[9] And it's how I ended up writing this for you. As you read on, perhaps keep this framework in mind, and consider what might be at the heart of your own diagram. The people interviewed in this book have all found their unique and important climate roles, which I hope will inspire you to seek or refine or double down on yours.

We'll start answering our "What if?" questions here in part one—Possibility—where we'll set the scene, introduce the stakes, and converse with climate scientist Dr. Kate Marvel to get a planetary perspective.[10]

---

9 That story is told in the "Changing the Rules" section.

10 Bios for the contributors are at the end of the book.

Part two—**Replenish and Re-Green**—explores ecosystems, food systems, forestry, and rural communities. This section, and each of the next five, opens with a set of stats: 10 problems (to ground us) and 10 possibilities (to give us direction). We'll talk with environmental journalist Judith D. Schwartz, farmer and organizer Leah Penniman, and farmer/forester and historian Brian Donahue.

Part three poses the open-ended prompt **If We Build It . . .**[11] with responses touching on cities, infrastructure, art, and technology, from landscape architect Kate Orff and design justice innovator Bryan C. Lee Jr., museum curator Paola Antonelli, and artificial intelligence pioneer Mustafa Suleyman.

In part four, we **Follow the Money**, from divesting to investing, from deploying tax dollars to transforming our financial system. We'll hear from activist and journalist Bill McKibben, climate investment mobilizer Régine Clément, and clean-energy financier Jigar Shah. Plus, executive K. Corley Kenna and I propose steps for corporate responsibility.

Part five deals with the fact that **Culture Is the Context** for all the decisions we make, from Hollywood to the news to activism. We'll get into it with producer Franklin Leonard and filmmaker Adam McKay, journalist Kendra Pierre-Louis, and youth activists Xiye Bastida and Ayisha Siddiqa.

Part six, **Changing the Rules**, is all about policy—the rules of the game—from the Green New Deal to a blue one, from the United Nations to the U.S. courts. We'll dialogue with international policy expert Kelly Sims Gallagher, domestic policy architect Rhiana Gunn-Wright, and, because the Earth needs a good lawyer, Abigail Dillen.

Part seven[12] puts **Community Foremost**, delving into emergency management, diasporas and home, and equity and power, through conversations with disasterologist Samantha Montano, attorney and climate justice leader Colette Pichon Battle, and Indigenous political/cultural strategist Jade Begay.

Part eight—**Transformation**—features a salty interlude on the future

---

11 Yes, this is a movie reference. And, yes, I hope they/you will come, will flock to climate solutions.

12 À la Prince, "They stand in the way of love and we will smoke them all."

of coastal communities with Bren Smith, a regenerative ocean farmer. Then, architect-philosopher Oana Stănescu and I offer up a climate redux of the Hippocratic Oath. And we'll end where we started, with possibility, converging on how we can each find our joyous roles in climate solutions. I'll leave you with my own "get it right" answer, and a musical parting gift.

———

Twenty years ago, my mother, one of my guiding lights, found her climate role. She retired from being an English teacher, sold my childhood home in Brooklyn, moved way upstate, and slowly created a small homestead farm. As I sit here, in the fullness of August, at a desk made from unfinished plywood, with calendula tea (dried from our garden) steeping in a mug made by my father's hands, as I look out at fruit ripening on trees my mother planted, while surrounded by books on civil rights, Indigenous knowledge, economics, and the glories of nature, as grackles squawk and roosters crow, and Mom feeds the chickens, and chipmunks and woodpeckers eat the cherries I had promised I'd pick today, as I gaze

past the dead flies on the windowsill to revel in every shade of green that summer offers, I want to make this moment last, this moment in the history of Earth when nature still often resembles what we as a species have known it to be for millennia: a world that nourishes and soothes us. That desire is the "why" for this book.

We find ourselves in a time of reckoning, at an inflection point for humanity. What we will inflect toward is not clear. It has not yet been determined how much global temperature will increase, how much sea levels will rise, how we will adapt to the inevitable and prevent the worst, or how we will treat each other amidst it all. Set aside your resignation and nihilism. There is a wide range of possible futures. Peril and possibility coexist.

A few things feel clear about this world we must build together: There can be enough for each of us. There can be a home for each of us. There can be a role for each of us. The imperative is transformation, and the goal is to thrive. Even if that's all we know for sure, it's enough to get started.

Now, *will* we get it right? I have no idea; it's a long shot. But we could . . .

Reality Check

Note: So that we're all on the same page about how catastrophically bad things are, how radical the change that's required is, and how urgently we need to get it together, here's a quick reality check. Then, after this brief wallop of bad news (this whole thing is !!!), we'll set aside the gloom and horrors (for the most part—context is important!), and focus on "what ifs."

———

At parties, usually late at night when inhibitions are long gone, people who know that I do climate work sidle up with their big question, often whispered: **"So, tell me the truth, how fucked are we?"** I usually answer, "We're pretty fucked, but . . ."[13] and then immediately pivot to solutions. But here I'll lay out the dire scenario. Okay, deep breath.

The Earth is hotter now than at any other point in human history. We spew greenhouse gases (aka pollution) out of more than a billion tail-pipes and smokestacks, creating a dangerously insulating blanket around the planet. And measuring atmospheric temperatures actually masks the true scale of climate change, because the ocean has absorbed the vast majority of the heat—in Florida the *water* surpassed 100°F (37.8°C), jacuzzi temperatures. Not good. Heat waves are more frequent and last longer. Hurricanes are getting stronger and wetter. Glaciers are melting faster than expected. The massive ocean currents that regulate our climate are slowing down, screwed up by excess heat and excess fresh water—water that was recently ice. Sea level is rising—two more meters (over six feet!) of water could be coming soon to a coastline near you, displacing hundreds of millions of people.

We have changed the pH *of the entire ocean*. It has absorbed so much carbon dioxide (CO_2) that it's getting more acidic. That sucks for ocean creatures trying to build a shell or skeleton, or just not crumble. Plus, we hunt fish using sonar, helicopters, nets larger than football fields, and tons of fuel—most fish populations are overfished or fished to the max.

---

13 I'm not usually one for cursing in print, but there's not another word that quite fits the bill here. So, yeah, I've been cursing more than usual in the past few years.

Meanwhile, many animals both in the sea and on land are making a one-way migration toward the poles seeking cooler zones, while corals and trees are stuck frying and shriveling in place. The Amazon rainforest is in danger of drying out. And then there are the bulldozers and saws. Every year, an area the size of nearly 20 million football fields is deforested globally, hugely contributing to climate change and to our biodiversity crisis. We are in the process of driving one million species extinct.

Simultaneously, we are on track to have more plastic in the ocean than fish, and the remaining fish are eating plastic. There's plastic in seafood. There's plastic in most drinking water—and in beer! There's plastic in blood and in breast milk and in semen. Plastic is made from fossil fuels. There's plastic in clouds. There's plastic in rain. There's plastic in glaciers, and glaciers are disappearing—and along with them disappears meltwater for drinking and for crops. In springtime, which can now arrive weeks sooner, snow melts earlier and flowers bloom earlier. Asynchronies in when animals emerge and when their food emerges are throwing food webs out of whack. By 2070, one-fifth of the planet could be as scorchingly hot as the (rapidly expanding) Sahara Desert. And already, around one-quarter of humanity (mostly in poor countries) is dealing with drought, which leads to famine.

Why is all this badness happening? Humans. We are burning ancient plants and animals (aka fossil fuels, not renewable) to jet around and wear fast fashion, and build highways and skyscrapers, and heat outdoor swimming pools in autumn, and shiver inside in summer, and convert lush ecosystems into sprawling and unwalkable suburbs with silly lawns, and commute alone in our cars to jobs that do not change this reckless status quo, and manufacture things we don't need (probably plastic things), and power the devices we're addicted to so we can "like" posts about biodiversity loss and climate disasters, and then proceed unchanged. We are fracturing rocks deep underground—causing earthquakes and polluting drinking water—to extract fracked methane (an extra-potent greenhouse gas) to light on fire to cook our food, food that is produced by dousing the soil with chemical fertilizers (made from fossil fuel) and with poisonous pesticides (derived from fossil fuel) that have been used for chemical warfare (how's that for a red flag). When it rains, these chemicals run from land down rivers to the sea, toxifying the water and causing low-oxygen dead zones that suffocate marine life.

We package food in plastics (and, heck, even package up water) to transport it thousands of miles, burning fossil fuels for shipping. We have thousands of plants and animals we can eat, but we cultivate just a few and in enormous monocultures so that they are susceptible to disease and drought, and we create a toxic cycle of increased pesticide use and increasingly exploitative conditions for farmworkers. And then, we throw away a third of the food we produce, which releases tons of methane as it rots in landfills. After all this, much of what we are eating is over-processed junk, which is probably making us sicker and sadder and dumber. Which is maybe why we do things like bulldoze the coastal mangroves and marshes that are the nursery for baby fish and could offer us better protection from storms than seawalls. Or maybe the "why" is just the usual greed and selfishness. Banks are after all still bankrolling all these absurdities to the tune of trillions of dollars per year. And then there's the basic air pollution. Burning fossil fuels puts sooty particles into the air, causing lung problems and heart problems and birth defects and cancer, and almost 9 million premature deaths every year. That's one in five deaths. Plus, as hot and wet habitat expands, so do mosquitoes and the range of diseases they carry.

To make matters even worse, when it's hotter people get irritated and aggressive and there's more violence. Meanwhile, the news media barely connects the dots between our changed climate, famine, unrest, war, and migration. Climate shocks are now the second-biggest cause of hunger, after conflict. Plus, our military considers climate change a "threat multiplier" that will increase the odds of wars—what a cycle. Holy hell. Unsurprisingly, inequalities are exacerbated by climate change—storms, pollution, droughts, and wildfires hit poor communities and communities of color first and worst, even though they have contributed the least to cause it all. The most brutal injustice. In the face of all this, governments and corporations are making weak-ass climate pledges (2050 is too late) and wealthy nations are not even coming through with the checks they promised to help save rainforests or help developing countries handle this onslaught.

So, yeah, it's too late to "solve" or "stop" climate change. We have already changed the climate. We have already frayed the web of life. The greenhouse gases are out of the bag, and we don't have a time machine. We are at the stage of figuring out how to *minimize* the damage, *mitigate* the impacts, and *adapt* to this unknown new world—while ensuring that those who are already marginalized and struggling aren't placed in yet more danger. Sure, space exploration is cool and all, but 8 billion of us aren't hopping on rockets to Mars anytime soon to frolic there for eternity. And even if we could, since we clearly haven't learned our lesson, we'd just destroy that place too.

End litany. Exhale.

This is not hyperbole. While it feels awful to rattle all this off, and overwhelming to see it all summarized, it is what the science tells us.[14] The scenario really is this extreme. In fact, all this is pretty much exactly what scientists predicted and have been warning us about for decades. They gave us ample time to prevent so much of this damage—even fossil fuel companies' own scientists warned in the 1970s that this was where we were headed. With their cutting-edge research and development (R&D) labs, they could have diversified, could have innovated renewable energy. We could have started to get our shit together half a century ago! Heck, Eunice Newton Foote figured out in 1856 that more carbon dioxide in the atmosphere would lead to planetary warming.[15] Eighteen. Fifty. Six.

These truths keep a fire in my belly. What could be more absurd, more perverse, than destroying the environment on this one planet that sits in the sweet spot at just the right distance from our sun, with ecosystems that support a nearly unfathomable diversity of life, and an atmosphere that allows us to breathe and romp? *Notwithstanding*, cruelly, fossil-fuel and big-agriculture executives—and the banks, lobbyists, politicians, and advertising and PR firms doing their bidding—repeatedly make choices that have us hurtling toward disaster. To be clear, *they* carry most of the blame for this planetary dumpster fire. All this enrages and motivates me because it simply did not have to be this way. But now, here we are. And given the gross precarity of it all, tweaks and adjustments ain't gonna cut it. Go big or lose home. Tick tock. This is it.

---

14 Shoutout to James Gaines, fact-checker extraordinaire.
15 Shoutout to the OG climate feminist.

And yet, there is good news too—lots of it. We have *many* solutions, and they are underway. Nature is incredibly resilient. Scientific research and our own eyes show us time and again that things can recover, can re-green. We can restore some balance to this magnificent planet, phase out fossil fuels, get carbon back where it belongs, salvage our ecological life-support systems. Like most things, this is not binary, not all or nothing. It's not planet destroyed or planet saved. Even degraded ecosystems can still support food security, economies, and well-being. There remains a huge spectrum of possibilities. And so many lives, livelihoods, and cultures hang in the balance. Instead of accepting the climate and biodiversity apocalypse as fated (or ignoring it), we can choose to dig our way out of this morass.[16]

As you can probably tell by now, in this book you'll find no sugar coating or downplaying of scientific projections or of the speed and scope of change required in order to birth a future where we have a chance to flourish. You also won't find wallowing. We are learning lessons and looking forward. Fundamentally changing how we interact with nature and each other certainly will not be easy. This transformation will be messy. It will require new ideas and technologies, and ancient ones. But we can do hard things.

Onward to the glorious mess. The only way out is through.

---

16 "Morass," originally a term for a swamp, is also used to describe a situation that's a shitshow. Apologies to swamps, and note that the way we use language is but another example of how we disrespect nature.

Earth Is the Best Planet

Interview with Kate Marvel

When I cohosted the *How to Save a Planet* podcast, we'd ask all of our guests the big question: "How screwed are we?" (I was told we shouldn't curse on air.) Guests responded with awkward blurts of laughter, lengthy pauses, and deep sighs. But Dr. Kate Marvel, my favorite climate scientist, immediately and firmly said, "We have a choice." That's what our conversation here is about. That and reveling in the wondrousness of our planet's interconnected systems and cycles.

Kate has spent the last decade conducting research at NASA and teaching at Columbia University. Her particular expertise is designing climate models, aka using the data we have about Earth's history, plus what we know about physics and chemistry, to crunch the numbers on what the physical world will look like if we get things right—or various permutations of wrong. A lot of that is focused on greenhouse gases,[17] but this conversation will convince you that the water cycle deserves way more of our attention. The atmosphere is, apparently, thirsty.[18]

The stellar science communication skills Kate has honed have guided many people into a deeper understanding of our climate crisis. She tells the world, in the plainest of terms, what the climate science is telling us. And makes sure we know it's not too late to do something about it. She's itching for us to turn science into action:

Kate: Who do we need to talk to? Who do we need to pressure? Because we can debate ideological frameworks until we're blue in the face. The atmosphere doesn't care. The atmosphere just cares about chemistry.

---

17 Greenhouse gases include carbon dioxide, methane, nitrous oxide, and water vapor, plus industrial fluorinated gases such as hydrofluorocarbons, perfluorocarbons, and sulfur hexafluoride.

18 Does this mean hurricanes are thirst traps?

Ayana: Does the atmosphere care if corporations have really lovely social media posts about maybe doing something someday?

Kate: Does not give a shit.

Straight talk. Always.

———

Ayana: One of the first things I distinctly remember hearing you say was that "Earth is the best planet." Please explain.

Kate: I have a PhD in astrophysics, which is the study of the entire universe, so I have an informed scientific opinion when I say the rest of the universe sucks. There may very well be life out there on other planets, but you know what? It's really far away and you would die before you got there. In terms of the planets we can actually reach in our solar system, they are terrible. Mercury, you would burn up. Venus, you'd get poisoned and burn up. Mars, you would freeze. And also Mars is so boring—it's just a bunch of red rocks. I'm glad we send rovers there. They're adorable and it's interesting to learn more. We have to learn about other planets because otherwise how would we know how terrible they are? I'm very supportive of research on other planets.

Ayana: You're hate-following other planets.

Kate: Totally. But these billionaires saying, oh, we'll just go live on Mars. Like, you won't even go to the Bronx, you're not living on Mars.

Ayana: Hahaha. That is so right.

Kate: The thing about Earth is that everything I have ever cared about or will ever care about is here. There is nobody anywhere else in the universe that I give a shit about. This is the only planet we know of that has civilizations or where anybody has a sense of humor. And I think that's important.

You hear it a lot in slogans: "There is no planet B." But it actually is profound that there's no way out of this, there's no escape plan. We can't just mess this up and go somewhere else. Or even if we could, it would be horrible. You would be hanging out in an oxygen tank with a bunch of tech billionaires on an inhospitable planet.

Ayana: Hard pass. So this is it.

Kate: This is it.

Ayana: Okay, please give us a reality check for planet Earth. How has the climate already changed and where are we heading if we don't get it together and shift this trajectory we're on?

Kate: What's going on right now is both really, really complicated and incredibly simple. The incredibly simple part is that the chemistry of the atmosphere is changing. Before the Industrial Revolution there were 280 parts per million of carbon dioxide in the atmosphere and now it's around 420. And that change is all due to human activities, mostly digging up dead stuff and setting it on fire, but also changing the way we use land, cutting down forests. So the atmosphere is fundamentally different now.

Ayana: It's wild that we have known about the greenhouse effect for well over a hundred years.

Kate: This is not cutting-edge, fun new science. This is stuff that we've understood for a long time. We know how changing the chemistry of the atmosphere works. And as a result of these changes, the planet is about 1.3°C [2.34°F] warmer than it was before the Industrial Revolution.[19] That doesn't sound like an enormous number, but it's a pretty big change when you put that into the context of the last ice age. Twenty-one thousand years ago, New York City, where I am right now, would've been covered with a massive sheet of ice. Back then, it was somewhere between 4°C and 7°C colder.

Ayana: A few degrees makes a big difference on this planet.

Kate: Exactly. And we're already seeing this 1.3°C of warming make a big difference. We're starting to see really extreme, record-shattering events. Things like heat waves that simply would not have happened had the Earth not been warmed by humans changing the chemistry of the atmosphere. It is really well understood that when you warm the whole place up, you get more extreme heat events.

*   Something else we understand really well is that warm air holds more water vapor. And more water vapor in a warmer atmosphere means

---

19 In 2023, for the first time, Earth's temperature was 1.5°C hotter than in pre-industrial times. However, the UN's target of limiting warming to 1.5°C is technically not yet surpassed (as of mid-2024), as that is calculated based on a multi-year average.

there is more to dump on us. So we're seeing an increase in really heavy rainfall and sometimes snowfall events.

We also understand where that water vapor is coming from. We know that warmer air is thirstier air and creates increased evaporation—just like when you get really hot your body covers you with sweat, so that liquid can be evaporated away from your skin and cool you. The Earth is literally sweating in the heat. We are seeing increased evaporation taking more moisture away from the surface of the Earth, which is creating more severe drought. For example, the southwest of the United States and northern Mexico is having its worst drought in at least 1,200 years. And that's not because of a massive decline in rainfall. What is driving that drought is much more evaporation of water away from the surface as a result of atmospheric warming.

Ayana: This is the first I'm hearing about this. Thirsty air. More evaporation is another dangerous side effect of warming.

Kate: And on top of that, in a warmer world, we get more rain as opposed to snow, so we're not building up the snowpack as much anymore. And that's changing spring runoff, which a lot of water managers in the West depend on. So we're seeing big changes to the water cycle—changes in droughts and downpours and floods and soil moisture and runoff.[20]

Also, the seas are rising. That's happening for two reasons. One is because of land ice. Ice sheets on Greenland and Antarctica are melting, so we get water that used to be frozen up on land going into the ocean. The other is because, as you may remember from fifth-grade science class, as things (including water) get warmer, they expand.

Ayana: Yeah, and ocean surface temperature overall is nearly a degree Celsius warmer now.

Kate: And those warmer sea surface temperatures, that's hurricane food—it fuels stronger hurricanes. The number of hurricanes is not necessarily changing, but we are certainly seeing more severe and more rapidly intensifying hurricanes.

Ayana: As you said, warmer air holds more water.

---

20 More on the glories of the water cycle in the following interview with Judith Schwartz.

Kate: The storms are dumping more rainfall. And because the sea levels are higher, the storm surges are more severe and reaching farther inland. We're seeing changes to storm systems in general, not just hurricanes. For example, an increase in severe thunderstorm activity, which is especially relevant in the center of the U.S.

Ayana: The heat the ocean has absorbed has protected us from an even more rapidly warming atmosphere. We've saved ourselves dozens of degrees of atmospheric temperature increase, because the ocean has absorbed over 90% of that heat we've trapped with greenhouse gases. Without the ocean, Earth would probably be around 36°C [almost 65°F] warmer, which would be unlivable.

Kate: Thank you, oceans. And, sorry.

Ayana: This planet has already changed a *lot*. What should we expect if we don't change our ways, if we don't make major changes in terms of burning fossil fuels and how we manage and protect ecosystems?

Kate: The way I like to talk about future projections is in terms of global warming levels. For instance, what the world looks like at 1.5°C [2.7°F] above pre-industrial temperatures, what it looks like at two degrees, three degrees, four degrees, and so on. And the short answer is: bad. We don't want to live in any of those worlds. In a 2°C [3.6°F] world, we're gonna see a lot more and more extreme downpours than we are in a 1.5°C world, for example. We're going to see stronger hurricanes, larger sea level rise, more severe impacts to ecosystems, more droughts, more floods.

The Earth system is incredibly complex. The reason we have deserts where they are is that the air that has risen from the tropics has shed all of its water, and now it's coming down to rest on the deserts. Then prevailing winds bring dust from the Sahara Desert to fertilize the Amazon. You don't have deserts without tropical rainforests. And you don't have rainforests without deserts.

Everything really is connected. And that's hard if you're a scientist, if you're trying to understand how a change in one system is going to cascade through all of the systems that are linked to it. That's why I always start with the physics, because if there is a physical principle that's clear, that you can write down in an equation, that gives you something to hang on to when stuff starts getting complicated.

Ayana: We all need something to hang on to in times of uncertainty. And I love that you recommend we grab a physics equation and hold on for dear life.

Kate: Well, ecosystems are complex. For example, the Amazon does a lot of work for us. The Amazon produces oxygen, but it also sequesters a lot of carbon dioxide because trees are made of carbon. And the more trees you have, the more carbon they're sucking out of the atmosphere through the magic of photosynthesis.[21]

Ayana: Deeply magical.

Kate: I wish I could say, "Here is the level of warming that we can reach where we will not risk the entire Amazon turning into a savannah and all the knock-on effects of that." But I can't tell you exactly where that point is. I can't tell you that if we limit warming to below level X, we will avoid a large-scale disintegration of the West Antarctic ice sheet and the resulting feet of sea level rise. And I can't tell you the safe level where we will guarantee that we won't see a shutdown of the North Atlantic meridional overturning circulation in the ocean.

Ayana: Which is that massive ocean current that keeps Northern Europe more temperate than it would otherwise be for how far north it is.

Kate: Exactly. There are a whole bunch of what we call tipping elements because they're essentially thresholds you cross where you see changes that are not reversible on any time scales that are relevant to human civilization. That's the thing that scares me.

Ayana: Right. Once the Amazon is gone, once the ice melts, once the currents have stopped . . .

Kate: And so we can't say what exactly will happen when warming reaches a specific point. We have some uncertainty.

Ayana: You can predict a range and likelihoods.

Kate: Yes. But the best response to the uncertainty is to say: Let's not do this experiment with our planet. Let's not go there. What we know for sure is that the more we limit warming, the better. But we can't ever say, oh, we're doomed if we exceed this value and we're saved if we don't. We just don't have that level of precision and we never will.

---

21 Photosynthesis: carbon dioxide + water + photons → glucose + oxygen. (Aka, magic.) ♥

Ayana: Another thing science has been clear about is that the cause of the problem is humans, it's anthropogenic. Can you give us a quick rundown of how we know that?[22]

Kate: A lot of people like to tell climate scientists, "Oh, don't you know the climate is always changing?" And we're like, "Yeah, we know, we're the ones who told you that." And, it's true, other things besides humans can affect the climate. For example, ice ages are caused by natural wobbles in the Earth's orbit. And it's true that if the sun were to get a lot hotter, that would warm up the planet. But if the sun got a lot hotter, we would notice, and it hasn't. Also, sometimes you hear people blame volcanoes. Volcanoes are an interesting one because they really can affect the climate. In 1815, Mount Tambora blew up in Indonesia and spread so much gas and dust all throughout the atmosphere that it actually made the Earth much, much colder. That was called the Year Without a Summer. During that year, Mary Shelley got trapped indoors with a bunch of insufferable men and wrote *Frankenstein*.

Ayana: Ha! Turning lemons into timeless, classic fiction.

Kate: The effects of volcanoes can be really big, but they don't last for very long. So, based on all the long-term data we have, we can rule out solar variations and orbital variations and volcanoes.

That brings us to humans, the clear culprit. We have been emitting carbon dioxide as the inevitable product of combustion, of burning fossil fuels. But that's not the only thing humans do to the climate. Humans change the land—we cut down a lot of forests and we replace them with grasslands or pastures or croplands. These surface changes affect how much sunlight is absorbed or reflected, which changes the temperature of the planet by changing how shiny it is.[23] On top of that, deforestation and other land-use changes release carbon and other greenhouse gases

---

22 I asked her to explain this because polling by Yale and George Mason universities shows that while only 11% of Americans are staunch climate-science deniers, 29% still think climate change is due to natural causes. (I say "climate-science deniers" and not simply "climate deniers" to make clear what exactly is being denied.)

23 The scientific term for this shininess or reflectiveness is "albedo." More on that in the next interview.

into the atmosphere. And we emit aerosols, which is basically filth—gas and dust that we put in the upper atmosphere that blocks the sun. Aerosols can have a fairly substantial cooling effect while greenhouse gases make it a lot warmer.

Ayana: And the net effect of all these variables has been warming. Greenhouse gases skyrocketing has trumped everything else.

Kate: There are *many* lines of evidence that show human beings are causing this warming.

Ayana: Is all this good science about this very best planet being put to good use?

Kate: Right now feels like a special moment. We spent years and years and years as scientists essentially screaming into the void. Climate scientist Dr. Jim Hansen testified to Congress in the 1980s. I testified to Congress in 2019 and said essentially the same things he did. We're on the sixth IPCC assessment report.[24] We're on the Fifth National Climate Assessment.

Ayana: And you led one of the chapters of that U.S. National Climate Assessment.

Kate: Yeah, the chapter on climate trends. We had the best team in the world working on it, just the most brilliant people. For so long it's been scientists crying out in the wilderness saying, "Hey, somebody pay attention. This is real. This is scary." And then being ignored over and over and over again. And then four years later you reconvene, you do the same thing, you write the report.

Ayana: And then it's, "Now we have even more data and, actually, it's still getting worse, and faster."

Kate: "That thing we've known about since the 1800s? That's happening, like we said it would." But about five years ago something changed.

---

24 Every five years, the Intergovernmental Panel on Climate Change (IPCC) of the United Nations releases assessment reports compiled by leading scientists around the world. Because of the lag between when research is conducted and when it is published, and because the report is a consensus document, it is often considered conservative as regards the severity of climate change.

Ayana: You mean when we met? When we became neighbors and everything changed?[25]

Kate: That was a catalyst. But I think it really was the IPCC 1.5°C report, which I'll be honest, I wasn't supportive of. I didn't think it was a good idea.

Ayana: Let's pause and explain this scenario. In 2015, there was an agreement, the Paris Agreement, that came out of the United Nations' climate negotiations.[26] Its consensus goal was to reduce emissions enough to keep global warming under 2°C. Limiting warming to 1.5°C was stated as the stretch goal.

Kate: Yeah, the High Ambition Coalition goal.

Ayana: The push for 1.5°C was initiated by small island states who were saying, "We're gonna be underwater at 2°C. We can't have that be the target." At the time, I was very surprised they got that more ambitious goal written into the agreement. And that they then successfully pushed for a scientific report on the difference between 1.5°C and 2°C to show how much it mattered.

Kate: But I was grumpy about this report because I was like, we're not going to limit warming to 1.5°C, so this is basically writing fan fiction. You're writing another report that's going to be ignored. And I was so wrong. That report came out and people for the first time were like, "Oh my God. 1.5°C is bad and 2°C is *really* bad. We gotta do something." At 1.5°C we are still going to be seeing massive disruptions to our nice, stable Holocene climate that all of our cultures have developed in.

Ayana: And right now we're on a path to . . . ?

---

25 I would like to take this opportunity to publicly thank Dr. Kate Marvel for keeping (most of) my houseplants alive in Brooklyn when I sequestered upstate with my mom for seven months in 2020.

26 The Paris Agreement is a legally binding United Nations climate treaty that was adopted by 196 states at COP21 in Paris in 2015. It set the goal of limiting global warming "to well below 2°C above pre-industrial levels and pursuing efforts to limit the temperature increase to 1.5°C above pre-industrial levels." COP is the Conference of the Parties, the annual meeting for all parties to the United Nations Framework Convention on Climate Change. COP21, for example, was the twenty-first COP. It is at the COPs that nations gather to negotiate collective action for addressing the climate crisis. For more, see: unfccc.int

Kate: Now we're on a path to somewhere between 2°C and 3°C, barring unexpected shocks to the carbon cycle, which I don't think we can rule out.

Ayana: That report landed and people organized around it. "1.5 to stay alive" was a rallying cry from small island states.

Kate: And you heard people saying, "Twelve years to save the world." That never resonated with me. Cause I'm like, no, you have negative thirty years, honestly.

Ayana: But that soundbite came from the part of that report that said we have to cut emissions by 45% by 2030, which was twelve years from 2018, when the report was published.

So, what would it look like if we got it together and took aggressive actions to limit warming?

Kate: Right now we have evidence from climate models that if we were to snap our fingers and from tomorrow on there were no more human-caused carbon dioxide emissions, then warming would stop. There would be a reduction in ocean heat uptake, which would increase the planet's temperature, and that would be balanced out by the biosphere absorbing CO_2, which has a cooling effect. Basically our best guess is that we would stabilize at 1.3°C. And that's great. And if methane emissions stopped too, the net result would be a cooler planet. If we could do that, I would be very supportive of that. But the way to really bring the temperature back down is to take the stuff out of the atmosphere. You can take carbon dioxide out of the atmosphere by allowing forests to grow back.

Ayana: Back to the magic of photosynthesis.

Kate: Yes, through reforestation. We let the forests grow back in all the places that used to be forest. We restore blue carbon habitats.

Ayana: Mangroves, wetlands, seagrass beds, kelp. My favorites.

Kate: There's also talk of afforestation, which is planting trees where there didn't used to be trees before. That is a more dicey proposition because if you're planting a monoculture of trees, what kind of habitat are you disrupting? And what are the effects on biodiversity versus benefits of carbon sequestration? Sometimes there's a trade-off. There's also soil carbon sequestration; allowing soil to regain its health sequesters carbon. These are "natural climate solutions."

That's all well and good, but that's not going to be able to remove the massive amount of carbon dioxide that humans have put in the atmosphere. We have put so much CO_2 in the atmosphere that it weighs more than all the animals and plants on the Earth. It weighs more than everything we have ever built. It's just incredible. In order to take that out, we are going to have to have a conversation about artificial <u>carbon dioxide removal</u>. That essentially means building machines that suck CO_2 out of the atmosphere (aka artificial trees). Or "rock weathering," which is spreading rock dust over cropland to enhance the geological carbon cycle, taking more CO_2 out of the atmosphere. There is a whole realm of interesting scientific carbon dioxide removal solutions.

This goes back to the *What if we get it right?* question. Getting it right means we stop using fossil fuels. We stop destroying forests, but we also remove CO_2 from the atmosphere. And the reason we need to do that is if you don't remove it, CO_2 sticks around in the atmosphere for a really, really, really long time.

Ayana: Like hundreds of years.

Kate: Hundreds, maybe even thousands of years. But when we talk about carbon dioxide removal, there is a justifiable fear that it is going to create a moral hazard, that it's going to be used to justify continued burning of fossil fuels.

Ayana: That's not gonna work.

Kate: No, because when we talk about carbon dioxide removal solutions, those are limited by physics. In order to separate out what is essentially a very powerful but trace gas in the atmosphere takes energy. There is simply not unlimited energy to do that. So when we talk about getting it
* right, we have to think about carbon dioxide removal in terms of removing what is already there, as opposed to allowing it to compensate for future emissions.

Ayana: Thanks for laying that out so clearly. Because some people are thinking of it like a get-out-of-jail-free card.

Kate: No, it absolutely is not.

Ayana: There are wildly different futures that we could potentially have. What needs to happen right now to set us up for having the best possible version of the future? To prevent three degrees of warming, how hard do we need to pivot away from the current status quo?

Kate: From a scientific perspective, the longer we wait to drastically cut emissions, the more drastic those cuts have to be. If we had started in the 1980s, we could have had a more gradual phase-out of fossil fuels. Because we have waited so long, those cuts have to be incredibly rapid. We need to start deploying the technologies we have. We need to start building <u>wind turbines</u>. Those wind turbines need to be offshore and onshore. We need to build <u>transmission lines</u>. We need to be able to get that electricity generated by those turbines to the people who use the electricity. We have <u>solar panels</u>. We need to be putting solar panels on a lot of stuff. We need great <u>public transit</u>. We have a lot of solutions that we can deploy right now.

Ayana: What do you see as the role of scientists, specifically, in making a better future become reality?

Kate: Our job is to generate the most objective science possible, to make our research relevant to more people than we have so far succeeded in making it relevant to, and to frame a lot of research explicitly around solutions. I would also love to see more engagement with possible futures.

Now, there is no such thing as purely objective science. It has always been done by humans. Humans always have their biases and their preferences.

Ayana: And those biases show up largely in which questions we even ask, in what types of research we even choose to do. Not bias in the sense of false data, but in the sense of which data we are even collecting or analyzing.

Kate: Exactly. And the most obvious way to counter that is to make the ranks of scientists look a lot more like humanity writ large. And there are changes to the structure of scientific retention and training that are necessary in order to make that happen.

Ayana: Diversify science to diversify the hypotheses we're even testing.

Kate: Diversify science to keep it alive. Non-diversity is an existential * threat to science because if only a certain type of people are asking the research questions, you're missing really interesting and important things. Diversifying science is not just a nice-to-have. Also there's a role for activists—they can help scientists ask better or more relevant research questions.

* Another thing scientists can do is acknowledge that there are always going to be trade-offs, and to think hard about how we present those to people. Because solar panels are not going to manufacture themselves. Wind turbines are made out of stuff. Electric car batteries are made out of stuff that has to be taken out of the ground, at least right now, because we don't have a circular economy for that yet.

Ayana: It's not like we switch to renewables and then we get to use as much electricity as we want. Because of this physical limitation on building things and the impacts of mining, et cetera, we still need to conserve energy.

Kate: The second law of thermodynamics says that there is not a perpetual motion machine. There's not unlimited energy. But at the same
* time, we have to be really clear that if we don't build renewable-energy infrastructure, the alternative is not no infrastructure. The alternative is the existing fossil fuel infrastructure.

You can't shut off your brain and say, well, science dictates this. The science may be obvious, but it doesn't necessarily prescribe particular policies, and there are always going to be questions about impacts on different communities. A better future is certainly possible, but we shouldn't pretend that it does not require any trade-offs.

Also, climate change is not pass-fail, it's not binary.

Ayana: This is such an important point. The health of the planet is not just total extinction apocalypse on one hand or perfectly pristine on the other.

Kate: Climate change is not something that all of a sudden happens and everybody dies. It's something that happens on the planet that we've built, and we can make changes. So even if we don't get everything right, even if everything doesn't go perfectly and we don't draw carbon down, even if we're looking at a planet that's two degrees warmer, that's not a planet that I particularly wanna live on, but it may be the planet we get.

It strikes me as not very useful to throw up our hands like, well, we're all gonna die, and therefore we don't have to think about how we live better lives on a worse planet.

Ayana: Giving up is not helpful.

Kate: Giving up is not helpful. Because, look, we're not going to go extinct at two degrees of warming. Life will be worse at two degrees and life will be worse for many people who did nothing to cause the problem. I'm not minimizing the effect of a two-degree world, but I am saying that there will be humans in that world. There will be people making choices, ♥ doing all the things that people do—falling in love, power struggles, family feuds. For me, that's helpful to remember so I can relate to people in the future.

Ayana: Kate, what do you think is the nerdiest or least sexy thing we need to do right now to start to turn the tide toward getting this better future?

Kate: Transmission. We need to build transmission lines. Fundamen- ✱ tally, an electrical grid that is heavily populated by renewables needs a different structure than the current electricity grid where we just burn stuff. And people who deeply care about equitable climate solutions need to have an articulate vision and strategy for how we build boring stuff that nobody wants built near them, like transmission lines.

Ayana: What are the top three things you wish everyone knew about climate science and/or climate solutions?

Kate: Number one: This is not a scientific problem. We know why these changes are happening and we know how to stop them. But science can't tell us what the best course of action is. We need to stop leaving the solutions to climate change to scientists.

Number two, it's not cutting-edge science that tells us that climate change is happening. We've known since the 1800s that carbon dioxide is a greenhouse gas. Nobody disagrees on this, just like nobody disagrees that gravity exists.

Number three, when you look at future climate projections, remember that they are always dependent on a particular emissions scenario. So if you read some scary news article that says the entire Amazon could be gone by the year 2100, or the West Antarctic ice sheet could completely disintegrate, or we could be experiencing 180 days per year of marine heat waves in the Gulf of Mexico, take a step back and think about the underlying assumptions that projection is making about human behavior and human society. The future we get is always dependent on the choices that humans make.

Ayana: Choices about how much fossil fuel we continue to burn and how we treat ecosystems.

Kate: The biggest uncertainty in climate projections, the wild card, is what humans will do. So if you don't like what an article is reporting about a possible future trend, you have the ability to help change that.

On Another Panel About Climate, They Ask Me to Sell the Future and All I've Got Is a Love Poem

Ayisha Siddiqa

What if the future is soft and revolution is so kind
    that there is no end to us in sight?
Whole cities breathe and bad luck is bested by a promise to the leaves.

To withstand your own end is difficult.
Anger against injustice makes the voice grow harsh yet.
The future frolics about, promised to no one, as is her right.
But if she leaves without us, the silence that will follow
    will be an unspeakable nothing.

What if we convince her to stay?
How rare and beautiful it is that we exist.
What if we stun existence one more time?

When I wake up, get out of bed, my seven-year-old cousin
with her ruptured belly tags along.
Then follow my grandmother, aunts, my other cousins, and
    the violent shape
of their drinking water.

The earth remembers everything,
our bodies are the color of the earth and we are nobodies.
Been born from so many apocalypses, what's one more?
Love is still the only revenge. It grows each time the earth is set on fire.

But for what it's worth, I'd do this again.
Gamble on humanity one hundred times over.

Commit to life unto life, as the trees fall and take us with them.

I'd follow love into extinction.

REPLENISH AND RE-GREEN

Nature works so well. And it is so beautiful.

—Mom

What if we look to nature for solutions?

What if smaller farms can thrive?

What if food systems are regional and regenerative?

What if we revive rural communities?

What if we revere biodiversity and photosynthesis?

What if we are good neighbors to other species?

What if we live within nature instead of on top of it?

What if . . . ?

10 Problems

- Approximately 1 million species are at risk of extinction due to habitat loss, exploitation, climate change, pesticides, and pollution.

- Up to 1 billion tons of carbon dioxide (a gigaton!) are released from degraded and destroyed coastal ecosystems every year.

- By 2070, one-fifth of the planet could essentially be too hot for human life. And around one-quarter of humanity (mostly in poor countries) is already dealing with drought.

- Over half of the world's GDP is moderately or highly dependent on nature and its services.

- Every year, an area the size of roughly 20 million football fields is deforested, resulting in roughly 17% of global greenhouse gas pollution. Agriculture is responsible for at least 90% of tropical deforestation.

- The industrial food system is responsible for 33% of global greenhouse gas pollution, with roughly half of that from meat and dairy production, which uses more than one-third of habitable land globally.

- Industrial agriculture has led to more than 133 billion metric tons of carbon in soil being released. Much of U.S. topsoil has lost 30% to 50% of its organic matter in the last 100 years.

- Despite technological advancements, farming productivity is 21% lower now than it was in the 1960s, due to the effects of climate change on weather patterns.

- Synthetic nitrogen fertilizer alone (112 million tons of it annually) is responsible for 2.1% of global greenhouse gas emissions.

- Residential lawn maintenance and landscaping in the U.S. uses nearly 3 trillion gallons of water, 70 million pounds of pesticides, and 3 billion gallons of gasoline annually.

10 Possibilities

+ Protecting and restoring ecosystems could provide 37% of the CO_2 mitigation needed to stabilize warming below 2°C (3.6°F) by 2030.

+ Mangroves and coastal wetlands can hold up to five times more carbon per hectare than tropical forests.

+ Reducing greenhouse gas pollution, ecosystem degradation, and biodiversity loss and restoring land could have an estimated $140 trillion of benefit annually—a third more than the entire 2023 global GDP.

+ 16% of land area, and growing, has legal protections. (The UN Convention on Biological Diversity goal is to protect 30% by 2030.)

+ Restoring 15% of croplands and pasturelands to natural ecosystems could prevent 60% of expected species extinctions and sequester almost 300 gigatons of CO_2.

+ Every $1 invested in restoring degraded forests can generate up to $30 in economic benefits.

+ Species biodiversity and abundance are on average 10% and 15% greater, respectively, inside protected areas than outside of them.

+ Improving farming practices could reduce greenhouse gas pollution from agriculture by 10%. Every 1% increase in soil organic matter could sequester roughly 8.5 metric tons of atmospheric carbon and hold 20,000 more gallons of water per acre.

+ Shifting to plant-based diets could reduce greenhouse gas pollution from agriculture by up to 80%.

+ In reforestation, mixed-species replantings hold 70% more carbon than monocultures.

First, Nature

Interview with Judith D. Schwartz

We'll get into policy and technology and finance and culture, but first: nature. The climate solutions that nature offers can comprise more than one-third of the CO_2 <u>mitigation</u> needed to hold global warming to below 2°C—while also providing water filtration, flood buffering, healthy soil, habitat for biodiversity, and other forms of climate resilience. Shoutout to ecosystems. Restoring nature's cycles—the carbon cycle, the water cycle, the nitrogen cycle—must be at the center of how we rebalance our atmosphere. Respect for these intricately elegant cycles should be the foundation upon which we make decisions about how to live on this planet harmoniously. Respect and love (biophilia) are a winning combination.

The quest to understand nature is a multi-generational, multi-millennial, never-ending task. So our respect must extend to the people who hold wisdom about how nature works—Indigenous peoples, scientists, elders, naturalists. Otherwise, our attempts to address the climate crisis will be misguided. We don't even know how many species there are on Earth (almost every time a submarine descends to the depths of the ocean, new species are discovered), but scientists estimate there are 8.7 million (±1.3 million). The vast majority of eukaryotic life[27] has yet to even be described.

But we don't need to know everything in order to make decisions that protect life. Right now, only about 3% of the ocean and 16% of land and inland waters have full legal protection from exploitation. We need to get to something closer to 50%, "half Earth" in the terminology of pioneering biologist E.O. Wilson. We've got a long way to go. And we are past the point where conservation is enough, where sustainability is enough. Yes, there are some nearly pristine places, but in general, merely sustain-

---

27 That's anything with a nucleus, in case you're rusty on biology terminology.

ing nature in its woefully degraded state would be pitiful.[28] We need to restore nature—to replant and revive; to end the deluge of poisons, of pesticides. Luckily, nature is at once fragile and powerful and resilient. We can destroy it (we are), it can kill us (it does), and it can replenish itself (#goals).

From Scandinavia to Texas to Jordan, science journalist Judith D. Schwartz has been examining environmental challenges and chronicling efforts at what she terms "Earth repair." I first learned of her work a few years ago when my mom gave me one of Judith's books, *Water in Plain Sight*. Mom then informed me that my Christmas present back to her would be reading and discussing it with her. I laughed, accepted this sweetest manipulation, and delighted in the collected stories of transformative ecological restoration and re-greening across the planet, and the breadth of climate solutions inspired by nature.

———————

Ayana: Maybe we can start with the basics. What are nature-based solutions?

Judith: What I tend to say these days is "nature-inspired solutions," which asks: What is nature doing? What are the processes that bring a given ecosystem into health, that support the synergies among different species? Because one overarching perspective that's been missing from the conversation about climate is the role that functioning, healthy ecosystems play in climate regulation.

Ayana: What would you say is your favorite nature-inspired solution? To give people a more concrete understanding of what we're talking about here.

Judith: I am endlessly fascinated by the extent to which wildlife, specifically animals, create the conditions under which they thrive. For example, reindeer help sustain the frozen environment. Reindeer, herded in vast numbers by the Indigenous Saami people in Norway, Sweden, Fin-

———————

28 I once heard someone say, "If you asked me how my marriage was doing and I replied that it was 'sustainable' you would be concerned. Sustainable is not enough."

land, and Russia, trample snow as they go. Fluffy snow acts as an insulator, protecting the ground from the frigid air. But when it gets pressed down, that insulating effect is reduced, which keeps the permafrost frosty, which in turn prevents the release of massive amounts of carbon that happens when permafrost thaws.[29]

And then, in the summer, the reindeer are brought into different areas to graze. With the increased warming, there's now a shrubification of the Arctic—plants that had not been able to get established in the far north are now doing so. But the reindeer are nibbling those shrubs, keeping them under control. Because those shrubs have darker leaves, when they get eaten, that lightens the color of the landscape, or to use the technical term, it raises the albedo.

Ayana: Albedo is a measure of how reflective a surface is. Ice has a high albedo and dark plants (and soils and rocks) have a low albedo and absorb more heat.

Judith: Exactly. Albedo is really significant. Reindeer are ensuring that those shrubs don't grow, that the darker leaves that would start absorbing more heat are managed, and that the native heath plants, which are lighter and therefore are more reflective, are still able to dominate. This helps keep that environment from warming.

The Saami work with the reindeer intimately—they know their grazing patterns, know what they're nibbling on, what nutrition they need, where to bring them to get that nutrition. These animals then interact ✱ with their environments in a way that perpetuates the health and stability of those landscapes.

Ayana: When we think about the ways nature and ecosystems are part of climate solutions, we're often just thinking about planting and protecting trees, about photosynthesis. Rarely are we thinking about these more intricate ecological connections that involve animals.

Judith: Right. Animals, including birds and insects, are managing plant ecosystems, which ensures biodiversity. Plus, if you think of a large ani-

---

29 Increasing herds of large herbivores in northern high-latitude ecosystems could prevent nearly 40% of all Arctic permafrost soils from thawing, by keeping soil temperature below −4°C (24.8°F).

mal, like a bear or a whale, the amount of carbon that is embodied in those creatures is extraordinary.[30]

Ayana: You've done a lot of writing about nature's cycles—the carbon cycle, the water cycle.

Judith: The <u>carbon cycle</u> and the <u>water cycle</u> are always interacting. A simple example is how healthy soil, with plants and roots and soil-dwelling critters, has carbon in it.[31] That organic matter creates a sponge-like aspect that absorbs water. But when the soil is not healthy, you get a situation like we saw in California in 2023, which was in a severe drought, then got a lot of very wet weather that led to massive flooding.

✱ **Ayana:** If we had healthier soils, more of that water could have been absorbed into soils, instead of causing landslides or eroding topsoil or leading to flooding.

Judith: And then of course a lot of that rain fell onto the built environment.

Ayana: All that concrete and asphalt, that's certainly not absorbing water. Highways and parking lots and all that.

Judith: Exactly. Those are all built to sluice water away, built with the belief that water is an inconvenience that must be dealt with, with dispatch. In so many instances, what could be a benefit becomes a problem if we haven't been managing our land appropriately.

Ayana: You contributed a beautiful essay to the *All We Can Save* anthology, and there's a line from it that I come back to often, which is:

> Basically what we need is life: life to transport water and, by extension, to regulate heat; life to seed the rain; life to slow down moving water so that it has a chance to infiltrate.

This idea that we simply need to support nature in doing its thing is powerful, because often when we think about climate solutions, we think

---

30 One whale is worth a thousand trees, in terms of carbon. When a whale dies and falls to the ocean floor, that's an average of 33 tons of carbon sequestered. Not to mention, while they are alive, all their pee and poo are nourishing photosynthesizing phytoplankton. Whales are climate heroes.

31 Soil holds approximately 75% of all terrestrial carbon.

primarily about control and technology in really dominating ways as opposed to enabling this flourishing of nature's cycles.

Judith: Yes. If we pause for a second and ask ourselves, "How does the Earth manage heat?," it becomes clear the powerful role that water plays in regulating temperature and generating our weather.

Ayana: Break this down for us.

Judith: So, the Earth has a small water cycle and a large water cycle. In the small water cycle, water falls down as rain, is held in the ground, and helps to grow plants. Plants release moisture through a process called transpiration, which is the upward movement of water through plants— from roots up to the leaves and out through leaf pores called stomata. And that moisture evaporates up into the atmosphere and, ultimately, it condenses and comes down as precipitation again. In some healthy ecosystems and areas like the Amazon rainforest, a raindrop may have five or six cycles of staying in the same area before ultimately it moves to another area. This evapotranspiration is a cooling process.[32]

Ayana: Similar to humans sweating.

Judith: Yes, that is how the plant regulates its temperature, by releasing or holding on to the moisture. Plants also help regulate the temperature of the soil, and dissipate heat from solar radiation.

Let's say you have the sun beaming down on some bare ground and there are no plants growing there. That heat is not dissipated, it just lands on the ground and heats up the surface. But if you step maybe ten yards to the left and you're in a nice, healthy meadow, then that solar energy beams down on the land and the plants are dissipating that heat, keeping that area above the ground moist and cool.[33]

Ayana: Like walking barefoot on a hot sand beach versus in the cool grass.

---

32 *Evapotranspiration* is the combined process by which water moves from plants, soils, and other surfaces into the atmosphere. Evaporation + transpiration.

33 Or, as Ciara Nugent wrote in *Time*, "Growing trees alongside crops is like installing an air conditioner and sprinkler system, or, in the words of one prominent farmer, 'planting water.'" On a sunny day, a single tree can transpire hundreds of liters of water, representing a cooling power equivalent to running two or more domestic air conditioners.

Judith: Exactly. And one other aspect of the small water cycle is holding that water in that local environment. Nature has all kinds of strategies to do that. One is having carbon-rich soil. Another is having soil-dwelling species like earthworms and dung beetles and prairie dogs. They're all making these passages in the soil that allow for water to be absorbed. And nature's engineers, like beavers, are creating dams to hold water in the area.

★ **Ayana:** I was reading recently about how beavers could help to manage wildfires in the western United States because their dams help landscapes absorb more water. It's such a lovely thought that beavers are part of the answer, not an annoyance.

Judith: Oh my, yes. Beavers are the answers to so many things. And then there are trees. It's only been a few years since our public conversations about forests have started to move beyond trees as sticks of carbon, to how biodiverse forests create their own rain, by emitting aerosols, which seed the rain.

Ayana: That is such a cool phenomenon. Cloud condensation nuclei, tiny aerosol particles that plants and fungi release into the air, are what water molecules condense around.

Judith: And again, the cooling and evapotranspiration is a huge factor in the management and conveyance of energy in the biosphere. It's really important to understand how vast an effect that has on climate.[34]

Which brings us to the large water cycle, the connection between the ocean and the land. This is where the theory of the biotic pump comes in. This concept was developed by two Russian physicists, Anastassia Makarieva and the late Victor Gorshkov, to answer the question: If our rain ultimately comes from the oceans, then how can land thousands of kilometers away from the ocean get moisture?

Essentially, their research suggests that it is the biology of forests, not just the physics of the ocean and atmosphere, that leads to more rainfall far inland. Transpiration creates a low-pressure zone as the water vapor

---

34 Globally, 40% of rainfall over land originates from terrestrial evaporation. Destruction of ecosystems is proven to reduce evapotranspiration. (E.g., Deforestation has reduced rainfall around the Panama Canal, greatly reducing its usefulness as a shipping route.)

rises and condenses, which generates winds that pull in moist air from higher-pressure areas, such as over the ocean. Almost like a tug-of-war between the ocean and the forest. In the past, when we had lots of intact forest in the Amazon, for example, the forest was winning as the dominant condensation zone, tugging in all that moisture from air over the ocean. But now, because of deforestation, that flow is a little shaky, so sometimes the rain falls over the ocean and you get drought inland, as we've seen several times in the last couple of decades.

So you want to make sure that there's enough concentration of healthy forest to regulate rainfall. It's unnerving to see how dicey things are, but let's be curious, because the more we understand how these systems work, the more we know how to work with them.

Ayana: Understanding the water cycle can be powerful when we're thinking about agriculture, because climate change is impacting our ability to grow crops in different places, and we need to be able to counterbalance drought conditions as arid regions expand.

Judith: I have an example of this from Texas. A few years ago I visited a couple, Katherine and Markus Ottmers, in the far west of the state, an unpopulated area. Very rough, very rugged.

Ayana: Like tumbleweeds Texas?

Judith: Absolutely. When I got there, it had been even drier than usual and people were saying, "If the heat doesn't get us the lack of water will."

This couple has a singular relationship to water. Katherine would talk about "moisture events." She would talk about nutrient-dense dew. She told me that she and Markus are so tuned in to nuances of the water cycle that they're like desert plants. The most extraordinary thing they had done was design their barn so that it captured condensation.

Markus told me how he was inspired by the Namib desert beetle in Namibia, by how it puts its bottom to the sun and moisture condenses and runs down its belly. That is what they re-created with their barn.

Ayana: He saw a beetle, ass in the air, drinking condensate, and thought, "I will design a barn based on this beetle pose." That is <u>biomimicry</u> at its best.

Judith: On the basis of this flash of inspiration, they angled in the metal roof of the barn so it would catch the most heat. And when the cool breezes (that no one else even noticed) were coming through and cooled the roof, that created enough of a temperature differential for condensation, and that water was collected in a tank. And a revelation for them was a time in early January, when they had not seen a raindrop in four months, when the water tank overflowed.

Markus couldn't believe it. He got up around four in the morning, the coolest part of the day, and saw water spilling over. He thought maybe there was a blockage or something, but he saw that the water was really flowing.

They are able to collect enough water this way for six people to take light showers every day. And the plants they grew would be put in a place that gets the morning sun as late as possible so they could hold on to the dew. It was remarkable. There are lots of examples of people dealing with extreme circumstances by working with natural solutions.

Ayana: Please do tell us another.

Judith: One that I wrote about in *The Reindeer Chronicles* is the Al Baydha Project in western Saudi Arabia. Neal Spackman is an American, a pioneer of hyper-arid agroforestries, who, because he had studied Arabic in college, was invited to be part of that poverty-reduction project. He wanted to bring in an ecological component, so he did a fast-track permaculture design course with Geoff Lawton of Australia, who famously created a food forest in the Jordanian desert.

The Al Baydha Project had about a hundred-acre demonstration plot, and the approach revolved around holding on to water. This area regularly exceeds 100°F, and it gets an average of only three inches of rain a year—but it might go several years without any rain.

Neal, working with a team of four local Bedouin,[35] started building swales, and troughs, and little earthen mounds, and they put in stones to steer where the water would go—all to slow the water down and hold it. This was a few men with their hands and a few simple tools. They were preparing for the rain. And when it came, they would rush to plant. Then they'd go back to shaping the terrain and preparing it to hold water.

---

35 The Bedouin are nomadic Arab tribes who have traditionally inhabited the deserts.

One thing that really struck me is Neal's definition of a desert. We tend to think that there's a certain threshold of rainfall that if you get less than that it's a desert, and if you get more than that, it's not. The way Neal defines it is that "a desert is a place where when it rains, it floods."   *

Ayana: Ah, it's a place where the water's not being absorbed and retained at all.

Judith: Exactly. Because once you create the conditions for water to be held in the soil, it ceases to be a desert.

Ayana: Because then things start growing.

Judith: This project continued for several years and things were working out well. And then in 2016, they ran out of funds, so they had to stop irrigating. And Neal's reaction was, "Oh, well the permaculture ideal is to create a self-sustaining system. So let's see how we did." Two and a half years and about a half an inch of rain later, Neal went back and found that for four of the twelve species planted, more than 80% of the trees survived, and between 100 and 200 native trees, including acacias that they hadn't planted, grew and had a 100% survival rate.

Ayana: Wow.

Judith: They had managed not only to grow particular plants, but to create a whole ecosystem that could withstand that long period of time between rains. In the longer term, most of those trees didn't survive without tending, but that project gave an indication of what's possible even in such a forbidding environment.

Ayana: It's exciting to think about these ways we can collaborate with nature's cycles to feed ourselves and re-green the planet—not only absorbing more carbon by having more things growing, but also having healthier soils to deal with climate changes that are going to get more extreme. If we get it right, we are working *with* the water cycle, *with* the carbon cycle, and we're thinking about soil in different ways.

Judith: And we'd be learning from Indigenous people, including pastoralists. Pastoralists with their animals are often marginalized because what they do is not always understood or appreciated. Thinking about the pastoralists reminds us that love of animals and love of all life forms and our love of nature itself can drive these short-term and long-term shifts we need to make, and can remind us how joyful it is to work with nature, to be in nature, and to be among other creatures.

Ayana: Yes, this is biophilia. It's very real.

Judith: Exactly. Listen to that and let that guide us. Sometimes all we hear about are the technologies. Because people are investing in them, because it gets into our stock market news. But in the meantime, quietly, nature is doing all of this work and we can ally with it.

Ayana: Can you paint us your vision of the future if we get it right with all of these nature-inspired solutions?

Judith: My vision starts with us opening ourselves up to being curious. By curiously observing, we open ourselves up to the possibility of getting back into what is termed "right relationship" with nature, experimenting and allowing our imaginations to flow.

Ayana: You're talking about this on an individual level, but institutions and governments can also be more curious and imaginative.

Judith: Yes, very much so. We need to be open to what's possible, and really consider: How might we live in harmony with nature?

*   One thing that would be different if we get it right is that we would all be more place-based so that no matter where we are, we are growing the food that is appropriate for that ecosystem—certain fruits and plants in dry areas, and different ones in moist and cooler areas. That is <u>bioregionalism</u>.

A permaculture teacher named Andrew Millison talks about how many problems would be solved if our boundaries, national and state, were determined by watershed.

Ayana: That's a nice idea.

Judith: There's something clarifying about that framework. And that reminds me of when I was in Hawai'i, I forget the Hawaiian term for it, but their land ethic was basically watershed-based.

Ayana: Oh, I remember this term, _ahupua'a_.[36] It's a ridge-to-reef framework—different chiefs controlled different watersheds.

---

36 Per the University of Hawai'i, "Each _ahupua'a_ is like a slice of pie that begins at the top of the mountain and goes down to the ocean. In old Hawai'i, people who lived in the mountains would barter with those who lived near the ocean. A complete balance in this land system is what kept the Hawaiian people alive."

Judith: Right. And that made sense because a community would have access to all the different parts of the ecosystem, and it could be managed in a way that appreciated that what happens upstream affects what happens downstream. But the way that we are currently organized is that, even if you're affected downstream, you have no say in what happens upstream.

Ayana: What are the top three things you wish everyone knew about climate solutions?

Judith: First, I wish people knew how important biodiversity is. All the animals and birds and insects, and the underground biodiversity, all of these creatures are driving ecosystem processes that are central to the regulation of climate. Second is the role of water and understanding how water flows and infiltrates, how transpiration and condensation work, and how that greatly expands our ability to manage heat in a direct sense. So, the water toolbox. And third, if each of us embraces a sense of agency and realizes that what each of us does matters, restoring the world's ecosystems and the cycles of life can begin right where you are, for any of us.

Ayana: This is a wonderful trifecta.

Judith: Where I live, I'm working on enhancing habitat for at-risk native pollinators. I can't tell you how joyful that is, to anticipate the increase in diversity of insects. And the beautiful plants I get to play with, and the colors, and my hands in the soil. That joy is infectious because other people see and sense your excitement and want to join in. You're happier. And the happier you are, the less burnt-out you are and the more curious you get. It's pollinators for me, but it will be something else that brings you that same joy. It all matters.

A Vision

Wendell Berry

If we will have the wisdom to survive,
to stand like slow-growing trees
on a ruined place, renewing, enriching it,
if we will make our seasons welcome here,
asking not too much of earth or heaven,
then a long time after we are dead
the lives our lives prepare will live
here, their houses strongly placed
upon the valley sides, fields and gardens
rich in the windows. The river will run
clear, as we will never know it,
and over it, birdsong like a canopy.
On the levels of the hills will be
green meadows, stock bells in noon shade.
On the steeps where greed and ignorance cut down
the old forest, an old forest will stand,
its rich leaf-fall drifting on its roots.
The veins of forgotten springs will have opened.
Families will be singing in the fields.
In their voices they will hear a music
risen out of the ground. They will take
nothing from the ground they will not return,
whatever the grief at parting. Memory,
native to this valley, will spread over it
like a grove, and memory will grow
into legend, legend into song, song
into sacrament. The abundance of this place,
the songs of its people and its birds,
will be health and wisdom and indwelling light.
This is no paradisal dream.
Its hardship is its possibility.

Go Farm, Young People

Interview with Brian Donahue

Consider this: "For more than a century rural America has endured a boom-and-bust economy that has supplied wealthy processors with cheap food, timber, and fuel, but has left the countryside impoverished, and nearly deserted. Rural America hasn't been 'left behind' in the march of progress—it has been systematically gutted."

That quotation is from a pamphlet titled "Go Farm, Young People, and Help Heal the Country," by historian, educator, and farmer Brian Donahue.[37] In it he frames the agricultural and economic predicament of rural America, unpacks how that's tied to our political one, and reminds us that it doesn't have to be this way.

Brian practices and advocates for sustainable and just agriculture and forestry, and we met in 2019 at a Green New Deal workshop in Washington, D.C., organized by policy phenom Rhiana Gunn-Wright,[38] back when that was a brand-new policy proposal.[39] I so valued the insights Brian shared in those discussions that when I received an email from him a few years later, I eagerly opened this pamphlet he attached. He describes a countryside with diversified agroecological farming, healthy soils, protected waterways, sustainable forestry, high biodiversity and carbon sequestration, renewable energy, significant land returned to people of color, and many wild places.

After I read Brian's vision, I excitedly called him to discuss it all, inspired to impulsively reorganize my entire life around his premise that "the relocation of a relatively small slice of the citizenry could first enable, and then fulfill, the vision of a just, sustainable world."   *

Our conversation focuses on the northeast U.S., where Brian has lived,

---

37 Available in full at newperennialspublishing.org.

38 Rhiana is interviewed in the "Changing the Rules" section.

39 At this workshop, Brian and I were in the subgroup on ecosystems, agriculture, and the ocean, and I was thinking, "How is this all one subgroup? It's the entire natural world!" Nature really doesn't get enough love in climate policy.

worked, and farmed for most of his life. He taught environmental studies at Brandeis University in Massachusetts for twenty-five years, and he's been growing food in the area for even longer. There are, of course, many specificities from region to region and town to town, but most of the insights and solutions he puts forward are broadly applicable. He made me realize that, while I've spent a lot of time thinking about agriculture, and a little time thinking about rural economies, I've not spent nearly enough time thinking about forestry or rural politics.

!   In the U.S., there are 4 million fewer farms now than there were in 1935, a 70% decline. And farm size has doubled—large-scale farms account for 3% of the farms but 47% of the value of production.[40] How did we get here? In New England, understanding the current socio-ecological-political scenario requires an appreciation of colonial history, so we started centuries back.

Ayana: As an environmental historian, can you lay out the transformation arc of the New England landscape?

Brian: For thousands of years there were Native people living here practicing widely diverse foraging systems, with some degree of agriculture. No one is sure how much farming, but anyway, taking good care of the place. And then European settlers showed up in the 17th century, mostly from England, a bunch of Puritans. Twenty thousand of them landed in the 1630s.

Ayana: I didn't realize it was that many. That's a lot of boats.

Brian: A lot of boats. And there were probably something like 100,000 Native people in New England at the time. But as we know, they suffered grievously from both the aggressive settlers who arrived and the diseases they brought with them. And five generations later, there were 2 million of these Yankee settlers.

Ayana: They had a lot of babies is what you're telling me.

----

40 Also, the average age of American farmers is now almost sixty years old, compared to around fifty in the 1950s.

Brian: Right—a lot of boats, and then a lot of babies. Families had an average of eight kids. And they kept moving and settling new towns, expropriating more land and putting in place an agrarian landscape that, by the middle of the 19th century, seven or eight generations in, had cleared about three-quarters of the land in central and southern New England. They were producing food and wood for their own use and for the growing urban markets of industrializing towns and the city of Boston and so forth.

And then, as the whole country was settled by White farmers, New England farming changed and specialized. They decided they didn't need a lot of the pasture land they had cleared, and that was now quite degraded in many cases. So in the late 19th and early 20th centuries, that cleared land started going rapidly back to forest. Even though agriculture was still doing okay economically, farmers didn't need as much pasture.

Then, in the 20th century and especially after World War II, New England farmers found it increasingly difficult to compete in an industrialized agricultural system. Today, little of New England is still farmland, something like 5%—it used to be more than half—and the region only produces something like 10% of its food. Southern New England—Rhode Island, Connecticut, Massachusetts—is now nearly 60% forested, and northern New England states are 80–90% forested. And we cut almost none of that wood. The trees just grow. So heck yeah, we ought to be able to produce more of our wood products here than we are.

Ayana: It's a remarkable transformation from forest to farmland and back to forest. What can we learn from this?

Brian: This story tells us a lot of interesting things about agriculture. One is that we had some very successful farming here. A hundred years ago, we settled into dairy, fruits, vegetables, eggs. Boy, that works well here. And it doesn't take a lot of the landscape to do it. The other thing is we got our forest back, after we had cleared too much. That was lucky.

Ayana: Every region in the country has its own story—and New England may have lessons for how the rest of us can rethink agriculture. What would New England look like if we get it right?

Brian: Well, put simply, we would be producing more food and wood in New England. Recently, a regional network of organizations, with

Food Solutions New England as the backbone team, spent three years co-creating something called A New England Food Vision.[41] I was a lead author on that and we figure that New England could grow half the food it needs right here in the region by 2060 by utilizing a little more of the reforested landscape.

Ayana: Half is a lot.

Brian: Yeah, half is plenty. We could reach that by increasing farmland from 5% to 15%. We don't want to clear all the forests again the way it was in the mid-19th century, but by clearing a small amount of forest and reconfiguring the things we grow, concentrating on fruits and vegetables, and increasing grass-based dairy and beef, which suits our climate, we could grow a lot more of our food here. We could expand urban and suburban production and bring back more agriculture into rural areas. You would see more small and medium-sized farms out in the countryside again.

Ayana: When we think about climate change we have this obsession with preserving forests, and rightfully so. That makes it sort of counterintuitive that cutting some of that down would be a part of the solution, but you are an advocate for sustainable forestry and a local wood movement. What might that look like here, where there are quite a lot of trees?

Brian: New England only produces about one-quarter of the lumber it uses, and about two-thirds of its pulp. A region that is about 80% forested is importing a huge amount of wood. Most American wood now comes from the American South, where there's intensive plantation forestry going on.

A group called Wildlands, Woodlands, Farmlands & Communities, that I'm a part of, is making the argument that the forest we have ought to be protected from development—some of it should be designated as wild reserves, but a large part of it can be sustainably managed through

---

41 "A New England Food Vision describes a future in which New England produces at least half of the region's food—and no one goes hungry. It looks ahead to 2060 and sees farming and fishing as important regional economic forces; soils, forests, and waterways cared for sustainably; healthy diets as a norm; and access to food valued as a basic human right."—nefoodvision.org

ecological forestry. That would allow the forest to give us environmental benefits and produce wood products at the same time.

We have an affordable housing crisis in this country. We don't have enough housing. We need to build both multi-family houses and single-family houses—especially smaller ones, for young families. And wood is a wonderful way to do that. It's much better than concrete and steel from an environmental point of view, as long as you're harvesting in ways that protect the environmental values of the forest. ✱

Ayana: What does sustainable forestry really look like?

Brian: Overall it looks like cutting very carefully. More specifically, there are a few key elements to what the people I work with call ecological forestry. First, you cut on long rotations. You leave the trees to grow for a nice long time so that you get the ecological values of a mature forest. One hundred and twenty, one hundred and forty years, something like that. Then, when you finally harvest that timber, you more or less mimic natural disturbance patterns, like how a storm might knock down a tree or patch of trees.

Next, as the forest grows back, there's a lot more young trees out there than will survive. You can go in and thin the forest periodically as it's growing to remove some of those small trees and utilize that wood and try to direct most of the forest growth into the trees that are ultimately going to be best for timber use.

We could produce all the wood we need in New England this way, and have plenty of wonderful timber to build houses with. But a lot of the forest in the southern and central part of New England is owned by families or nonprofit organizations whose first instinct is to leave it alone, protected as a beautiful forest. The good news is there are ways of managing forests that people can live with. I've been part of ecological forestry for decades in the Weston town forest in the suburbs, and we do it now on our own farm. It definitely doesn't mean you cut everything down and have to go up there and weep for the next thirty or forty years.

Ayana: That's the problem. Forestry often brings to mind these horrible images of a clear cut, where the landscape looks like a bomb went off and you just see all these sad stumps and wood chips. But you're describing something completely different where you would still have a forest there. It would be carefully, selectively logged.

Brian: If you are going to let trees grow to their harvest age of over a century, that means you're cutting 1% or less of the forest per year. So even if you do a clear cut—which you might want to do in some cases, to regenerate trees that need a lot of light—that would still only be 1% of your forest.

But mostly, you're going to harvest a little patch of trees here, and a little patch of trees there. That more selective logging will let in less light, but be good for shade-tolerant trees, like sugar maples. It will vary from spot to spot. This is totally different from what's happening in the American South, for example, where entire landscapes of young trees are cut at once, and all the industry cares about is maximizing the amount of sawtimber and biomass those landscapes can produce within fifty years, and it's unattractive.

Ayana: Why isn't ecological forestry the norm now?

Brian: The same reason we do little of our agriculture in small pieces: because it cannot compete economically with the larger enterprises that are pumping out the commodity in the cheapest possible way. As a society, we have to make a choice to produce these things, food and wood, in ways that keep the land beautiful and keep it ecologically whole. That will make things more expensive, but it's a small part of our economy overall and it will create a landscape that people want to live in. So it's worth it—the added benefits outweigh the costs.

* **Ayana:** Sustainable forestry and agriculture could potentially create many more local jobs, as well.

Brian: Oh, absolutely, it would, and that's why it would be more expensive—and because we're taking all those costs that are usually externalized (like environmental damage) and internalizing them. It then becomes a problem of who has access and means to afford these sustainable products. Making these products accessible is a necessary part of the same equation.

Ayana: This gets to the question of who holds the political power for crafting the policies that would enable this kind of shift, which is one reason I wanted to interview you. Your pamphlet, "Go Farm, Young People, and Heal the Country," made a big impression on me. What prompted you to write this manifesto of sorts?

Brian: Well, the big picture is that in order to flourish, rural America ✳ needs more people. The peak of farm population for much of rural America came in the early 20th century. The 20th century is a story of rural depopulation with agriculture moving toward a model of very large farms with operations that involve big machines and very few people, and that produce a huge amount of commodity crops at a very low price. This industrial agriculture creates a lot of environmental problems and also creates places that can't sustain communities.

Ayana: Why can't big ag sustain communities?

Brian: There's nobody there. There aren't enough people to have a school or a doctor. These large operations create an empty landscape. If you've ever been in rural Kansas, or any place like that, it's full of these little towns that have dried up and almost completely disappeared, places that people loved and wanted to live in, but now have very limited work available. If we want to imagine an alternative model, it has to be one that not only involves farming more sustainably, but has more people doing it. The driving question is, How in the world would you accomplish that? How would you get people back into these places?

Ayana: That would be a huge transformation.

Brian: It requires a massive change in the way we do things. But, relative to our entire population, it wouldn't actually involve a large number of people because rural America is so depopulated. We're not talking about a mass migration, but a relatively small percent of the urban and suburban population returning to rural areas.

Ayana: What are we talking about, like a million people across the U.S.?

Brian: A couple million actual farmers. A few million more to build out communities. Really, a small percentage of our total population—but an enormous increase for rural America.

Ayana: You make the case that so much of the fate of our country, our energy, our food system, our politics is in the hands of rural America.

Brian: There's a raw political question that runs alongside the environ- ! mental one, which has to do with the massive political power held by rural parts of the country because of the way our Electoral College and Senate are arranged. For instance, a disproportionate part of the Senate comes from rural states with small populations. We might want to pass progressive policies like the Green New Deal and have universal health-care and a living wage and all the rest of that great stuff, but rural America often votes against all of it. That's a political battle that's very hard to win. And you can't change that system precisely because you don't have the votes, in the right places, to change the system. So, to state it in a rather naive way, why don't more progressive people just move to those states?

Ayana: It wouldn't take many people moving to rural areas to shift voting power.

* **Brian:** The kind of world we want to create—sustainable farming and forestry, sequestering carbon in the soil, and taking good care of the countryside, thriving rural communities—is completely achievable. We know how to do this. But the very problem of rural depopulation and alienation prevents passing the policies that would fix it. It's tied up in that knot. Now, on the flip side, repopulation would build the political power to pass sweeping climate policy, which would greatly benefit those places.

Ayana: And rural folks are figuring out more climate-friendly ways to farm. Can you describe some of that work that's happening?

Brian: So much of our cropland today is used to grow feed for livestock.[42] In terms of climate impact, there's no doubt in my mind that we need to be consuming less meat.[43] But it's also true that we could be growing meat in different ways—raising cows less on grain and more grazing on grass.

And as Wes Jackson and the whole crew out at The Land Institute in Kansas have been pushing for decades, we can produce grain crops from
* perennial plants so that we don't have to plow as much, so we don't dis-

---

42 Nearly all of the alfalfa, most of the soybeans, and about 40% of corn (three of the four biggest crops by value) go to livestock feed. Irrigating cattle-feed crops accounts for 23% of all water consumption nationally—and 55% along the Colorado River.

43 Global meat production has more than quadrupled since 1965.

turb the soil and release its carbon.[44] We can move toward perennializing the land again. That is, not have as much land devoted to annual crops—corn and soybeans and so forth. More in pasture, and legume rotations, and agroforestry, and eventually perennial grains.

Something that we're working on at The Land Institute is to try to use farming techniques that keep the soil permanently covered with perennial crops, which prevents erosion and builds the soil's potential for absorbing carbon dioxide from the atmosphere.

Now, getting carbon back into the soil and keeping it there is not easy. And so this is not gonna solve our climate problem all by itself or some miraculous thing like that. But it could, if widely practiced, make a positive contribution by simultaneously rebuilding healthy soil with lots of organic matter that holds water and nutrients better, while taking down some of the carbon dioxide that's in the atmosphere. That can be done.

Ayana: More broadly, how do you imagine our food system would be changing if we get it right?

Brian: We would re-localize and re-regionalize our food production. Not all of it, but significant parts of it. In particular, fruits and vegetables could be grown closer to where people live.

Ayana: Maybe fewer strawberries will be eaten in the Northeast in the winter.

Brian: I'm not against strawberries, but we do need to rely less on the Central Valley of California and northern Mexico. About half of the fruits and vegetables for the whole country come from there, and how sustainable is that in a warming world? There are parts of the food system that it makes a lot of sense to bring much closer to home. We can have fresher vegetables and nice small-scale farms.

But we can't grow all of our food that way. In a place like New England with a lot of people and not much cropland, we can grow some grain and beans and meat and so forth, but nowhere near all of it. We need the Midwest for that. We also need to imagine how the Midwest can

---

44 Perennials are plants that come back every year (as opposed to annuals). Deep-rooted perennial grains like Kernza have roots that can extend down ten feet, helping to restore soil carbon and nutrient cycling.

continue to produce surpluses of those things, and do it in more sustainable ways. It's not that each region has to be totally self-reliant, but they can become *more* self-reliant and reap a lot of benefits that way.

Ayana: When you talk about all this, it's not theoretical for you. You've been working on farms since the 1970s.

Brian: Yeah, I dropped out of college and co-founded Land's Sake Farm, a nonprofit community farm. It's on land owned by Weston, a wealthy town outside Boston. We grew fruits and vegetables. The program's still there, steaming right along almost fifty years later.

That work got me thinking about how community land—forestry and farming and working with kids and the schools and so forth—could work in other places, like in the many other suburban towns that have conserved a lot of land. And how it could work in towns that weren't so affluent, too.

The thing we did that was a little more unusual than farming was starting a forest management program on a couple thousand acres of town forest. We were taking kids out and splitting wood and actually logging in a forest owned by a suburban town. The idea was to get people acquainted with the concept that if we really want to have a sound relationship with forests, we have to use them to some extent. You can't just lock 'em up and say, "Oh good, we're preserving nature," when all your resources are coming from some other part of the world where they're being extracted. Of course, we couldn't produce much of the wood that Weston consumes, but at least we made some noise!

Ayana: When did you start your own family farm?

Brian: My wife, Faith, and I since we were both kids had wanted to go live in the country. We have a friend and his wife who wanted to do the same kind of thing, and now we co-own a place with them. Farming takes a lot of time, so you need to do it with other people, especially when you've got other full-time jobs. After years and years and years of looking and thinking about this, we bought this place that we could eventually retire to, essentially. If you can call running a farm retirement.

Ayana: I love that you pooled your resources to buy a farm together with another family. We need more of this communalism. How many acres are we talking?

Brian: We're talking about 170 in western Massachusetts. Maybe fifty acres of grass, a hundred acres of woods, twenty acres of beaver meadow and swamp.

Ayana: That's a good mix.

Brian: Oh yeah. It's got everything. And I should say that we had the means to buy this place because we had reasonably well-paying jobs, and had inherited a little wealth from our parents, the kind of thing that upper-middle-class Americans have access to—and a lot of other equally deserving people don't. This was in 2007, when we were in our fifties. So still young and energetic.

But land is still so expensive that we couldn't have bought it if it wasn't land protected by the Commonwealth of Massachusetts. It's what they call APR land, Agricultural Preservation Restriction, where the state, usually working through a local land trust, acquires the development rights. So the land cannot be developed. It's permanently farmland. And it really has to be farmed—the state actually checks. We raise beef cattle, pumpkins and squash, and a few pigs. Sort of a small-scale, semi-commercial operation. And we manage the woods as well, so we make timber sales, which has commonly been an important part of New England farms.

Ayana: If we want more folks to be able to have farms in this country, we certainly can't rely only on the people who can inherit a down payment. That's not gonna work. And having jobs off the farm to make ends meet implies the need to do things more collectively. And you're managing a scenario with multiple crops. It gets complicated real fast—very different from the vision of a farmer with a large field of corn and a tractor.

Brian: It's a very different model. It looks a lot more like the way we were farming a hundred years or so ago. There are still people doing mixed-husbandry agriculture on a medium scale, but it takes a powerful community to support you to do that. It's important to note that most farmers in America are part-time farmers.

Ayana: A lot of them are also truckers or teachers or contractors or something.

Brian: Yeah. Not just the quasi-hobby farmers like we are, but even commercial large-scale farmers often have other jobs to help support themselves. And they're usually a couple. One part of the couple typically has a full-time job as well. Farming can be a range of things for different kinds of people.

Ayana: On the topic of who owns the land and how people who want to farm can afford land, you're an advocate for land trusts. What is a land trust? Why do you think they're a good idea?

Brian: A <u>land trust</u> is a nonprofit organization that's organized to protect land, and there are different ways that can be done. One is that the land trust, alone or together with the state, can own an easement on private property that prevents the property from being developed.

This is the case on our farm. They say, "Gee, we'd love to keep these small farms and woodlots, and we're okay with the idea of them being in private hands, but we don't want them to be developed." So the land trust, working with the state, will purchase the development rights. Through that you get an increasing part of the landscape that cannot be developed.

Ayana: And that helps defray some of the cost for the people who purchase it?

Brian: It makes the land somewhat cheaper for the people who purchase it. Although, conservation through a land trust can create places that are so beautiful and nice to own that they're still going to cost a lot of money, even with the easement.

So we get to the second part of the problem, which is how we can protect land but make it available to more people, who don't necessarily have a lot of money. We can imagine a world in which much more of the land would be owned by local land trusts, and then leased long-term and at affordable rates to people who wanted to farm it. An organization called Agrarian Trust is working on this, along with many others. All they need is boatloads of more money.

Ultimately, to do any of this at scale is going to take a lot of federal and state money, which again requires political power. One potential funding source is states who have climate plans—they have some incentive to get farming and forestry right.

Ayana: One of the things that I worry about when we think about <u>rural repopulation</u> is that, as a Black woman, going into some of these places feels uncomfortable or even risky. And beyond race there are social tensions around outsiders, in general. Plus, in many cases, people who can afford to move to the country and buy land are going to have more resources than the people who are already there, so there's a class dynamic too. It feels like we're setting ourselves up for a clash. You've suggested previously that we can start by actively being a good neighbor. And that sounds nice, in theory.

Brian: On the racial side, I don't know what that feels like. I'm an old White guy.

Ayana: I see a lot of Confederate flags in upstate New York and New England and I can tell you it doesn't feel great.

Brian: For sure, people need to go where their heart tells them they are reasonably safe. But the countryside isn't monolithic. Yes, there are some Proud Boy types out there who are genuinely scary. But then there are a lot of culturally conservative people who vote hard Republican because they're pissed off. I think they're a huge obstacle politically but for the most part you can get along with them and work with them as neighbors—of course I haven't had to do that as a person of color, but I suspect in most cases it wouldn't make that much difference. These folks respect hard work, and once they see you out there working and taking care of business, you've made a connection.

There's a sizable minority of progressive people out in rural areas— I know this from living in Kansas. They've been fighting these fights for sustainable farming, revitalizing small village art centers, what have you, for decades. Wherever someone feels comfortable going, there's already good work and good people out there to latch onto. So what I'm saying to people who want to move to the countryside is, find your people, whoever that might be, or bring your people. Don't feel like you're supposed to go it alone. This requires community.

And if we look at the country as a whole, there are a lot of people of color out in rural areas,[45] including lots of immigrants from Central America and Mexico who provide a lot of farmwork, who are part of this

---

45 According to the Brookings Institute, "24% of rural Americans were people of color in 2020. The median rural county saw its population of color increase by 3.5 percentage points between 2010 and 2020."

solution. It's not as though these places are currently completely White. I think that needs to be said.

To the other point you raise, there are elite White people gentrifying the countryside. They may just want to have a nice country place. And there's gonna be tension and resentment, but again, that's happening anyway. And so to me, the question is how do we utilize that force of urban people who have some wealth, who want to live in the country and say, okay, go live there, but try to make positive changes where you go.

Ayana: Contribute something.

Brian: Well, contribute a lot. Don't put your house on top of the hill, but think about building it closer to the road and helping protect that nice land up on top of the hill. Work with local land trusts. Be part of ensuring that there's affordable housing so that it's not just wealthy people who can live in these places.

I don't wanna be naive. The country is in such a polarized place right now that these relocations are not necessarily going to be happily received by folks who are feeling alienated. But we need more people living in country areas. We need more progressive values for the country as a whole. I can't back away from those kinds of ideals.

I have a certain amount of faith that if people move to rural areas and basically bring economic growth, and that is done in a way that does not drive out the people who live there now, but creates good jobs and there's still land available for working-class people, that on balance it will be received as a good thing.

Ayana: What is holding back this change, the scaling of this approach and solutions?

Brian: Market capitalism.

Ayana: Getting right to the heart of it there.

Brian: It's hard to say, "Oh, I'm going to go live in the countryside," if it's super hard to make a living out there.

Ayana: Are there places where your vision of rural transformation is already unfolding?

Brian: Well, Vermont, to be cliché about it. Around thirty or forty thousand hippies went back to the land in the 1970s, into a rural state that was rock-ribbed Republican. And have you looked at the senators from Vermont recently? It's not as though Vermont has become a kind of a rural paradise. It's still tough to make a living out there. There's still plenty of tension.

Ayana: So, this is a <u>generational project</u>. That's what I'm hearing you say.

Brian: Oh, for sure. Like life itself.

Ayana: What would you say are the top three things you wish everyone knew about climate, climate solutions?

Brian: We totally have to get off of fossil fuels. That's number one. There's no shortcut around that, no magically sucking it all back into the soil or whatever.

Second, it's not like climate is the only thing we have to fix. And it's more than just a technological change. If we get off fossil fuels, but keep the same extractive consuming economy we have now, we're going to create a whole bunch of other problems—the amount of mining we do, or the amount of trash we produce.

Third, there's the justice part of this. The people who are suffering most from changes in climate are the people with the least means to deal with it. And so whatever solutions we put in place in terms of both mitigating and adapting have to be aimed squarely at that.

Ayana: How would you summarize your vision for rural America, if we get it right?

Brian: More people live there again. There are thriving communities. ♥ The agriculture and forestry that are being practiced are sustainable—people grow healthy food and harvest wood in ways that also maintain thriving ecosystems and don't pollute. The people doing this work are reasonably prosperous, live decent lives, and are having a good time.

Give Us Back Our Bones

Erica Deeman

With *Give Us Back Our Bones* (2022–23), artist Erica Deeman envisions a decolonized future that returns to Black and Indigenous belief systems and practices of tending and caring for the land and sea. She finds inspiration from her maternal family's connection to the Jamaican land they farmed before emigrating to the UK in the 1960s as part of the Windrush Generation, as well as contemporary farmers and coastal fishing communities who have been struggling to sustain their livelihoods in the face of climate change. Conceived as a contemplative portal that allows viewers to consider their roles in potential climate solutions, the installation features hundreds of fragments that have been hand-molded from a dehydrated form of gypsum and cast with seeds sourced from Black farmers working in the United States—for instance, callaloo, okra, and rice—as a reference to the Black diasporic journey. These fragments cascade from biodegradable fishing lines, a material that was recently used in grassroots efforts to restore Jamaica's coral reefs. Erica has created a space for reflecting on what we bring into our climate future, what new traditions we might need, and what we should leave behind.

Seeds and Sovereignty

Interview with Leah Penniman

Over the last two decades, my mother has devoted herself to creating a small homestead farm in upstate New York. I've had a front-row seat to the tribulations and glories of raising a large flock of chickens, starting an orchard, and cultivating an expansive vegetable garden. Up with the sun, endlessly shoveling shit, endlessly weeding, endlessly trying to outwit pests and predators, endlessly praying for rain but not too much at once. (Clearly, growing food is hard enough without also having to deal with climate change, with the changes in seasons and the extreme weather it brings.)

Not exactly a chill retirement for my mom, but certainly a gratifying one. She is enamored with the concept of <u>climate victory gardens</u>.[46] It's amazing how much food you can grow on a single acre. We have (in my biased opinion) the world's most beautiful chickens gallivanting among fruit and shade trees, eating bugs and comfrey and our leftovers and whatever vegetables they can pilfer from the raised beds, fertilizing the soil with their poop, and laying perfect eggs.

Like many, I descend from farmers and fishermen when you go back a few generations. Back to when food was real, local, and fresh by default, when there was no need for the term "organic." Today, instead of diverse crops and animals, integrated and nourishing each other and the soil, with flourishing microbiota, we have a high-pesticide, long-distance, ultra-processed, mostly-monoculture scenario. This is unhealthy for us and for ecosystems. Nutrients in some vegetables are up to 30% lower

---

46 Per the nonprofit Green America: "During the victory garden movement of World Wars I and II, Americans planted gardens to feed and support both their local communities and troops overseas. These efforts were wildly successful. By 1944, nearly 20 million victory gardens produced 8 million tons of food—around 40% of the fresh fruits and vegetables consumed in the U.S. at the time. This incredible show of grassroots organizing and community efforts are the inspiration for today's Climate Victory Garden movement." For tips on how to start your own, see: greenamerica.org/climate-victory-gardening -101

compared to the 1950s, because the soil is so depleted. (You have to eat three times as much broccoli to get the same amount of calcium!) The ocean side of things has headed in a similar direction, with overfishing and aquaculture that destroy habitats and have horrific records of human rights abuses.

This is in part a cultural issue. Compared to people in other rich countries, people in the U.S. spend the least on food (as a percentage of income) and the most on healthcare. We expect food at bargain-basement prices, subsidize all the wrong things, and growers are in a race to the bottom to the detriment of farmworkers, eaters, and biodiversity alike. Globally, our food system is the source of 33% of greenhouse gas emissions.

We are told it has to be this way, that agriculture has to be super industrial, in order to feed the nearly eight billion people on the planet. But that's not true. Done right, farming (of both land and sea) can absorb tons of excess carbon, pulling it from the air (where it exists in massive and still-increasing concentrations) and into plants and soils (where it very much belongs). Clearly, there is ample room for improvement.

Before we settle into this next conversation, a few words about soil, the magnificent substrate that can absorb gigatons of carbon from the atmosphere—but only if it's alive, teeming with microorganisms, webbed with roots. To restore our agricultural soils, we need to re-embrace use of regenerative organic practices. That includes keeping roots in the ground, always—never bare and prone to erosion—by reducing tilling,[47] and planting cover crops and perennials. We need to grow a greater diversity of crops, rotate those crops, use compost to replenish the nutrients in the soil, and eliminate chemical pesticides and fertilizers. Re-embracing these practices of regenerative agriculture will nurture healthy soil, which makes for healthy plants that collect atmospheric CO_2 through photosynthesis. This CO_2 is used to both build their tissues

---

47 In no-till farming, seeds are planted through the remains of previous crops, as opposed to tilling, which overturns up to a foot of soil before planting new crops. A no-till approach allows soil structure to remain intact, improving soil's ability to hold on to water and carbon, reducing erosion and runoff, slowing evaporation, and fostering soil biodiversity.

and feed carbon to the microorganisms (aka microbes) at their roots that deposit carbon into the soil. Along with worms and other critters, these microbes (such as bacteria and fungi) allow more absorption of water (instead of runoff), decompose organic matter, and help transfer minerals from the soil into the plant roots. Good for the water cycle, good for the carbon cycle, good for biodiversity. A virtuous symbiosis readying soil for future seeds.[48]

Some version of "back to the land" has to be part of our climate solution: more people farming again. Leah Penniman and her family started Soul Fire Farm in 2006, and they have been manifesting such a vision, growing food in a way that is grounded in community, justice, and food sovereignty. I first learned of Leah in December of 2016, when I got this email from my mother:

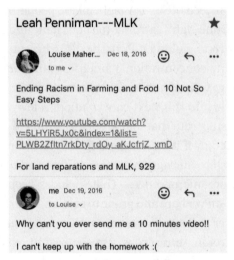

The hour-long video, which I ended up watching in full, was of Leah speaking at a Northeast Organic Farming Association conference. When I interviewed her for this book, she said that it had been her first-ever keynote and she was so nervous, clutching the sweaty paper and feeling like she was all over the place. But even so, she helped me see farming in new ways and brought wisdom, context, and humanity to the question of how our food system can and must transform.

---

48 For more, see: regenorganic.org

Ayana: This seed-braid concept struck me so deeply the first time I heard you speak. Can you share that story? It seems seeds are such a center of your work. Perhaps it's silly to say that to a farmer, because of course seeds are important, but they seem to have a particularly large space in your consciousness.

Leah: Even as you say that, I'm feeling this energy, this thing bubbling up in my heart thinking about how much I love seeds. It's winter now. This is the time of year when the seed catalogs start coming in, and every farmer thumbs through the catalog with this reverent anticipation. The seed is a bundle of genetic promise. The seed is 9,000-plus years of peo-   ✳ ple who pay attention trying to figure out how we coax more food and more joy and more life from the Earth.

Seeds can also be metaphors. Our ancestral grandmothers in West Africa were surrounded by war, kidnapping, enslavement—and finally were forced into these transatlantic slave ships with no report-backs. And amidst this chaos and this horror they somehow found the audacity to gather up the seeds they'd been saving for generations—their okra, molokhia, Levant cotton, black rice, melon, Amara kale, basil— and braided these seeds into their hair as insurance, because they believed against odds in a future on soil.

They were braiding hope into their hair. They believed that their de-   ♥ scendants would exist to inherit the seed. The seed was this precious legacy representing generations of selective breeding and microclimate adaptation and cultural cohesion. This was what they brought and passed on, and what we have inherited and have a duty to steward for our future generations.

Ayana: I love this. It's inspiring because it raises the bar for our responsibility, to each other and the future.

Leah: When I'm faced with hopeless, challenging situations and all that we're up against, I call into mind this powerful story because if these ancestors, facing horrors I can't imagine, still had the hope to carry seeds, then I sure as hell better not give up on my descendants.

There was also ecosystemic and cultural knowledge about being in "right relationship" with land that came with our ancestors. So many of the practices we think of as good farming, good stewardship—things like no-till, semi-permanent raised beds, making pyrogenic <u>compost</u>, terracing, or planting <u>polycultures</u> and <u>cover crops</u>, <u>mulching</u>—these technologies have roots in Afro-Indigenous wisdom. So the seed to me is also the representation of that agrarian genius that is passed down in our lineage. This what-if-we-win, what-if-we-get-it-right future has a lot to do with how we honor the seeds and all of the wisdom that comes with them.

Ayana: And there are traditions of Indigenous farming practices and seed-keeping on every continent. The reason this story moves me so much is that it feels like the same scenario I find myself in when I think about climate. Even though, as a middle-aged person, I'm not going to be around in 2100, wouldn't it be great if all these beautiful people were still eating melons and okra then and enjoying life? Let's act as if, let's bring the seeds.

Leah: We do this climate-beneficial, carbon-sequestering agriculture for the long game. In the meantime, these same practices stabilize soil aggregates, help water to infiltrate, increase biodiversity, make the air cleaner, make the water cleaner, make the environment more beautiful. There's a real-time win even if there's not an eventual guarantee. Plus, as I always say to my children, people with hope who act in a pro-social way are just a lot more pleasant to be around. So I wanna be pleasant to be around compared to a cynic who's given up.

Ayana: Overall, how does our food system work now, and how would it be different if we got it right for people and justice and the climate, from seed to table?

Leah: Today the food system is fundamentally based on stolen land, exploited labor, and a profit motive. Farmworkers, arguably one of the most important professions in the world, are not protected under the Fair Labor Standards Act or the National Labor Relations Act. They don't have a right to overtime pay, to a day off in seven, to <u>collective bargaining</u>, to protections against wage theft or child labor.

Since the beginning of the agricultural system in the U.S., which was based on chattel slavery, workers have always been exploited. There are the huge disparities in terms of land ownership and access, and ownership is rooted in a legacy of attempted genocide and land theft from

Native people and expulsion of Black people from their land holdings starting in the early 1900s.

On the consumer side of things, we live in a system of food apartheid, where your zip code and the color of your skin can determine your life expectancy, your access to fresh food, your disproportionate burden of diet-related illnesses like kidney failure and heart disease. One in four children of color often go to bed hungry. The food system is not based on taking care of people or taking care of the Earth. It's based on extracting a profit for the owners.

A reenvisioned food system, a democratic food system, would have the interests of all the members of our sacred community in mind. Land would be apportioned fairly and owned collectively. Workers would be treated fairly, with dignity, and have an ownership stake in their cooperative. Everyone would have access to culturally appropriate, life-giving, affordable foods that nourish and prepare us for civic life and contributing the best of our brilliance and capabilities. Everyone gets to eat. Regardless of the color of your skin or where you were born, you have enough calories and nutrients and vitamins and minerals. And the soils would be healthy—growing, not washing away. It's got, in our region, at least 10% organic matter and a rich microbiome. The seed is a heritage seed that is not genetically modified, that is passing on a story of a people. The carbon would revert its trajectory and be headed back into the soils. And we would be continually innovating around the question of: How can we do better in serving our human and beyond-human kin?

Ayana: Yes, please. I am into that future.

Leah: My daughter, Neshima, when she was really little she said, The food system is everything it takes to get sunshine onto your plate.

Ayana: Ooh.

Leah: I love this because it makes me think about all the stops along the journey that sunshine takes. The soil, the seeds, the workers, the plants, the processors, the cooks, the eaters. Sunshine is on this whole journey. And if we could infuse democracy and justice and ecological humility into every single step, that's what getting it right looks like.

Ayana: Let's take a step back. You didn't grow up farming, and you were formerly a high school biology and chemistry teacher. You've clearly become not just a farmer but a wisdom keeper, gathering wisdom from previous generations and being deliberate about passing that on. How did this become your work?

Leah: A little bit by accident. I've always had a close relationship with wild things. We grew up three Brown kids in an almost-all-White school district in Massachusetts. And to say that the other children bullied us would be an understatement. Racialized bullying was rampant. So we spent a lot of time in the woods. We befriended the oak and pine trees. We befriended the beautiful Lake Watatic near our house. We spent time with wild blueberries and deer and turkeys, and this friendship was so deep and devoted.

At age six or seven, my sister and I started a Junior Ecologist Kids Club. We'd go around and pick up trash. And we would put our bodies in front of the loggers that were honestly just trying to do some thinning in the State Forest. But we didn't have an analysis of ecosystems; we were doing our best. We had our little blue Huffy bicycles and we did pollution patrol. We had this book, *50 Simple Things Kids Can Do to Save the Earth*. We did all of them. Then we got the sequel, *50 More Things You Can Do to Save the Earth*, and did those, too, including handwriting letters to every address in the phone book of our town of 1,000, telling them what they could do to save the planet.

Ayana: Okay, this is all clicking for me now.

Leah: Yeah. When I was old enough to get a job, my mom found a flier at church about work on a farm, and I thought, "Awesome, that's an Earth thing to do."[49] So I got a job at The Food Project in Boston working with urban and peri-urban farms, growing food, and serving it in low-income farmers markets, and I totally fell in love.

What is cooler than planting a seed and then watching it grow and getting to feed somebody? It made everything clear. Being a teenager is pretty complicated and confusing, but there was something elegantly simple and pure about cultivating a seed and feeding the community. That was the anchor I needed to make it through adolescence.

Ayana: And you did eventually get your own farm, with your family. Tell us about Soul Fire Farm.

♥ 49 When in doubt, ask yourself: Is this an Earth thing to do?

Leah: We are a community farm in upstate New York. There are about fifteen of us on the team at this point. Half of us are working the land, the other half do our national advocacy work. It started in 2005 when our family moved to Albany, New York. We were living in the South End, a neighborhood under food apartheid with no supermarkets, <u>farmers markets</u>, or accessible <u>community garden</u> plots.[50] My spouse Jonah and I were struggling to get fresh vegetables and fruits for our young children. And our neighbors were as well. So when they found out that Jonah and I both had farming backgrounds, it became a running joke. They'd tease us like, *When are you gonna start a farm for us? When are you gonna bring food to us?*

Long story short, we wedded ourselves to these eighty acres of degraded mountain land, about thirty minutes outside of Albany, and got to work building a home, repairing the soils, and in 2010 opened Soul Fire Farm with one program: a doorstep-delivery solidarity share that brought low- and no-cost food to our neighbors in the South End.

That remains our longest-standing program. But over time we started a nonprofit and a co-op and educational programming, became a community farm, developed our advocacy work, all kinds of other things. But foundationally, the farm was started in order to feed folks who deserved good food but couldn't afford it in Albany, New York.

Ayana: What seeds are you dreaming about for next year?

Leah: I love dry beans. I love staple crops very much. There's a solidity to them. I think often of Fannie Lou Hamer, who is well known for her work with the Mississippi Freedom Democratic Party, but not as well known for her incredible work as a farmer and co-op organizer. She said, "When you've got 400 quarts of greens and gumbo soup canned for the winter, nobody can push you around or tell you what to say or do." She said it's not just about the vote, it's about land and food and water and home, and making sure those things are community-controlled because, if you're starving, you will put down your ballots to get a piece of bread. Bread is the power. And that has shaped this fundamental understanding that undergirds Soul Fire Farm: We gotta feed ourselves to free ourselves.

---

50 There are more than 29,000 community gardens in the largest 100 U.S. cities, a 44% increase since 2012, and more than 12,000 schools have edible gardens.

Ayana: There you have it.

Leah: Food is a liberatory tool. We're not dependent on a system that's designed to kill us or extract from us. We're dependent on our own community labor, our own sweat, and this land that we have mutual regard for in order to feed ourselves. And that jazzes me up. That self-sufficiency tip is really exciting.

Ayana: That is exciting. And it's from you that I learned this term, "<u>food sovereignty</u>."

Leah: La Via Campesina, which is a peasant movement, the largest in the world made up of people who are closest to the Earth, came up with this concept of food sovereignty. It's the idea that every person has the right to democratic participation in the food system at all levels. It's a vision for care of the ecosystem as a family member. In a material sense, land is the source of food, the place where we build our shelters, where we gather, where we bury our dead, where we enjoy scenic beauty and healing from nature. But on a metaphysical level, there's something even more going on. I've experienced this firsthand when people come out to Soul Fire Farm for our farmer training programs.

I've gotten proud of my soil chemistry workshops that I teach, and folks walk away understanding cation exchange capacity. But in their evaluations, people who visited Soul Fire Farm kept talking about healing from trauma, kept talking about being able to let go of addictions, leave toxic marriages and dead-end jobs, make different decisions.

And I might be going out on the limb a little bit here, but the land was the scene of the crime, right? It's where all this really jacked-up stuff happened to people of color—chattel slavery, sharecropping. And so it makes sense that our ancestors got the hell outta Dodge and left the red clays of Georgia for the paved streets of Pittsburgh.

But there was something left behind more than intergenerational wealth or deeds. I will put forth that a bit of our souls, our culture, our spiritual understanding of self, our connection to higher power was left in those soils. And the ache of that gap has been eating away at us for generations. The Earth has been crying for the footsteps of her children to come home. And when folks finally taste it, a little bit of it, a flood of remembering comes back. *I am somebody. I'm worth something. My life matters. I have choices. I have options. I belong here.*

There is a deep yearning to belong to land as home without which I ♥ don't think we are complete as people. Land is everything.

Ayana: You open your first book with this quote from Malcolm X: "Revolution is based on land. Land is the basis for all independence. Land is the basis of freedom, justice, and equality." I have to admit, I was surprised these were his words. I bet others will be too. But I certainly agree.

And I remember visiting your farm, walking up this dirt road, through the trees, and then coming into this clearing and seeing the house that you and your partner built, and seeing rows of luscious vegetables and seeing your butterfly gardens and the ways in which you are supporting the bees, and having you show me around and introduce me to your plants. To me it felt like, *Ohhh,* this *is the ultimate luxury. This is what it* ✳ *means to be rich.* Standing at your front door and looking out along your rows of food gave me a sense of what it could look like to be free.

Leah: You're making me realize that there's an ironic truth that the freedom in belonging to land is a freedom that comes with being bound and in service. That is ultimate freedom. It's something that grows inside of you, year after year, as you bind yourself up more and more deeply and inexorably with the land that has claimed you and that you have claimed. You are bound, you are wed, you are married, and you're also free at the same time.

Ayana: You are creating a way of doing things that is in deep contrast with the way our food system is set up right now. What is it that you are working to subvert or invert?

Leah: I will say two things that need to be inverted. One of them relates to the tragic irony that you need to be certified in this country to be organic. You pay money and you have someone check up and make sure that you're nurturing the soil and not adding harmful pesticides and herbicides and that you're treating animals with dignity.

Whereas you can go ahead and ravage the atmosphere, putrefy fresh ! and salt waters, destroy topsoil, exploit your workers, and poison biodiversity with pesticides—and there's no certification required. There's no taxation of your externalities.

Ayana: And worse, that's now called "conventional" agriculture. That really gets my goat.

Leah: Right. And further, the U.S. Department of Agriculture will mightily subsidize your operation because you're providing commodity crops and because you're at scale. The entire system is set up for "go big or get out." Instead, imagine a world where you had to pay and get special permits to trash the planet—industrial agriculture would have to go through all of these hoops to justify why you need to spray poison all over the place or destroy the organic matter in your soil, and the subsidies and the supports would go to those of us who steward the public trust, those of us who are creating pollinator habitat and sequestering carbon and protecting watersheds . . .

Ayana: And preventing erosion of topsoil . . .

Leah: Exactly. That's one thing that needs to be inverted. And that would benefit smallholders, Indigenous people, Black and Latine farmers who disproportionately are doing that type of agriculture, growing what the government now relegates to the category of "specialty crops." That's what they call all the healthy food.

Ayana: Where's the carrot lobby? Let me join the carrot lobby. I will show up in D.C.

Leah: Exactly. Where's the collard greens lobby?

The other thing that needs to be inverted is land ownership. According to the 2017 USDA census of agriculture, over 95% of the agricultural land in this country is White-owned. Which means that every other racial group is sharing less than 5% of the pie, of the land that feeds our nation.[51]

Ayana: You created an incredible book, *Farming While Black*, and your introductory chapter is called "Black Land Matters." That term seems to be an organizing principle for your work.

Leah: Absolutely. I was talking to this fantastic young Black organizer and writer, Brea Baker, who's writing a book about Black land.[52] And she told me something that I can't get out of my head. She said in her family they have a large land holding, and the Black elders who live on that land

---

51 Meanwhile, 85% of farmworkers in the U.S. are Hispanic or Latine.

52 This book is now published: *Rooted: The American Legacy of Land Theft and the Modern Movement for Black Land Ownership*, by Brea Baker.

live, on average, ten years longer than the family members who moved to urban industrial areas. Not only living ten years longer, but she went down to interview some of her uncles and aunties and they would say, *You want me to chop you some firewood? Go harvest something from the garden for you?* Like active, healthy elders.

This is the reason why the National Black Food and Justice Alliance, of which Soul Fire Farm is a member, is working on a Black agrarian commons to figure out how to reclaim the 16 million acres that were taken from Black people, so we can realize the benefits that come with access to green space and a long-term secure homeland that has eluded our communities for generations.

Ayana: Tell us more about that 16 million number. It astounds me that that much land was owned by Black people in the first place. How was it acquired, and how did it get taken away?

Leah: Somehow, *somehow*, our ancestors, in spite of debt peonage and convict leasing and racial terrorism, figured out that they could work two, three extra jobs, and save money over generations to purchase these scrappy, marginal, smallholder plots. And those eventually added up to 16 million acres. The peak of Black land ownership in this country was in 1910, with 14% of the nation's farms being under Black control. ✱

Ayana: And what are we at now?

Leah: Oof. Around 1.5%.[53]

Ayana: I better buy some land.

Leah: Buy some land. And for folks who have land in their family, don't let it go. But the reason that there's so little land left now is because of violent backlash to that era of peak land ownership. The Klan, the White ! Citizens' Council, the Whitecaps lynched Black land owners, burned down their homes, and drove them off their land for the audacity to leave the plantation. It was a threat to the White Southern way of life, so many Black land owners were murdered, lynched. We know the names of about 4,000 people who were murdered in Southern states between the 1870s and 1950, but there were many more. And that was a push factor

---

53 For context, 16 million acres is about twice the size of the state of Maryland, or larger than the state of West Virginia—a lot of land when all those plots were added together.

for the Great Migration, when 6 million Black people fled the South to urban areas in the North and West.

Ayana: It wasn't that Chicago seemed fun.

Leah: Yeah, that's a big risk, right? To take your whole family up to the cold. But the other major factor was the U.S. Department of Agriculture's discrimination. The USDA is responsible for shoring up the food system, so they provide grants, low-interest loans, technical assistance, crop allotment, and more to farmers. Over generations, White farmers were getting these goodies and Black farmers would be delayed or denied so systematically that the U.S. Commission on Civil Rights published reports in the '60s and '80s that said the decline of the Black farmer was on the hands of the USDA. And this was so terrible in the 1960s that if a Black farmer was found to have registered to vote or joined the NAACP or signed a petition, they would almost certainly have their loan applications denied.

Ayana: Part of your vision of the future is about how we start to bring that ownership level back up. How does that transformation happen?

Leah: I'm tantalized by this question because I get to dream into it. I think we give Indigenous land back first, actually. We have to recognize that this entire beautiful continent of Turtle Island was violently wrested from the hands of Indigenous communities. Most are unceded territories. We have to figure out what that return to tribal sovereignty looks like, and forge collaboration between Indigenous, Black, and other communities of color in reimagining land stewardship.

The good news is it's already happening in pockets. You see it in the work of organizations like Sogorea Te' Land Trust, the Land Back movement, Eastern Woodlands Rematriation Collective, the Native Land Conservancy, and First Light. Individuals are coming together to say, "We're done with this whole idea of private property and the idea that a single individual can control what happens to the land from underground minerals up to the column of air above. That doesn't make any sense."

And so the land trust movement is working to address this. Here I have to shoutout Baba Charles Sherrod, who recently passed away, and Shirley Sherrod. They started the first-ever community land trust in the United States, as Black farmers in the '60s. They brought together 500 Black families on 6,000 acres to figure out how to hold the land as commons but still have equity to seed intergenerational wealth. Brilliant, genius work.

Ayana: Wow.

Leah: Now there are hundreds if not thousands of community land trusts. But that first one, the one the Sherrods developed, was attacked. Talk about cruelty: The White population's backlash was intense—they killed their hogs, they diluted their fertilizer, they poured water in their gas tanks. It was a whole mess.

But focusing on solutions, I'm super excited about the land trust ✶ movement. We were one of the founders of the Northeast Farmers of Color Land Trust, which is an Indigenous and Black collaboration that's working on land return across the Northeast.

Ayana: Where is that land coming from that's being transferred back?

Leah: The right thing to do would be wholesale land reform, where the government steps in and does some redistribution. The closest we've gotten is a provision in the Justice for Black Farmers Act of 2021, introduced by Senators Cory Booker and Elizabeth Warren, that we were able to help draft along with hundreds of other farmers. The Act provides a fund to purchase land off the market and put it into the commons. That legislation is held up in court by charges of reverse racism because of provisions for debt forgiveness for Black farmers. In the meantime, what we're doing in the private sphere, the community sphere, is encouraging good-hearted people who know what's right to donate their land or to will their land upon their passing to these land trusts or directly to tribal governments.[54] And lots of folks are stepping up.

Ayana: When I think about the way that agriculture in the majority of this country is happening right now, it's this large-scale industrial approach that's heavily reliant on fossil fuels for machinery, for fertilizer, for shipping and processing. It's a significant part of the climate problem. And we know that on the flip side, agriculture could be a significant part of the solution. Sequestering carbon instead of emitting it, supporting more jobs and more biodiversity, reducing erosion of topsoil. There's so much potential.

---

54 A directory of Indigenous-run land trusts: oregonlandjustice.org/indigenous-land
-trusts

* **Leah:** It's all connected. Fundamentally, the philosophy that allows us to oppress human beings is the same philosophy that allows extraction from the Earth. When you designate something as "other" or "non-kin," that gives permission to ravage that thing. And when we re-embrace people of all cultures and faiths and races and backgrounds as kin and simultaneously re-embrace the soil microbes and the hawks and the white pine trees as kin, it is impossible to ravage. And I'm hopeful. I'm hopeful based on evidence. It's a grounded hope.

When we got to Soul Fire Farm in 2006, we signed the purchase papers, we put a shovel in the ground and heard a loud clink because the topsoil was almost completely washed away. We were at hard-pan subsoil, leached clay. It doesn't smell like anything. It forms a brick in the sun. We literally built our house out of it. It's great building material for a straw bale cobhouse.

When the USDA agent came out, because we were applying for our farm number, they ranked our soils. We got the lowest ranking. That actually makes you ineligible for a lot of funding. And they said, you can't farm out here. You can't grow vegetables out here. The best you can do is cut your forest for timber and then wait thirty years and come back. There's really no hope. But we had been urban farmers working in areas with lead contamination and on top of concrete, so we knew about remediation, we knew about these Afro-Indigenous technologies. And if you reach your hand into our soil now, it is a rich, black, dark, beautiful humus that smells like a rainstorm in the forest.

Ayana: Black gold.

Leah: It's black gold. Our vegetables are beautiful. The USDA agent was wrong. We can make things better. Humans don't have to be a blight on
! the Earth. But right now agriculture is responsible for around one-third of the greenhouse gas emissions, 70% of freshwater withdrawals, half the land use, and 78% of the eutrophication.[55] It's a hot mess.

*   Instead, if everyone were to farm regeneratively, for every 1% increase in soil organic matter, that would absorb around 8.5 metric tons of at-

---

55 Eutrophication is defined as excess nutrients in a body of water, which is often caused by fertilizer runoff from industrial farms, resulting in algal blooms and dead zones—not to mention pesticide runoff screwing up marine ecosystems with toxicity.

mospheric carbon per acre. That adds up. That carbon gets absorbed out of the atmosphere, back into the soil where it belongs—that could be a hundred gigatons of the carbon solution in this coming generation. That's not insignificant. As Larisa Jacobson, one of our former farm managers, said, our job as farmers is to call the carbon and call the life back into the soil. That's our number-one duty as farmers, to call the carbon home.

Ayana: I love that, because we've sort of vilified carbon, but carbon is the building block of all life. It's just in the wrong place. It's in the atmosphere as methane and CO_2, causing the greenhouse effect. We need it in the soil and in the bodies of living things.

Leah: Exactly. Carbon is life. We need it in our roots and in the nematodes and in the fungi and bacteria. We need it back in the soil ecosystem where it's doing immense good, feeding us and stabilizing the soil when the waters come. That's what carbon is up to. Making these sticky exudates and aggregates that make everything healthy and yummy and smell good.

Ayana: It strikes me that so much of this is not in fact new. My mom always says, "We need to go back to the future"—she's never seen the movie, she has no idea that the phrase has a whole other meaning. But she means we need to go back to nature and take some ways of doing things from the past and leapfrog that over whatever mess we're in now to get to the future that we want. And so I'd love to hear if you have a version of that, of combining old and new.

Leah: Brilliant. I wanna hang out with your mom.

There is a catching up to our ancestors we need to do. But these practices are not frozen in time. If we were just doing old things, we would have teosinte and not maize. Our ancestors were plant breeders, they were innovators. They were constantly creating new techniques, like African black earth, which is a sophisticated compost generation method.

We want to base our innovation on these values of ecological stewardship, of care for the workers and care for the Earth, and not on the warped, in my opinion, values of maximizing profit and extracting resources as quickly as we can to get maximum yield in the short term. Those are the wrong values. And right now there's an under-investment in resources in trying to figure out how we do sustainable better.

Ayana: To make all this possible, what needs to change in terms of policy, in terms of how corporations operate, in terms of activism? What needs to happen first?

Leah: We need to short-game and long-game it. I'm going to go metaphorical and then specific.

Ayana: I love it. Bring it on.

*   **Leah:** We think about change in terms of this butterfly of transformative social justice. This butterfly has four winglets—butterflies cannot fly with one, two, or three; they need all four.

One of the winglets is "Resist." This is directly confronting oppression. That's the blockades, the <u>strikes</u>, the protests, the <u>boycotts</u>. That's necessary. We need to get in the face of oppression. You will not put this pipeline through my community; I will chain myself to it.

Another winglet is "Reform." This has to do with policy change, getting inside of our institutions and changing the narrative; making change from the inside. This is a lot of the slow, painful bureaucratic work.

And then we have the winglet that I situate myself on. That's "Building." Building alternative institutions that try to model our higher values. The freedom schools and co-ops and land trusts and community farms and free libraries.

Ayana: The seed banks.

Leah: Yes, exactly. And the final winglet is "Heal." Because there's no way we can go through 500 years of this BS and not be completely traumatized. We need art and therapy and <u>ritual</u> and spirituality and collective healing. And we need every farm to sign up to be <u>food justice certified</u> for workers' rights.

Ayana: There's an actual certification?

Leah: There's an actual certification. Just like organic, where you get inspected and make sure you're not trashing the planet, there is a certification to make sure that you're obeying labor laws at minimum, and treating your workers fairly.

Every farm needs to be food justice certified, and institutional purchasers should only be supporting farms that are certified. We also need farmworkers protected under the Fair Labor Standards Act and the National Labor Relations Act. Like immediately. The only reason that there is an exclusion for farmworkers is because when that legislation first passed in the 1930s, Southern Democrats would not vote for it if it benefited people of color, who were the farmworkers. It was overtly racist and we somehow haven't figured out how to un-racist that yet.

Ayana: Racism, still holding us back.

Leah: Another thing we need to do is fix SNAP, the supplemental nutrition assistance program, which provides food aid to low-income people.[56] Right now it is underfunded and too onerous to apply for and the allocations are small. We need to make sure that food is a human right in this country. We need to fully fund SNAP, and also make sure that our daycares and prisons and elder-care centers have fresh, healthy food. If we treated junk-food companies like tobacco companies and started making them put warning labels and pay extra taxes for trashing people's health, that'd be a step in the right direction.

Ayana: The carrot and collard lobbies need to get on this.

Leah: Exactly. Another thing we need to figure out is land reparations and rematriation and giving land back and giving stolen resources back. There is a bill that has been languishing in Congress for a long time called HR 40 that would authorize studying reparations. We could start by being courageous enough to have the conversation about what it would be like to give back the trillions of dollars of stolen wealth to Black communities.

Ayana: And HR 40 is named after 40 acres and a mule, right?[57]

Leah: That's right. And other pieces of legislation that need support are the Justice for Black Farmers Act, the Fairness for Farmworkers Act, and the Urban Agriculture Act, which would provide some resources to start to correct some of these historical wrongs.

And the last thing I'll mention is around ecology. Maybe one of the

---

56 SNAP is also sometimes known as Food Stamps.

57 As Leah relayed it to me, "The post–Civil War idea of 40 acres and a mule originated in the Black community. Reverend Garrison Frazier and other clergy met with Union generals to plan Reconstruction and essentially said, 'What we need are homes and the land beneath them so we can plant fruit trees and say to our children, these are yours.' And that became the basis of this idea of 40 acres and a mule, a land reparation. A little bit was given, all was taken back. And in fact, where reparations were given, they went to the so-called masters for so-called loss of their 'property' in many states."

only good pieces of work that the USDA does around the environment is the Environmental Quality Incentives Program. If we could make that most of what the USDA is doing for agriculture—stop subsidizing commodity crops, industrial ag that trashes the planet, and instead give a lot more resources to incentivize regenerative agriculture, like cover crops—we'd be in good shape.

Ayana: I love this concrete list of things that need to change now.

Leah: The good news is that since the food system, as we've said, is everything it takes to get sunshine onto your plate, there are a lot of right answers to the question of what people can do to help.

Ayana: Are there cultural shifts that we need to see as well?

Leah: I'm going to tell a story and then I'll bring it around to the cultural shift. One of my all-time heroes is ancestor Dr. George Washington Carver, who's famous for peanut and sweet potato inventions. Apparently he also discovered at least two species of fungus. This guy was super cool. He's the father of regenerative agriculture in this country. But something I learned more recently about him is his deeply spiritual nature when it came to his relationship to the forest. He would go out every morning before dawn, in the dark, into the forest, to listen to what the trees and flowers and soil had to say to him.

Ayana: Damn.

Leah: When he was asked, "How'd you come up with these thousands of peanut inventions and ideas?," he said, *Oh, the peanut told me what to do.* He thought of nature as God's radio. In all seriousness, he said, "I love to think of nature as unlimited broadcasting stations, through which God speaks to us every day, every hour."

Ayana: Tune in.

Leah: Tune in. Listen to learn again the language of the Earth. So I set myself to the task of talking to people like you and others who still remember the languages the Earth speaks. The ice cores are language. The birdsong, or lack thereof, is a language. The voice that we hear whispered in our dreams is a language. And the Earth is the primary source. As environmental expert Audrey Peterman puts it, we've started to play this dangerous game of telephone with the Earth where there are only a few people who hear what the Earth has to say. And then they tell someone else who tells someone else, who tells someone else. And like five thousand people later, we have a scrambly message about what it is we're supposed to be doing. And a lot more of us need to get literate again in the language of the Earth. That's the number-one cultural shift. ✳

But it's happening. There are so many folks, so many grassroots, homegrown, earth-listening, community-accountable people who are remembering the lessons of those seeds that were braided in our ancestors' hair, about how we take care of the Earth and how we take care of each other. And it's profound, but it's also simple: Treat all beings like family. Full stop.

Ayana: What are the top three things you wish everyone knew about this intersection of food and climate and justice solutions?

Leah: Thing number one is that the Earth is madly in love with you and ♥ you need to participate in this romance. We need to get back with the program of our relationship with the Earth.

Ayana: Hug a tree and mean it.

Leah: And mean it. Breathe with the tree. When I was a kid, I learned about photosynthesis and respiration. I had this idea that if I exhaled on the tree, the tree would directly take that in and give me back some oxygen. I had these whole breathing sessions with trees as a child.

Ayana: I wish we had been friends as children. I would've been like, *Let's go breathe on trees together.*

Leah: Oh my goodness, yes. You would've been a member of the Junior Ecologist Kids Club. For sure.

Thing number two is during the pandemic, society labeled farmworkers as essential. But it's not just the work that is essential; it's the lives, it's the people. And we need to immediately figure out how to have equal rights and dignity and justice for those who grow our food.

And number three, as Queen Quet of the Gullah/Geechee people says, the land is our family member. It's our kin. And the struggle we are in right now is to get our family back. And so the project of rematriation of land for Indigenous people, of reparations and land return for dispossessed Black and Brown farmers, is a spiritual as well as a material imperative.

Love the Earth back, treat farmworkers fairly, return the land. Those are the top three orders of business.

IF WE
BUILD IT . . .

If it can't be reduced, reused, repaired, rebuilt,
refurbished, refinished, resold, recycled, or
composted, then it should be restricted,
redesigned, or removed from production.

—Pete Seeger

What if climate adaptation is beautiful?

What if we value culture as a form of resilience?

What if all infrastructure is climate-smart?

What if we deconstruct instead of demolish?

What if technology is a tool instead of a mode?

What if museums are hubs of R&D for society?

What if we dream at landscape scale?

What if . . . ?

10 Problems

- Buildings and construction account for 38% of CO_2 emissions globally— cement manufacturing alone accounts for 8%.

- The construction industry is the world's largest consumer of resources and raw materials. 35% of construction and demolition debris is disposed of directly into landfills.

- Development in hazardous flood-prone areas has increased by 122% globally since 1985, outpacing the growth in flood-safe areas.

- In urban areas, extreme heat exposure has tripled since 1983, with warming 29% faster than in rural areas due to the dense concentration of dark pavement and buildings.

- Low-income neighborhoods in the U.S. have on average 15% less tree cover (62 million fewer trees total) and are up to 4°C (7.2°F) hotter than higher-income areas.

- Currently, no single U.S. state has a sufficient supply of affordable rental housing.

- In the U.S., transportation is responsible for the largest share (28%) of greenhouse gas emissions, with cars and light-duty trucks accounting for 57% of those emissions. Globally, transportation is about 20% of emissions.

- Manufacturing fossil-fuel-based plastics generates 3.4% of global greenhouse gas emissions, and rising. From 1950 to 2022, annual plastic production grew from 2 million tons to 400 million tons, and is projected to triple to 1.2 billion tons by 2060.

- Globally, only 7.2% of all materials are reused, and only 9% of plastic waste is truly recycled—actually turned into new things.

- Data centers in the U.S. consumed about 2.5% of the country's total electricity demand in 2022, and that is expected to triple to 7.5% by 2030.

10 Possibilities

+ A circular economy (based on the principles of using less, longer, and again) could reduce the demand for virgin material extraction by around 30%, reduce associated greenhouse gas pollution by almost 40%, and create 6 million jobs globally.

+ About 75% of the infrastructure that will be in place in 2050 has yet to be built. Building for climate-resilience adds about 3% to costs, but the benefits outweigh the cost by about 4:1.

+ Nature-based solutions received only 0.3% of overall spending on urban infrastructure, despite costing on average 50% less than gray infrastructure alternatives.

+ Designing steel structures for disassembly and reuse could reduce demolition-associated energy use by 70% and associated greenhouse gas pollution by 80%.

+ Building with wood instead of concrete and steel could reduce associated fossil fuel consumption by up to 19%, and carbon emissions by up to 31%.

+ Installing green roofs in urban areas could lower indoor temperatures by up to 18°C (32.4°F), and retain up to 88% of stormwater runoff.

+ High-speed rail is 3 times faster and 4 times more energy-efficient than driving, nearly 9 times more energy-efficient than flying, and can reduce emissions by up to 90% when powered by clean electricity.

+ Use of EVs cut oil demand by about 1.8 million barrels per day in 2023—equivalent to roughly 2% of the global supply—and could cut demand by 12.4 million barrels per day by 2035.

+ Currently 80% of solar-panel materials (glass, aluminum, silicon, copper, and silver) and over 95% of metals in batteries can be recycled.

+ Efficiency measures could reduce U.S. energy use and greenhouse gas emissions 50% by 2050 and save more than $700 billion.

Neighborhoods and Landscapes

Interview with Bryan C. Lee Jr. and Kate Orff

Walking through the streets of New York City with my architect father shaped how I see the world. He would point out the shoddy construction and cheap materials of a new building. He would point out the Chrysler Building as a paragon of design: "Look at the strong shoulders on that one." I was fascinated by the blueprints on my dad's desk. I was curious about construction sites. And I was surprised that when a street was dug up to repair a pipe, there would be red-brown soil under the asphalt— I expected it would be concrete all the way down. Even though I grew up with a vegetable garden in my inner-city backyard, making mud pies and playing with worms, it somehow didn't compute that we build cities on top of nature. Now, decades later, I find myself pondering what it means to live not on top of nature, or with it, but *within* it.

When I think about great architecture of the past, what comes to mind for me are stark, dramatic, imposing, and often egotistical monuments. Alternatively, what we need now are projects where you can't even really see the sophistication behind them—designs that sequester carbon, provide passive temperature control, absorb water and withstand flooding, and offer a home for bees and other biodiversity.

Architecture and design carry significant blame for our disconnect with nature, and for maladaptively turning extreme weather into disasters. They must also, of course, be part of our reintegration and resilience. Landscape architecture in particular could help us restore our relationship with nature at the same time as we restore ecosystems. As Kate Orff has put it previously, "Mending our ecological infrastructure, and pairing it with renewable energy, should become the highest priority for the profession of landscape architecture."

So, how do we make climate <u>adaptation</u> beautiful and unobtrusive and egoless? How do we design and build for the future, for the next hundred years or the next millennium, in line with scientific projections for heat, flood, fire, and sea level? How do we design for community cohesion and re-entwining with nature? How do we build enough green and affordable housing to meet the massive current gap? How do we create what people actually need, from materials that make ecological sense, instead of just building something cool-looking?

Let's have some foresight and some humility. Let's also embrace opportunity. The world's building stock is expected to double by the middle of this century.[58] About 75% of the infrastructure that will be in place in 2050 has yet to be built, and 55% of the global population, and growing, currently lives in cities. We can transform the built environment in climate-wise ways. And the shift is beginning—building codes for new construction are starting to ban gas lines and require good insulation. In the U.S. alone, there are 140 million existing homes that will need to be retrofitted—a gargantuan task that will be made a bit easier with tax credits to homeowners via the Inflation Reduction Act.

When contemplating these design challenges of where we live and how we live there, my mind often goes first to the coasts, my personal frame of reference. Eight of the ten largest cities in the world are coastal. There, gray infrastructure has replaced more than half of the natural shoreline ecosystems, aka green infrastructure. Coastal cities are largely unprepared for the climate changes that we know are coming and the ones that are already here. This is not a "coastal elite" issue. Twenty percent of Americans, 65 million people, live in coastal cities, and 40% live in coastal counties. Building ever-higher walls to fortify against rising sea levels is not viable. We simply cannot hold back the entire ocean.

In October 2022, while the tail end of Hurricane Ian doused New York City, I hosted an event to discuss the future of coastal cities. At Pioneer Works, an arts institution in Red Hook, Brooklyn (in a building that was flooded with several feet of water during Superstorm Sandy), I sat down with landscape architect Kate Orff and design justice pioneer Bryan Lee, and an audience of several hundred, to dissect the role of architecture in caring for communities in the context of our changing climate.

Kate and I met when we served together on the board of the Billion

---

58 Mostly due to their buildings, cities consume about 75% of world energy and are responsible for 70% of global carbon emissions.

Oyster Project.[59] I was enamored with her MacArthur genius award-winning visions for coastal restoration and the work of her landscape architecture firm SCAPE, which aims to "create positive change in communities by combining regenerative living infrastructure and new forms of public space." Bryan and I met at a TED conference in 2018, where he led a design justice workshop and I learned about his innovative, New Orleans–based design firm, Colloqate, which exists to "design spaces of racial, social, and cultural justice."

I've gotten to know Bryan and Kate over the last few years as they graciously accepted my invitation to serve as advisors to Urban Ocean Lab, the think tank I co-founded. This conversation was the chance I had been waiting for to delve into their expertise and insights. It was the first interview conducted for the book, and the only one conducted with a live audience, so you might sense a slightly different vibe.

––––––

Ayana: Kate, can you paint us your vision of the world if we get it right on landscape architecture?

Kate: For me, getting it right would mean regions around the United States rallying in support of transformative, connective climate-infrastructure landscapes, and working together to create them. For decades, we've been chopping down the forests that clean our air and absorb water, and plowing down protective dunes that buffer us from storms. Now it's time to reverse that and rebuild these landscapes that protect us. Healthy landscapes are climate infrastructure.

For instance, we could create a beautiful, just, and protective American Shoreway along the Eastern Seaboard. Or reconnect and give room to the Mississippi River as a park. Right now, the Mississippi River has levees along most of the lower way, and in New York Harbor the 500-plus miles of shoreline are largely bulkheaded. Instead of building all these walls and hard edges, if we got it right we would build con-

---

59 The Billion Oyster Project has restored 122 million oysters (and counting!) to New York Harbor—aiming for 1 billion by 2035. For context, in the 1600s the harbor is thought to have contained several trillion oysters—half the *world's* oyster population. See: billionoysterproject.com

nective, transformational, water-driven projects and, taking our cues from dynamic coastal landscapes, try to knit ourselves back together and revive some of these waterways at the same time.

Ayana: This is an exciting vision for American landscapes. We don't usually talk about projects on the scale of the entire Great Plains, of the Rocky Mountains, of the Eastern Seaboard. The work, the visions, seem to get caught up in jurisdictional boundaries as opposed to representing ecological sensibility.

Kate: Yes! You can imagine that each of these regions would begin to come together and coalesce around some key regenerative landscape initiatives. This is the work of our generation. We have to rally in support of the natural world before the landscapes that have sustained us are lost, leaving us even more vulnerable to climate change, food systems collapse, erosion, extreme heat, and so on. We've inherited the work of a very different generation, and I'm not willing to accept the world as it is now as a given. We can repair and heal what has been lost.

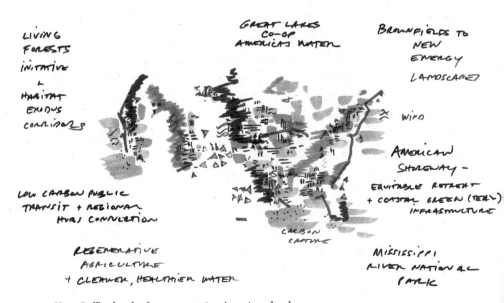

Kate Orff's sketch of a regenerative American landscape.

Ayana: Bryan, I'm curious to hear your answer to this same question. When it comes to "design justice," what does the future look like, the future that's full of good possibilities?

Bryan: One way we explain our work is to establish that design justice is ♥ what love looks like in public spaces. It's an active display of care for beloved communities. The sentiment is an adaptation of the Cornel West quote stating "Justice is what love looks like in public."[60] While Kate is looking at a regional scale, our work considers the scale of the city block, the neighborhood. Our approach is to start with envisioning dynamic development of healthy blocks and neighborhoods and then scale up to healthy cities, healthy states, and so on.

Ayana: This challenge of integrating the scales at which we live, from neighborhood to landscape, is fascinating because we need to be a part of ecosystems but many of us are based in urban places. This reweaving and reopening of the possibilities for nature to thrive in cities is, perhaps, a microcosm for what needs to happen at the national level.

Bryan: Yeah, and the way we've structured society lends itself toward entropy, lends itself toward us tearing apart at the seams and losing connections to one another. Many of us have fewer and fewer friends, we see them less and less.[61] So we need to think about how we stay in proximity ✱ to the things that we love and care for and, ultimately, need to live a fulfilling life.

Ayana: Which is critical in the context of climate change. Who's gonna come for you when a hurricane hits? Who's gonna be the first person to help you? It's your neighbors, it's your family. Emergency management expert Dr. Samantha Montano says one of the best things you can do to prepare for a disaster is to bring your neighbor a basket of muffins, because you have to get to know them before the storm.[62] Building community cohesion and resilience is important.

Bryan: This word "resilience," just like "sustainability," everyone gets tired of it. But it's important to recognize that resilience is just culture. ♥ Culture is the habits, tendencies, patterns, and routines that we collectively share; it's the consequence of persistent circumstance and prevail-

---

60 More fully: "Justice is what love looks like in public, just like tenderness is what love feels like in private."—Cornel West

61 Close friendships among Americans have declined over the years, from 3% saying they have no close friends in 1990 to 12% in 2021.

62 Stay tuned for an interview with Samantha in the "Community Foremost" section.

ing conditions. It's crucial that we design and create spaces where culture can thrive.

Ayana: I know both of your answers to this "What if we get it right?" question have evolved over time. Can you share a little bit about that evolution?

Kate: As someone living in New York, and loving the watery environs of the city, Superstorm Sandy was a major inflection point for me. In 2012, when the storm came, I had a lot of preconceived ideas about how it was going to play out and what we were going to work on afterwards—mostly along the lines of physical projects that I had always wanted to do on various parts of the shoreline. But then, as part of this post-Sandy reck- * oning, I had a huge shift toward this notion that infrastructure is other people. I realized that our social lives and our connectedness to each other are going to do as much to protect us from whatever risks we face as any physical project.

All these other things—behavior change, doing less with more, look- ing out for your neighbor—that seem to be not the thing are actually the main thing. That's been the biggest shift for me personally. I guess the old adage applies: If you're a hammer, everything is a nail, right? If you're a spatial designer, everything has a spatial solution, but I don't think like that anymore. My work aims to combine social movements with built- landscape transformation.

That's reflected in our Living Breakwaters project, which is under construction out in Raritan Bay, off the coast of Staten Island. The origin of the project was in the trauma that followed Sandy. That storm stopped New Yorkers in their tracks and made us realize that we live in a coastal city. After Sandy an idea arose that we should build a barrier across Ver- razzano Narrows and spend billions of dollars walling off the water. But I wondered, is this just? And what exactly does it mean to "wall off the water"?

After a series of research projects, my team at SCAPE concluded that part of what we need to do is reset our relationships to each other and the water. What if we approached the coming decades starting with the idea that we have to reconnect?

A <u>living breakwater</u> is basically a half-mile linear chain of break- waters that slow and clean the water, and help replenish the shoreline.

The idea is that these structures reduce risk and also bring back some of the subtidal and intertidal marine life that once thrived in these waters. These constructed tide pools with multiple scales of intricacy mimic the biocomplexity of an actual oyster reef. And we're seeding these break-waters with oyster larvae. So they'll protect the coast, but also be a big draw for harbor seals, fish and crabs, et cetera. That's the concept.

Ayana: When we think of New York City's waters, we should think of seals and seahorses and whales and all the cool stuff that hangs out here. The harbor is actually an incredible restoration success story—it's cleaner now than it's been in a hundred years, and aquatic life is replenishing, thanks largely to the Clean Water Act.

Kate: Yes, and the other part of the story here is that beyond this physical infrastructure, we're helping to re-energize communities on shore. We're helping create a network of schools and science teachers, and a shared curriculum around habitat restoration, bringing people to the shore with the Billion Oyster Project to restore oyster reefs that had disappeared.

Normally, physical infrastructure, programming, and social infrastructure are all segregated, but in the Living Breakwaters project we're aiming to loop them all together. Helping people in the surrounding communities to understand their immediate physical environment and each other may be the most important thing we can do to reduce risk. This is very different from the concept of protection that we've inherited from engineers of a different era—like 14-foot sea walls that were products of top-down planning.

Ayana: Bryan, as someone based in New Orleans dealing with not just the fast-moving disasters of hurricanes, but also the slow creep of sea level rise, what do you make of the prospect of climate-driven relocation, often called "managed retreat"?

Bryan: It's always going to be difficult for people who are rooted to the land, any land, to move by force of nature or capitalism. New Orleans is a place where if you ask anybody about moving, they'll tell you they are not leaving. Seventy-seven percent of people who are born there, die there. They are not going anywhere. So environmental justice organizations working in the city are not addressing that relocation question in the same way that national climate activists might.

Kate: And there's this framing of "equitable managed retreat" and no matter how you say it, it doesn't ring right. What I've seen is that this process can be incredibly slow and painful. I've also seen that although finding a path of action and relocation is difficult, doing nothing is even more unjust.

We have yet to put together the legal, financial, landscape, social, and spatial tools to create a process and a national framework in which people can equitably and voluntarily opt in to relocation. And we don't have a landscape vision for what would happen when that potential retreat would occur. We need to weave together multiple kinds of tools to be able to create an enabling environment for the people who want to move, who say, "You know what, this is the third time my house is flooded, so I need a different opportunity." This is a profound project.[63]

Bryan: One way I see this is through the work I do with some houseless communities across the country. What you see if you talk to folks in that particular condition is that people will offer them housing or what is perceived as a way out, and it'll be turned down. Because the primary source of capital they have is social capital, and social capital is relative to location. The same thing can be said of any community that faces the daunting task of surviving the ever-present placelessness of capitalism and

* climate change. Black folks are not going to leave a place facing climate risks if all of our community, all of our social and cultural capital exists in that space.

Ayana: What's important for us to consider as we go about making this climate-landscape-design-justice transformation we need?

Kate: Time. We need time to do the work of climate adaptation. And hopefully this work, while hard work—digging the holes and planting

* the forest and seeding the oysters—can be fun. There's so much to be said about the green-blue jobs of the future being enticing, being better than what we have now.

Ayana: Future jobs will not just be email. We'll be building the future together.

---

63 More on this in the interview with Colette Pichon Battle, in the "Community Foremost" section.

Kate: We're creating, we're making, we're building, we're interacting, we're reading, we're doing. Not conveying that is a weak spot in how we talk about this physical transformation. Coastal cities of the future should be much more tied to a joyful experience with water, with each other, and with our relationship to work.

Ayana: Bryan, same question to you.

Bryan: The first thing I would say is: Find ways to organize and vision together as much and as often as possible, to think about the radical vision for what a future looks like for you.[64] Whatever position you're in, try to do that. Everyone can have a voice in this space.

The second thing I would note is that systems rely on the biases of their operators. For instance, if you see a request for proposals, an RFP, that comes in from the government, they're often templates. Getting the template itself changed could change which projects get funded. We need to understand systems well enough to be able to understand where   ✱
you can make changes, where we can hit them hard enough to dismantle them.

Ayana: Indeed. Now, tell us what happens when you two collaborate and combine your superpowers.

Bryan: We're collaborating on designing The Launch School, an environmental justice school.

Ayana: Where is this geographically?

Kate: Floyd Bennett Airfield, on Jamaica Bay in New York City, one of my favorite landscapes.

It's an exciting project, because rather than just saying to the community, "Here's the building, and you're welcome," there was an intensive questioning phase: "What do you need? What would be a physical environment that can support all of this learning? How can we be safe from rising waters, but still embrace the bay?" It's a small-scale site plan, but one of many things that hopefully can be part of a larger shift to hands-on learning, learning by doing, which is a big shift we need for the future.

---

64 Some tools for sparking and holding such conversations are available from Colloqate, Bryan's design firm: colloqate.org

Bryan: Absolutely. We worked with young people throughout the process to drive not just the programmatic vision, but the experiential vision of the space. They became co-designers with us. The parents and the teachers were given an opportunity to voice what they envision. And what they wanted to do is maintain the level of community they have in their current school, right in the location that it currently exists. There will also be an outward-facing part of the school that is a community space.

Ayana: Bryan, I think of you as a pioneer of design justice, a term I didn't know until I heard you share it.

Bryan: Because I made it up.

Ayana: Aha! Coiner of "design justice"!

⭑ **Bryan:** It's not enough to have a pretty building; you have to have a process that doesn't just include people, but is so thoroughly invested in the needs and the outcomes of those communities that it expands the possibilities that are there.

Ayana: Mm. Kate, do the concepts of interweaving space and story, or maybe ecosystems and people and stories, also influence your practice?

Kate: So much of what we're doing as designers is having one foot in the world as it is now, and carrying these stories from the past, but also trying
⭑ to project an alternate future. That's the crux of it. Drawing the future together with people seems to be the most important thing.

I'm a landscape architect, so I'm always seeing the world not as something found but as something "made"—by the choices that Bryan's talking about. Whether that's a choice of the Army Corps of Engineers to make a 30-foot-tall levee, or the choice to redline[65] a community. This is a world that we have made—and we can't just accept it as it is. There's this constant pulling and stretching, not accepting the status quo, and always imagining that there could be an alternative way. So that resonates with me—how to project forward but also pull everyone through these kinds of processes and have that future be visualized, designed, and co-located together. There are transformational landscape projects yet to be discussed and conceived that can serve as that physical and so-

---

65 "Redlining is an illegal practice in which lenders avoid providing services to individuals living in communities of color because of the race or national origin of the people who live in those communities."—U.S. Department of Justice

cial fabric we've lost, that can help us reinvent landscapes in a much more just way.

Ayana: Okay, we're going to take a few questions from the audience.

Audience Member #1: As a landscape architect who works with communities, what do you do if a community doesn't come to a consensus?

Ayana: Consensus is really hard.

Bryan: Aiming for pure consensus as the goal is an insane thing to do. Instead, to me, one lens is to do no harm. Does this one thing that one part of the community is asking for perpetuate harm? If you get past that, then we often hire community members as part of the design team. You honor the primary users or the primary impacted people first. Then you have secondary users or people who have an investment in the space, but may not be impacted on a daily basis. Then the third, and clients don't like to hear this, is the operators. And developers are almost never in the equation at all, because they're not in the building. To make sure we have these priority perspectives in the room, we've hired more than seventy-five community organizers, community design advocates across the last few years.

Kate: Something I've learned is that there is no project that has no conflict. And that's okay, and that somehow relieves you of this dilemma. For example, we have a project that was built in Lexington, Kentucky, where on every account one would think it's a win-win-win project. There's a giant new stormwater system, a multi-use trail weaving through the city, et cetera. But there are still people who are like, "There used to be twenty-four parking spaces here and now there are fourteen, and I hate this project." So there's literally no scenario where everyone would be like, "Great, I love it." And I'm okay with that.

The other thing is that the world we've inherited has caused different forms of harm or injustice. We're working in this environment that has layers and layers of complexity. And it's not okay to just say that the status quo is the answer. Often, it's the easiest thing to do, but that status quo does not have design justice embedded in it.

Ayana: And the status quo doesn't work for the new world that we live in. I would add that consensus takes time. I'm an impatient person, by nature. I'm like, "We gotta go, we gotta solve all these problems, it's urgent." But to have any chance of consensus, we have to bring people along. We have to, as I've heard community organizers say, move at the speed of trust.

Audience Member #2: I'm a master's student studying sustainable environmental systems. And I feel like so many of the conversations that my cohort is having are on the precipice between climate doom and excitement to jump into this work. What do you tell yourself in order to not be exhausted when you think about the world as we hurtle toward 2050?

Kate: There's this swirl of mixed emotions that contain both doom and joy. I get emotional, let's be honest. Being with you, Ayana, and Katharine, with the *All We Can Save* cohort—that book and community was like a climate-grief therapy session for me during early COVID. Being connected to these other amazing people made me think it's going to be okay. This person is working on electrification, this person on soils, so I'm going to do my thing and have trust and understand that there are a lot of people out there who are doing just the right thing. That helps me not get overwhelmed.

And then I also get up every day and do the work. And figuring out what that work is, is a constant reinvention. But knowing that it's there to be invented and to be passionately pursued helps.

Ayana: Bryan, how do you keep moving forward?

Bryan: I think about it like a second line in New Orleans. It's a celebration of life and death. If you come to New Orleans and you do a second line—a brass band, thousands of us in the street, we dance, we sing, we drink, all of those things—it's a beautiful, beautiful experience. If you don't know that Black people weren't allowed to bury the dead in the city and had to carry bodies outside of the city, then you don't fully understand why that joy exists. Or if you don't know that in the 1940s, '50s, and '60s we couldn't get insurance to bury our dead within the boundaries of the city—so we had to raise money through these parades. And so, for me, it's the realization that culture is a collective coping mechanism. When we know that foundationally those traumas are always going to be with us, then we create connections with people to survive. To see culture vibrantly around us in New Orleans, around this country as a whole, that can do nothing for me but bring me joy and bring me inspiration to continue to push forward.

A Note from Dad

Archibald Frederick Johnson

A page from my father's college notebook.

Design for a Changing World

Interview with Paola Antonelli

A small part of my dad's legacy is that I love design. I revel in elegant, functional objects. I want our futures to be safe, but I also want them to be beautiful. I appreciate the (often hidden) roles that design serves in how we experience and move through the world, and how—from fine art to technology to furniture—it influences what we value and how we interact. In New York City's Museum of Modern Art (MoMA), my favorite gallery is the one displaying iconic household items. From kettles and toasters, to chairs and lamps, to telephones and video games, these everyday objects shape our days and our lives.[66]

Design is fundamental to shaping our climate futures. This includes buildings, urban planning, appliances, the internet, and materials. And intangibles, too, like design of communications and processes.

This conversation, exploring how design can help us get things right, is with Paola Antonelli. She is MoMA's senior curator for architecture and design as well as its founding director of research and development. Paola, from Milan, Italy, is a trained architect who describes herself as a *"pasionaria* of design." We met in 2019, at a science event at Pioneer Works in Brooklyn, the arts institution where, inspired to follow in her footsteps, I curated the *Climate Futurism* exhibition in 2023. We were introduced by Cyndi Stivers, a curator at TED, a cherished mentor who has coached my public speaking. When I later emailed Paola to discuss where there might be synergies between our projects, she replied, "I would be delighted to collaborate with you on anything. . . . We can meet on Skype and talk for hours." Who sends emails like that to a relative stranger?! But that perfectly captures her sparkling curiosity

---

66 My father took me to MoMA when I was little and I was totally baffled to see what I thought were our very own living room and kitchen chairs (iconic ones designed by Marcel Breuer and Charles and Ray Eames) on display. What the heck?! I could not understand how he snuck them into a museum, and I assumed it must be some elaborate and off-timed April Fool's joke. I have those same chairs from my childhood home in my home today. Good design lasts for and is appreciated by generations. That too is sustainability. Thanks, Dad.

and the fact that she somehow, amidst all her projects, magically, has time.

It's a joy to witness her curator's mind. Always looking, zooming in to parse, zooming out to contextualize, and reliably, joyously, spotting *that one thing*. This is perhaps the broadest ranging of the dialogues in this book. It arcs from recycling infrastructure, to rewilding, artificial intelligence, the role of museums, and the climate action value of a good list.

———

Ayana: What do you mean by the term "design emergency"?

Paola: I believe design is like an enzyme that takes revolutions that happen in science, in politics, in technology, in history, and metabolizes them into life. In this way, design influences society, and changes the world. If you look at things from this viewpoint, since there is always an emergency of some kind in the world, there is always a design emergency.

Many people think design is about embellishing things or restyling them at the last minute, but it's not. Design starts at the very beginning of a process, of a transformation, and comes under many different guises, not only objects. There's an act of design when it comes not only to the planning of a neighborhood or to the production of a new subway train, but an act of design can also be the spark that can help change behaviors to cope with the environmental crisis. That's what design really does. It does not only solve problems or give form to functions, it changes behaviors.

One of the examples that I often use is the internet. At the beginning, it was lines of code, and only people who knew the code could use it and could communicate with each other. And then, in 1993, the designers of the Mosaic interface all of a sudden introduced buttons, windows, hyperlinks, a whole intuitive interface that allowed everybody to use it. Design is about the translation of new, important discoveries for everybody's use, via an object that can be physical or digital, or even an infrastructure or system.

Ayana: Can you share an example of how design changes behavior as it relates to the environment or climate?

Paola: Take the design of infrastructures for recycling, which are so different in different locations. In New York, it's quite disastrous. There are not nearly enough recycling bins, and we know that the plastic that we try to recycle doesn't really get recycled.

In Germany or Italy instead, the infrastructure includes recycling centers and trucks and receptacles that are visibly designed to communicate an attitude that flows from the citizen to the municipality. You feel a different sense of responsibility. In my family's home in Italy, there are five bins—for paper, plastic, glass, food waste, and landfill.[67] All of a sudden, you start being mindful about what you bring home. You start pushing back on packaging at the supermarket. You start saying, I want less of this.

In the example of the recycling infrastructure, design happens at many different scales—from the scale of the individual citizen, to a family's home, to the neighborhood, and then of the city, all the way to the scale of the country. Design happens not only with objects but also very much with interfaces and policies. Design lives in each component of these large infrastructures and ecosystems, so we can deploy it in all these different ways when we try to address the environmental crisis.

Ayana: In the early days of COVID, you launched the @design.emergency Instagram account to highlight inspiring examples of how, as it read in the account bio, "Design can be a social, political, and ecological tool that can foster positive change at a time when we urgently need it."[68] There is so much room for creativity in how we approach all these climate solutions. Tell us a bit about your curation for this project.

Paola: Design Emergency started in May 2020, and it came from my longtime friendship and alliance with Alice Rawsthorn, the great design critic. Alice and I share the belief that design is a powerful agent for change, and a tool for progress in all fields—social, political, and environmental, among others.

The rationale for Design Emergency was to take the opportunity of a momentous and worldwide crisis, so universal and devastating everybody's life, to explain the power and importance of design to as wide an

---

67 Italy recycles 51% of the waste it collects, the most of any European country. The U.S. recycles around 32%, and the global average is nearly 20%.

68 What started as an Instagram account and Instagram Live interview series spawned a published book of essays and interviews, and is now also a podcast.

audience as possible, by highlighting examples of designers working to help humanity cope with it. We interviewed U.S. architects working on emergency hospital wards, Afghan teenage girls building stopgap ventilators, Belgian fashion designers turning their ateliers into PPE production and distribution centers, and more.

As the crisis ebbed, we decided to keep the platform going and to focus on how design can help improve the whole of society, of human life and other than human life, and therefore we also began talking about the environmental crisis. We brought in several examples of how to tackle it, from rewilding, to building self-sufficient communes, and architecture built with local materials.

Ayana: Let's pause on that last one.

Paola: Yes, we interviewed Vinu Daniel, a super interesting architect who works in Kerala, in India.

Ayana: In that interview, Vinu talks about meeting Laurie Baker, a British-Indian architect a generation older than he is (who he describes as a prophet) and how Baker took a whole day to talk to him and show him around and invest in sharing the wisdom he had. In Vinu's words:

> There was no reason for a legend like him to literally pluck tiles off his roof to educate us, but I think he did it because he believes that we are the last generation of architects who matter. We are living at a time when architects can still do something about the big problems in front of us. We are the generation that needs to put into practice the techniques that have been taught for generations, whether earthen techniques, passive cooling techniques, salvaging techniques, or traditional craft techniques. We need to think about what we will give to our children. What happens to the people who come later? You should enact your vision, but can you reduce something? Can you salvage something? Can you not cut that one tree? That is a step towards sustainable building.

What a powerful paragraph. Wonderful questions.

Paola: Vinu is arrestingly clear and lucid in the way he speaks, and in the way he understands what happened before him, including the Gandhian idea of building with the materials that you find within a five-mile radius.

His work shows how tackling the crisis is not about suffocating ego and creativity, but rather about seeing things from a bird's-eye (or moon's-eye) perspective and understanding the ecosystems we live in, and that sometimes doing less is doing more. Not in the modernist architects' way, but in the sense that when we are designing and acting in a whole ecosystem, that ecosystem might need less of us. It's a true vision. And his work is gorgeous, also.

Ayana: It really is.[69] This reminds me of an early lesson I was taught by my father, who was an architect, which is that having constraints and limitations can be a way to unleash creativity. We probably need to bring a lot of that vibe right now to living within our ecological means.

Paola: Yes, there are many examples of this all around us. For instance, sometimes I feel that what Gen Zers are recuperating is basically my grandmother's culture! I love them because it's not only rediscovering point-and-shoot cameras, it's also mending and repairing, recycling and upcycling, jumping with abandon on second-hand, third-hand clothing. It's how people used to live before the system of consumerism set in.

Italy after World War II was a poor country. And I remember my great-aunts, who lived through it, making risotto with the lung of the cow. It was about using every single part of everything, no waste, about being economical. Because "economy" literally means to make something out of what you have in your home. An economical attitude is the attitude of many great designers.

I feel lucky to have architecture and design as the ground under my feet, with their ethos and their pragmatism, and the degrees of freedom within. Sometimes I feel bad for my colleagues who deal with art because it's so much harder for them. They do not have a litmus test—Does it hurt or put somebody in danger? Is it making somebody's life better?—or a Hippocratic Oath that enables you to know whether something is good or bad.

Ayana: I like the idea of an oath like that.[70] For Design Emergency, you also covered the idea of "rewilding."

---

69 See: wallmakers.org

70 Inspired by this conversation, I co-created a "Climate Oath," which you can find near the end of this book.

Paola: Alice interviewed Isabella Tree, a well-known British author. For more than twenty years, Isabella has been working on Knepp, a castle and estate in West Sussex, England, together with her husband Charlie Burrell, a conservationist. They found rewilding to be the best solution to bring back the land from a complex situation of mis- and overgrowth—not to mention an invasion of creeping thistle. They left this beautiful property to the animals and to nature, seemingly intervening as little as possible. If you look at pictures of their estate you see wild pigs roaming happily, white storks that had not been seen for centuries, native ponies. . . . Altogether, it's considered a paragon of environmentalist success. Alice wanted to highlight this as one of the possible paths toward a more responsible attitude.

Ayana: On the flip side of rewilding, there are high-tech approaches. Artificial intelligence is playing such an enormous role in how the future will be shaped—the design of algorithms shapes the physical world, too. How do you think about the role of AI, and about what boundaries or limits on it are needed in order for it to be a good addition to this future that we're creating?

Paola: It's an excellent question and such a hard one, because AI is so pervasive. It's everywhere in our life already. AI is in some of our home appliances, in our cars, in many other objects that are mainstays of our life. And, like any other tool that is so widespread, it can be used well, abused, overused, and anything in between. So it's hard to put boundaries because it's a matter not only of legislation, but also of individual agency. Certain types of AI are becoming more and more available to individuals for free, and I think we need to evolve with AI and its new applications and strengthen our moral and intellectual core so that we can cope with it, push back if necessary.

Ayana: I recently had a chance to visit you at MoMA and we walked through an exhibition you curated titled *Systems*, which features AI.

Paola: The anchor piece of the room is *Anatomy of an AI*, a massive visualization that shows all the systems of extraction—labor, data, and resources—that exist behind an Amazon Echo. This piece was made in 2018 by Kate Crawford (an artist, academic, and researcher who's been working on the ethical, social, and environmental ramifications of AI for a long time) together with her Serbian colleague Vladan Joler. It's a big mural, an intricate, detailed diagram showing you the earthly implications of an AI—it is incredibly resource-intensive. That piece becomes a way to talk about ecosystems and infrastructures.

Citizens[71] need to understand that AI is not only digital or metaphysical, but also very much physical. And it has implications that concern its implementation, the human rights abuses that it is based on and can lead to, and the biases it can reinforce.[72]

It's important to debunk myths, to alert citizens to what could go wrong, but also to show the positive potential of AI as a tool, which is immense. If we got it right, AI could be a tool to improve conditions not only for all humans, but also for all other species; to create and find balances that we didn't know existed. But sometimes I feel that we are running after a wheel that runs faster than we can.

Ayana: We live in this world where everything is possible—and tech can be used in dangerous ways.

Paola: We need to be one step ahead of technology—not in terms of computational power, we would lose that game every single time—but in terms of goals, morals, and imagination. What do we do with all this power? We can talk about it for days or we can try to regain our spine. We're in charge until proven wrong. It is up to us to educate and tame technology.[73]

What is important to me as a curator is not to tell people what's good and what's bad, but rather to stimulate people's own critical sense. That's the effect that I want to have, either as an irritant or as an inspiration, whatever it takes. And to do so, it's important to present many different strategies, especially when talking about environmental responsibility.

---

71 Paola explained her word choice: "I say 'citizens' because every other word doesn't work well. I dislike the term 'users,' I hate 'consumers' even more. So, 'citizens' will do."

72 For example, per a ProPublica analysis, "black defendants were far more likely than white defendants to be incorrectly judged to be at a higher risk of recidivism, while white defendants were more likely than black defendants to be incorrectly flagged as low risk."

73 The following interview, with Mustafa Suleyman, is on AI.

Ayana: This brings us right to your *Broken Nature* exhibition at the Milan Triennale in 2019.[74]

Paola: The whole idea behind *Broken Nature* was that, as humans, we will become extinct. It's going to happen at some point, but we have some control over the *when* and quite a bit of control over the *how* and over the legacy that we will leave behind us. We don't want the next dominating   ! species to remember us as complete morons.

Ayana: Ha! I suppose that's good motivation, not wanting to be remembered as a moron.

Paola: No, really, even as individuals, when we start thinking we will die, we try to do better, don't we? Same as a species. It was about designing our own legacy, and having a more respectful attitude toward other species and the environment. It was about presenting great examples of how many ways there are to live life. Not in a righteous or sanctimonious way, but in a more satisfying way. You can be a better person and you don't have to renounce the pleasure, the delight, that comes from great design, from beauty and elegance.

The exhibition included biodegradable pregnancy kits, pristine office furniture made of electronic waste, and also several objects and diagrams that managed to give you an expanded sense of time, that forced you to consider reverberations of our action that last well beyond an individual's lifespan, beyond two to three generations, which is another important strategy to foster a more responsible attitude toward the environment.

Some of the pieces showed how a design project can be an ongoing strategy and a distributed approach. For instance, at the LUMA Foundation in Arles, France, there is a lab that studies algae, which are a renewable resource and also often the byproduct of the pollution of our waters. Their Algae Platform initiative gives workshops in different cities, teach-

---

74 "*Broken Nature: Design Takes on Human Survival*, highlights the concept of *restorative design* and studies the state of the threads that connect humans to their natural environments—some frayed, others altogether severed. . . . *Broken Nature* celebrates design's ability to offer powerful insight into the key issues of our age, moving beyond pious deference and inconclusive anxiety."—Paola Antonelli, brokennature.org

ing people to transform their local algae into biomaterials that can then be 3D printed to create traditional objects. The idea of digging out a new memory from pollution is beautiful.

There were also some works in the exhibition by Neri Oxman, a remarkable architect whose oeuvre is emblematic of an approach to design based on working with nature. Two projects in particular, one about co-creation with silkworms, and the other experimenting with melanin as an architectural element.

There was an example of a green burial system, *Capsula Mundi*, by Anna Citelli and Raoul Bretzel. I remember a Milanese lady, circa seventy-five years old, well-coiffed with an impeccable suit and a perfect bag, wearing elegant shoes. She recognized me as the curator and addressed me, "You need to tell me more about this pod because that's how I want to be buried. I want to tell my family."

Ayana: Wow.

Paola: If you can inspire people to picture what they see in a show in their own life, position it in their context and see if it works, well, that's a successful exhibition.

Ayana: For the show and in the accompanying book, you highlighted this concept of restorative design. What does that mean to you?

Paola: My idea of restorative design means being respectful of the environment, being responsible toward other beings and ecosystems, without however being forced to do away with all the beauty that comes with human creation and design, the sensuality, the pleasure. And when I say beauty, I never mean classical symmetry—it can be punk, it's more about formal intention. In other words, there is no need to self-flagellate and atone in order to be responsible toward other humans and the rest of the environment. That was the idea. In Italian, "restorative" translates into *ricostituente,* which when I was growing up meant a tonic, a medicine that you would take to give you a boost.

Ayana: Oh, like it reconstitutes you. I like thinking about each of the pieces you curated coming together to give us a boost as we expand our vision of what's possible. Paola, what does the future look like if we get this right?

Paola: If we get it right, designers and artists and scientists and politicians will all sit together at the table, in the mayor's office or president's office. And corporate shareholders will not be the most important creatures in the world, but rather they would want to share their shareholdings with other creatures.

Design is about reaching certain goals using the means at your disposal—concrete, e-waste, code, earth-boring machines, marquetry, or even policymaking—in the most elegant and economic way possible. To tackle our environmental predicament, a revision of goals is key, as is a revision of means, which we have focused most of our energy on so far. Ultimately, I think goals are the most important object that we have to ✱ design. And then everything else will flow, and we will all be co-creating and collaborating toward these goals.

Ayana: You're so deeply familiar with New York City. Can you imagine what it might look like if it were transformed in this climate-solutionist way?

Paola: During the pandemic, the fact that more bike lanes were introduced was major progress. But let's move into complete utopia, let's imagine what Olalekan Jeyifous depicts: Brooklyn becoming a layered garden city where both a formal and an informal economy and life come together, with many layers and modes of interaction.[75] To me, the dream is less legislation and regulation, and stronger, more imaginative initiatives by citizens.

Another idea hailed by environmentalists is the fifteen-minute city, but in truth, that's how we've always lived, and how many still live.

Ayana: That means everything you need is within fifteen minutes of you.

Paola: Exactly. This "revolutionary idea" always makes me smile because most arrondissements in Paris are like that, and so are some smaller Mediterranean cities. Many of the possible solutions are about unbuilding and reverting a little, learning from the past but using the technology of today.

Ayana: And what about buildings? How do you imagine they would be different experientially or practically?

Paola: New buildings should be built with different methods, goals, and metrics in mind. And old buildings should be retrofitted as much as pos-

---

75 Olalekan is the artist whose salvage punk, retro-futurist images appear in this book.

sible to be more energy-efficient—it's happening in many cities even if it's not as easy as it sounds. What I find outrageous is that so many buildings get demolished to make room for new buildings before they've finished their life, for reasons of profit or expedience. That's senseless waste.

★ **Ayana:** It seems there's a design challenge there. If you do need to take a building down, there should be a process for separating the parts so they can be used to make new things. As far as I understand, we don't have the infrastructure yet for doing that on a significant scale. Like you described with recycling and all the different bins, we need some version of that system for construction materials. I feel like many more people would love to use reclaimed wood and fixtures and local materials when they build or renovate, but right now it can be onerous for the average person to be able to do that—it can cost twice as much or take three times as long to source everything. It's starting to become more accessible in a few specific places, and definitely needs to be expanded and replicated.[76]

Paola: Yeah, there are many architects working on these issues. Vinu Daniel is one. Or David Benjamin from The Living, an architectural office in New York, who has been experimenting with new-but-old materials and techniques and talking about the concept of embodied energy.[77] At MoMA, after building the new extension that opened in 2019, we measured the embodied energy of both the old and the new building. It matters a lot, for instance, if you get the stone from a quarry in New York or Italy. We need architects and citizens to start thinking of buildings in terms of embodied energy and their whole life cycles—from where the materials are born until where they will be at the end of their life.

Ayana: As senior curator at MoMA, you are working in this serious, physical institution. What do you see as the role of museums in helping us to address the climate crisis?

---

76 In 2016, the city of Portland, Oregon, became one of the first U.S. municipalities to adopt an ordinance to require certain homes to be deconstructed, rather than demolished. Milwaukee, Palo Alto, San Jose, Pittsburgh, and San Antonio have similar initiatives.

77 "Embodied energy" is the energy consumed by, and the associated greenhouse gas pollution of, the production of a building, from mining and extraction of resources to delivery.

Paola: Museums need to really find a way to talk more clearly and influentially about all crises, not only the environmental one. In a survey about retention of museum staff in the United States, once again Gen Zers stood out. For them, it was much more important that museums dealt with diversity, equity, and inclusion, and with anything that has to do with environment and social justice, than having gravitas or artistic reputation. They have a moral stance; maybe even moralistic.

I've been mulling this over for many years. In 2008, for instance, I was itching to capitalize on the fact that the financial sector had shown its true colors at last, and that society should begin putting its trust in the cultural sector instead. So, I started the R&D department at MoMA, to show that museums can be the R&D of society. I believe that by ✳ strengthening education and the importance of culture and the role of creatives in a society, we can do a lot.

That's what most museums should do to remain relevant. Because of course we have great collections, great art, but we need to make sure that art is relevant to people's lives. It cannot be only higher education or escapism; it has to stimulate each visitor's own ideas and critical sense, so they can contribute to the betterment of society.

Ayana: Culture needs to change in order to make a lot of other things possible. When you think about this needed societal transformation around climate, what's holding us back?

Paola: I think it's unmoored capitalism. The pursuit of shareholders' interests is what's holding us back, and all the backchanneling and politics that go with it. As human beings and as citizens I think we're ready, but we're held back by ideology and by an idea of society that is unsustainable.

Ayana: And you've mentioned consumerism.

Paola: Consumerism is less of a problem because it can be redirected. There are other ways to entertain our instincts of hoarding, collecting, showing, having new things, because that's what consumerism also is. It is a way to express ourselves, but we were taught that to express ourselves, we should buy.

We can still have different types of "shopping therapies." We can still go on a shopping spree of upcycled clothes or have wonderful clothes swaps like the ones I used to have with my girlfriends in Milan. We can completely redirect that energy. Instead, the basic goals right now are producing more and moving value around. It's a perverted system.

Ayana: What is the nerdiest, most esoteric thing we need to do to turn your future vision into reality?

Paola: Well, once upon a time I was part of a group that was studying pie-in-the-sky ideas for the environment. We were dreaming of being able to mandate the display of water and carbon footprints on all products, similar to the nutrition facts label you see all over the world. It would be important to be able to clearly express, in a legal and legislated way, what the environmental impact is of different products.

Ayana: Oooh, yes. That kind of transparency enables or forces accountability, and empowers consumers in a useful way. A lot of people want to make better choices, but how would we even find that information?

Paola: It's completely obfuscated by, once again, the shareholders' empire. We need mandatory and legislated transparency.

Ayana: I'm in. What are some things we need to leave behind in order to have this better climate future?

Paola: There are so many. Single-use plastic is one. I feel horrible every day about it. I put it in the bin, and I know it's not going to be recycled. It's a real problem. Solving that could give us a sense of satisfaction and hope, and hope is a propellant. It's about being more vocal and pushing for better solutions, pushing legislators and companies to leave behind excessive materials and misguided solutions.

Ayana: And it sounds like you were also maybe alluding to leaving behind being passive about our role as citizens.

Paola: That's a beautiful way to put it. We need to leave behind complacency for sure. So, marching more, even though it might seem completely useless, making our voices heard, boycotting, whatever it takes to push back.

Ayana: How can people in general, or designers in particular, be more a part of the solutions that we need? Is there a call to action?

Paola: The call to action is to really be better humans. I don't know how else to put it. Be better humans by understanding that we live for others. Otherwise we don't have much of a reason to live. And when I say "others," I mean also the rest of the environment, all creatures and things. Love is the answer.

Ayana: I wish it didn't sound so corny to say that, because it is true.

What are three things you wish everyone knew about climate or climate solutions?

Paola: Number one, I wish everybody knew that the powers that be are not doing their job, yet. They're not punishing companies that pollute.

Number two, I would like everybody to know that they can make a difference. And I would like them to always make lists. People like making lists, so just make lists of how you can make a difference. Internalize and realize the potential that every single citizen has. ✱

And number three, it's incredibly important to communicate the urgency. That's paramount. I would like everybody to read and learn about the environment, not as a chore but as a pleasure.

Ayana: You are one of the people who influences my vision for what the future could be. Who is influencing your vision for the future?

Paola: I have to say I'm like a vampire and I draw blood from everybody.

Ayana: Ha!

Paola: No, it's true. Seriously. I'm in the right position because whenever I get depressed or down, I find somebody creative and I'll [*slurps*] just suck the creativity.

Ayana: This is the consummate curator's answer. And it's a place that I find myself more and more, with great joy. It's an inspiring, generative, and restorative place to be.

Paola: A *ricostituente.*

Does it need to exist?

—Virgil Abloh

The AI Deluge

Interview with Mustafa Suleyman

Technology pervades our lives, for better *and* worse. Technology brought you this book—from the research and typing, to the (AI-powered) interview transcriptions, to the graphic design, printing, and distribution. Human ingenuity has created everything from plows to plumbing, bicycles to boomboxes, and windmills to wi-fi. While I'm into my solar panels and geothermal HVAC system, I'm not into the technology that got us into this climate mess—the mining, drilling, fracking, combustion engines, plastics, the whole petrochemical industry.

I worry about who gets to design our technologies, who decides which problems we should solve and which solutions we should pursue. And I am concerned about a humanity addicted to social media and gaming, disappearing into tiny screens and losing track of reality, missing the delightful susurrus of trees and streams, diverted from the imperative of making a new world together. A tenuous and teetering scenario. In the words of E.O. Wilson:

> The real problem of humanity is the following: We have Paleolithic emotions, medieval institutions and godlike technology. And it is terrifically dangerous, and it is now approaching a point of crisis overall.

Arguably, of all technologies, the transformative power of artificial intelligence is unprecedented. Its emergence on the scene feels like a beast has been unleashed—and not a beast with a shyly charming Snuffleupagus vibe. The release of generative AI goes against all my training to adhere to the precautionary principle;[78] it seems to have outpaced our

---

78 Per the UN's 1992 Rio Declaration, "In order to protect the environment, the precautionary approach shall be widely applied by States according to their capabilities. Where there are threats of serious or irreversible damage, lack of full scientific certainty shall not be used as a reason for postponing cost-effective measures to prevent environmental degradation."

moral frameworks, and certainly our regulatory ones. What would it look like to, as a society, weigh the costs and benefits of how we proceed, as opposed to having tech executives decide for us? Since AI is here, and almost certainly here to stay, how might we tame this beast, and even put it to work on climate solutions? Technology is a tool; we decide how to use it.[79]

I met tech executive Mustafa Suleyman in 2019 at a gathering at MIT. He was then head of applied AI at DeepMind, the pioneering AI company he co-founded that had been acquired by Google, and he served as founding co-chair of Partnership on AI, a research nonprofit focused on responsible AI. He would soon after become Google's VP of AI product management and AI policy before leaving to found Inflection AI, which in 2023 released an AI chatbot named Pi (short for personal intelligence). Mustafa has been at the forefront not only of AI tech but of the discourse on AI ethics—and goodness do we need to bring some deep morality to bear.

Through our conversations in the last few years, which often quickly veer to the philosophical, it became clear that he not only sees the world differently than I do, but also sees a completely different world. When he looks, he sees the layers of technology behind it all, and the layers that will be there soon, pervading in ways that change the structures of interactions and power. For me, someone who's squeamish about big tech and tends to want to run away and live in the woods, his perspective, steeped in cutting-edge tech and Silicon Valley and venture capital, can feel alien. But it is critical to get our heads around it.

Before working in AI, Mustafa co-founded a conflict resolution consultancy, and in that role found himself at the heart of climate policy. So, he has some understanding of my world, and we'll now get a better understanding of his. If you, like me, find yourself thinking, "WTF is AI and is it going to destroy society, and can we just put it back into the box, and is there even a 'get it right' on this???," well . . . buckle up. And consider: Are the risks worth the rewards?

------

79 I mayyybe named my phone "AEJ Helper," to stay clear on who is in charge.

Ayana: Before you were deep in the AI game, you did some work on climate negotiations. Are there any lessons from that earlier phase of your work that you wish other people knew about?

Mustafa: Yeah, I co-facilitated one of the tracks at Copenhagen in 2009 in preparation for the UN climate summit, on reducing emissions from deforestation. And one of the hardest things was trying to get everybody in the room—who all came from very different scientific, technical, activist, and organizing backgrounds, and spoke many different languages—aligned on a coherent negotiating position among themselves. Everybody had a different view of what the solution should be, and quite often a different view of what the problem was.

Ayana: That's the hardest, when people don't even agree on what the problem is.

Mustafa: Climate is a collective action problem. It requires consensus, it requires compromise and major concessions. Everybody is partly responsible, some more than others, and that makes it easy for no one to be responsible. If we succeed with climate change, we will all collectively have contributed to improving the possible outcomes. No one's gonna be the sole "winner." But if we don't address it, there will be many, many individual and collective losers. It will be the success of our species if we were able to collectively organize and compromise and overcome.

Ayana: Worth a shot.

Mustafa: Right.

Ayana: As someone who is on the cutting edge of AI, can you tell us what AI has to do with climate change?

Mustafa: Well, let's first take a step back from climate change to think about what AI is and then you can see how fundamental it is not just to climate, but to every area of society. What we are trying to do on the quest to develop artificial intelligence is take what has made us, as a species, capable and effective, and try to reproduce that more cheaply and spread it more widely.

If you think about it, everything around you is a product of human intelligence—or a part of the natural world. Everything we have manufactured, engineered, or organized is a product of our ability to look back at past ideas, recombine them, and then predict what is likely to be useful

* or valuable or rewarding in some way in the future. Experimentation and guessing at what might be useful are the most fundamental characteristics of humans. They are what give us our ability as a species to predict that if we chip away at this piece of stone, it'll get sharper and then we can use it to go and hunt some deer.

Sometimes we think of intelligence as consciousness or as what it means to be human, these much looser concepts. But really it's about how we efficiently use what we have to produce new constructs, whether they are materials or organizations, or services or goods or ideas. That is the main thing that has made us productive as a species. And that's the thing that we are automating.

Ayana: And then there's something called "generative intelligence," which is a subset of AI more broadly.

Mustafa: The first phase of AI has been about classifying the perceptual environment. That means taking raw pixels that represent images, or audio spectrograms that capture sounds, and trying to cluster those chunks of pixels or sounds into patterns that represent phonemes in the case of speech, or elements of images in the case of pixel classification. And that's the first step. As a child, your first job is to process perceptual information.

Ayana: This is a tree, this is a puppy, this is my mother's voice.

Mustafa: That's exactly right. And then, as in the second phase of AI, once the child has built an understanding of trees and puppies, you can then say to the child, "How about you try drawing a new puppy that has never been seen before? You can make it pink and you can give it five legs if you want." Or it could be that you can rearrange LEGO bricks to create a representation of a puppy. And that's what is known as the generative intelligence that we now have. You can see that as kind of a prediction engine, generating new examples of those ideas that it has learned through observing billions and billions of examples. We've gone from classification to generation.

Ayana: Big leap.

Mustafa: And the next phase, the third phase, is that AI can not only generate new examples, but can interact with humans. So the AI can say: Was this the kind of puppy that you were looking for? Do you want me to produce a new one that is blue and has six legs? And you can give feedback.

That interactive feedback is incredibly helpful for producing the kinds of output that you actually want as a human. Because now the AI doesn't just wander off and produce some random stuff, it produces something in response to the guidance you've given it. We now have interactive generative AI that will operate across many domains—not just images and speech, but also arbitrary sequential data.

Ayana: Computer code, all languages, all forms of writing.

Mustafa: Exactly. And even more abstract data, like time-series data, telemetry[80] data. Or all the data people have been collecting about the climate for a long time. We are now getting much, much better at building climate models to simulate patterns in the environment and so on.

Ayana: When we hear about AI in the press, it's often about generating images or writing a term paper or an email. But the ability for AI to work with scientific data sets is something we don't hear much about.

Mustafa: Many sources of climate data are too limited for generative AI or large language models (LLMs) to work with. The data sets are quite small. The weakness of these AI models is that they depend on very, very large amounts of data. But now that we have, over the last ten years, built these foundational models, you can add small bits of new data that the AI model hasn't seen before, and it can do a pretty good job of predicting what comes next.

Ayana: Does that hold when things get erratic? Because of climate change, the past is no longer a good predictor of the future, so how can AI make that leap into the unknown?

Mustafa: Well, it's difficult to say. I wouldn't say that AI can select which future is most likely, but it could produce a wide range of possible futures, and you could then have a human decide which one is more or less likely. But in some ways, that's not the kind of prediction I'm getting at. What

---

80 Telemetry is the automatic collection and transmission of data from remote sources to a central location for monitoring and analysis (e.g., in manufacturing, aerospace, etc.).

AI tends to do is take systems that currently work well and make them much more efficient. That's what it's going to, I think, mostly deliver over the next five years.

Ayana: Can you give us some examples of how people are right now already using AI to address climate change or to implement climate solutions?

Mustafa: One example is using it to make cooling infrastructure more efficient. This applies to HVAC systems in buildings but also in data centers and large industrial facilities. AI is useful here because you have a clear objective, which is that you want to continue running your facility.

Ayana: At a stable temperature.

Mustafa: Yes, and you have a bunch of inputs and control mechanisms for which area you decide to heat and which area you decide to cool and how. There are many different ways that you can adjust the type of heat extraction—all different-sized fans and cooling equipment, plus how you handle the night times and lighting and all sorts of things.

What the AI models have been good at doing is finding the optimal combination of control settings to efficiently use the resources that are available to maintain the target. We did this with DeepMind back in 2016, where we basically trained an algorithm to predict what the cooling infrastructure should be set to every fifteen minutes across Google's data centers.

These data centers are campuses. Each data center hall can be the size of a football field, and there might be twenty of them on a campus. And they consume a lot of energy—some of these campuses can be 200-megawatt campuses. This is seismically large in terms of power consumption.[81] And if you go inside one of these football-field-sized halls, what you'll see is rows and rows of cabinets, which contain very powerful computers, which, when they are turned on to stream video so that you can watch Netflix, for example, produce vast amounts of heat.

If you extract the heat efficiently, then the computers last longer and they run better and they fail less frequently. But it's not obvious where

---

81 In 2022, data centers consumed at least 460 terrawatt-hours of electricity, over 1% of the global demand—which could double by 2026.

heat builds up. It sits in pockets around the data center because it's a dynamical system. There is no linear one-to-one mapping between if you do this, this will definitely happen, so these predictions are incredibly complicated and take decades of expert intuition to develop.

Ayana: Gotcha.

Mustafa: So given our knowledge of the incoming load—demand from YouTube and Google searches—

Ayana: And demand from sending so many emails that don't need to be sent.

Mustafa: Yeah, exactly, lot of boring stuff. With that knowledge, we could predict how to control the infrastructure, and we reduced the cost of cooling the data center by 30%.

Ayana: That's a lot.

Mustafa: A significant financial saving, yeah. And this is billions of dollars of infrastructure.

The AI model that we were using would make a recommendation to a human data entry controller to say, okay, you should reduce the pressure of this valve and you should speed up the fan in this aisle. And then what we found is when we put the model into production, the control-center operators got so confident in the AI's judgment that like 95% of the time they were just implementing the AI's recommendations.

And once the operators got really comfortable, we set the system to make adjustments automatically, without the human in the loop, every five minutes. Obviously a dynamical system is going to do better. We only ! introduced the fifteen-minute thing just to give the data operators comfort that they were in control. But then once we initiated the real-time control set points, we increased the saving from 30% to 40%, which is obviously super significant. And then that got rolled out across all of ＊ Google's data centers, and we saw an average energy savings of 30% across them all.

Ayana: And so, more broadly, there's an opportunity for AI to be used in energy efficiency improvements across industries, perhaps.

Mustafa: Totally. There's a bunch of startups now which have launched doing exactly this not only for data centers, but even for old-school industries like concrete manufacturers and aluminum smelters.

Ayana: Cement uses a lot of energy.[82]

Mustafa: Yeah, all of these things do. Absolutely. Part of the challenge is that while there are financial gains for the companies, the retrofitting requires up-front cost. It also requires a company's culture to shift toward one that cares about data, from depending on mechanical-control engineers to depending on software-control engineers. So culturally there's always a bit of a clash.

Ayana: Yeah, I can see that. Unrelated, I was mapping my driving route earlier today, and Google Maps told me that if I took the scenic route, I would save 33% energy, which was pretty cool.

Mustafa: Oh, that's super cool.

Ayana: I assume that's AI?

Mustafa: That's definitely AI.

Ayana: When I've hung out with you, you've pointed out the things that AI is already involved with, and explained that it's already gathering data and training and running all these models around us that we don't even think about. AI is such a big part of our present, not just our future.

But it doesn't seem like a lot of its capability is dedicated to addressing the climate crisis. Why do you think that is? Is it just not lucrative enough? Or is AI not yet sophisticated enough? Beyond energy conservation, it seems like AI could be helpful in public transit systems or waste management, for instance.

Mustafa: The areas that I've worked on are large-scale systems optimization where I think there's vast savings to be made. And I know that many companies and startups are doing that too. Where there's a profit incentive, I think things will scale very quickly.

* On the renewable side, the challenge AI can help with is building stable grids that can absorb and distribute the capacity created by renewable energy sources. Assuming we're not going to have multi-gigawatt storage at the scale that's needed anytime soon, we have to figure out how to manage real-time redistribution of energy to meet demand. Dirty energy, like coal, is fast and easy to spin up in hours, just by firing up a

---

82 Concrete manufacturing alone accounts for 8% of global greenhouse gas emissions.

power plant. But renewables, like solar, can take longer to spin up. That's part of the challenge.

I have seen applications of AI to help with simulations and predictions to help improve grid-load management. DeepMind certainly tried pretty hard on that a few years back. I had an energy and climate team there that was trying to apply AI to all of these contexts, which is why we did the data centers project.

We also did a project with the National Grid in the UK to try to help them with load balancing and grid management. It was not quite effective, but mostly because, unfortunately, they had narrow areas of control. It wasn't possible to make large-scale trade-offs for different parts of the system. And it's also hard to get access to large amounts of historic data and use that for training models. National Grid has started projects with AI again now, so I'm hoping that five years from now they'll have more data to be able to build these kinds of models.

Ayana: That seems promising. Let's zoom out to AI more broadly. You're someone who thinks about the future quite a lot and how things might ripple and evolve. If we get it right, if we are using AI in all of these good ways, what does that future look like? What should we be expecting?

Mustafa: It's unclear, for example, how we avoid the worst climate outcomes without making material progress on carbon capture and storage. It's unclear how we avoid the worst possible outcomes without making progress on batteries. Those two things are basically hard scientific problems. And in a way, my quest in developing AI is to create a tool that everyone in the world can access to help them be smarter about the problems they want to address.

We will use AI to address our big social challenges, from transport and healthcare to sustainable food to water desalination to renewable energies to carbon capture and storage. It's going to help us make progress on all these problems over the next 20 to 30 years.

At the same time, it's going to mean that many people are not going to be able to compete in the labor market, and their skills just won't be sufficiently valuable in 20 to 30 years. We have to face that. We have to collectively embrace that reality and not call that doomerism or pessimism. We have to confront that reality in a responsible way and ask: ✱ What is it going to take to carry people through that transition and be respectful of their livelihoods? And we should start with <u>progressive tax-ation</u>. We don't have to jump to <u>universal basic income</u>, because that's easily dismissed, and it's hard to see how that gets funded. But we can certainly start with massive taxes on the wealthy and massive subsidies for those who aren't able to contribute their skills. And those should be focused around education and retraining and community welfare and supporting people who are already adding value to society but don't nec-essarily get paid for it, like in the way that we care for our elderly and people who have disabilities.

Ayana: So there's a good version of the AI-filled future, where we're using AIs to come up with creative solutions that help make the world better. But for some people, like me, the specter of what AI might do in the fu-ture is unsettling. There are a lot of risks associated with unleashing this new potential in the world.

Recently there was a statement on AI risk from the Center of AI Safety that you signed. And the whole statement is one sentence, which I ap-preciate: "Mitigating the risk of extinction from AI should be a global priority alongside other societal-scale risks, such as pandemics and nu-clear war."

How do we get from chatbots writing term papers to the extinction of humankind? That seems like quite a leap. Where does this risk of things going so horribly wrong actually come from?

Mustafa: I think that the extinction side of things is a little bit overstated, and I think in some ways focusing too much on it is a bit of a distraction. It's really hard to predict what's gonna happen in a hundred years' time.

! Technically, extinction is conceivable if you create an intelligent sys-tem that has the ability to pursue its own goals, to acquire its own re-sources, that can act autonomously, and can process every kind of data that we have developed sensors for. Imagine it being able to see in ways that humans can't see. Hear in ways that humans can't hear. Imagine it being able to control physical infrastructure. Humans can drive cars and planes; machines will be able to control all kinds of highly complicated physical infrastructure.

And so the general fear is that, when you create something that can learn, has autonomy, and is smart, where is the limit on its ability to get ! smarter? And so as a thought exercise, can you control and contain something that is smarter than you? Have we ever done that before? I think generally the answer is no.

Like, the reason why we are able to keep animals that are stronger physically than us locked up for entertainment in zoos or as pets or elsewhere, is because we have intelligence. We are clever. The source of the AI fear is that if you're building something that's clever, how do you stop it from outwitting you? That's the challenge. But to me, at the same time, we're building something that's clever because we've reached, I think, or we're approaching the limits of what we are able to do as a species.

The world has become too complicated. It is changing too quickly and there is just too much information for us individually, or even collectively as groups of the smartest people, to make meaningful interventions. And so we have these runaway forces, whether it's climate or the way that culture is evolving or the way that we're developing AI. There's all these hyperobjects that are evolving faster than we are able to digest and understand as a species.

Ayana: I feel that tumult. For sure.

Mustafa: Right, so there are good reasons to try to build these kinds of AI systems and grapple with the challenge of containing them. Rather than say no to the potential benefits, I think it's better that we try to focus on limiting the harms of these technologies as best we can.

Ayana: I know you are concerned about the risks of AI that is not contained—this is the focus of your book, *The Coming Wave*. Of course, between here and extinction, there are a lot of things that could go wrong, from unpleasant to dangerous. Yet you're developing AI, investing in AI. Help us understand that balance of risk and reward.

Mustafa: I think there are lots of more near-term things we have to focus on. Like, how do we make sure many people in the world, or everyone, ultimately gets access to cutting-edge technologies? How do we make sure we don't end up getting inundated with synthetic media and losing sight of the truth?

Ayana: Like deepfakes, you mean?

Mustafa: Yeah. It'll be easier for anybody to create deepfakes. I think that's something we will learn to adapt to pretty quickly, but it will cause instability and potential near-term harms. People will believe false things.

It'll make it easier to spread misinformation. And ultimately it'll be easier to personalize persuasion and manipulation.

Ayana: And change the outcomes of <u>elections</u>, potentially.

Mustafa: And we've already seen that, right? The election process, particularly in the U.S. but elsewhere as well, is such a chaotic circus that
! small stories sway outcomes in significant ways. So being able to automate the production of those nudges that intervene in electoral processes is going to cause a lot of instability. But I think net-net, it's worth trying to battle through those side effects, those harms.

Ayana: Hmm . . . maybe. You've described getting it right as "a narrow path." And, as you put it, "containment [of AI] is not possible." How would you describe traveling this path from where we are now, where AI is essentially unregulated, to having it be safely managed?

! **Mustafa:** Well, if these tools are left to proliferate without restriction over the next twenty years, then everyone in the world is going to have state-like power. They're going to have the ability to persuade and broadcast and manipulate at the scale of religions, TV shows, newspapers. Take your pick, right? These AI tools amplify power. They make you smarter, more capable, more persuasive, more efficient. Wherever there is a motivation to spread an idea, these tools will ultimately help you to do that.

Ayana: And they won't all be dedicated toward minimizing our greenhouse gas emissions.

Mustafa: That's the problem.

Ayana: Sadly, I think that many selfish applications are in the works.

Mustafa: Well, the problem is that everyone thinks their work is good, right? No one sets out to be evil. Well, very few do. Ninety-nine percent of people, whether you disagree with them or not, genuinely think they are on the side of doing good in the world according to their definition of good. People are highly motivated to persuade other people that they are doing good. That's going to cause a lot of instability because it will mean our ideas and our interests will clash even more than they have done in the past.

! Likewise with the spread of synthetic biology, it will be possible for smaller and less-well-resourced and less-well-trained groups of individuals to engineer pandemic pathogens, which are way more transmissible and way more lethal than we've seen before with COVID or anything else. And so because these tools proliferate, they lower the barrier to entry, which is great because—

Ayana: Which is great?!

Mustafa: Well, lowering the barrier to entry is good because you get more people involved in the innovation process and more people creating and producing value.

Ayana: I don't know that more people with the capacity to cause pandemics seems great to me.

Mustafa: Well, that's the challenge, right? Some aspects of this have to be contained. The problem is that these tools are dual use. The same thing that could be used to make a pandemic could potentially be used to produce a more efficient crop.

We have to chart this narrow path between not completely preventing all innovation in this area and centralizing it. Because centralizing runs the risk of a dystopian, authoritarian surveillance system. But at the same time, complete openness potentially could lead to catastrophe.

Ayana: How do we walk that line? What are the guardrails we need in place?

Mustafa: Unfortunately, the boring answer is that we need competent regulation.

Ayana: I love a policy answer. This will not be boring to me.

Mustafa: Well, the downside of this is that we have to trust our govern- ✱ ments. We have to have competent, well-paid, educated regulators who understand the system, who are deeply embedded in the technical details, who are able to move quickly, make mistakes, get it wrong, change course, not get completely roasted in the media and by companies whenever they do make mistakes.

We need to give these regulators a lot more love and we have to give them support and encouragement to experiment, because they can't sit around chin-scratching for five years, because the field would've moved on during that time. And as soon as the regulation is written, it will need to be tweaked and changed, if not fundamentally course-corrected. We need a much more dynamic interaction between companies and regulators.

Ayana: Do we need something like a nuclear disarmament pact for AI? Is there anything that could be a model in terms of policy?

Mustafa: We do need international agreement and cooperation, which is super difficult, as you well know from climate.

Ayana: Yeah, I'm familiar with that morass.

! **Mustafa:** It's even difficult to get agreement on lethal autonomous weapons, which have been a known quantity for decades. We've so far not reached any international agreement at the UN on them. So it's hard to see exactly how it's going to happen on AI.

I do think that it requires an international agency that has audit rights and can oversee current large-scale AI development projects. The European Union's proposed AI Act is a good start.[83] There's a lot of regulation in that act, like requirements to disclose training data, and all sorts of bias and fairness requirements. I think that policy is generally headed in the right direction.[84]

Ayana: Is the EU policy a model that other countries should be adopting?

Mustafa: Definitely. There's lots of white papers by different countries looking at AI regulation, but the EU is furthest ahead by a long way—they've been drafting the policy for several years.

Ayana: There's the role of government and then there's the role of businesses. What do you think corporations need to do to put us on this good path? And what are you doing at Inflection AI to be on the solution side of things instead of the problem side?

Mustafa: Well, there's a new crop of AI tech CEOs that has been much more proactive than the last round of social-media company CEOs in raising the alarm and encouraging people to think critically and talk critically about the kinds of models we're building.

With DeepMind, we've been at the forefront of pushing the language around AI ethics and AI safety for almost fifteen years now. Our business plan when we founded the company back in 2010 included the tag-

---

83 The EU passed the AI Act in March 2024. For details, see: artificialintelligenceact.eu

84 63% of voters, across party lines, agree that regulating AI should be a top priority for U.S. lawmakers.

line "Building safe and ethical AGI,"[85] and that shaped the field. Both OpenAI and then Anthropic, the two main companies that came after us, and now my new company Inflection, all of us are focused on the question, *What does safety look like?* That means designing the AI models so that they understand human values and they can be evaluated with respect to their adherence to a behavior policy.

Ayana: Now that we have all this context, let's revisit this big question. If we get it right on AI, how would that benefit humankind and help avert climate catastrophe?

Mustafa: I think its benefit will be in giving everybody access to intelligence. The last wave of software gave billions of people access to unprocessed information on the web, like blogs and books and videos and podcasts. There's now about 5 billion people connected to the web, all of whom can, in various forms, get access to what others are thinking.

In the next decade, I think billions of people are going to get access to an intelligent agent or an AI that will help process that information. You're just going to have a natural-language conversation with your AI about anything that you are interested in or that you want to learn about, and it will adapt to your style, your expertise, your tone. I think that is going to unleash the greatest productivity gains we've seen in human history, because everybody will get access to the smartest expert in their pocket.

I think that's going to do a tremendous amount of good because if you're motivated, like we were, to make our wind farms more efficient at Google, a project we did in 2018, you are now going to have an intellectual aid that knows everything there is to know about wind turbines—where they sit and how they work and what their tensile strength is and what their weaknesses are and how to place them. And you'll be able to get into a dialogue with that AI expert as you're thinking about the question of whether I can advocate for a wind farm in my local area or in my region, and what considerations I need to factor in before I can make a proposal like that.

---

85 AGI = artificial general intelligence

Ayana: What is the nerdiest, least sexy, most esoteric thing we need to do now in order to make this good vision of AI possible?

Mustafa: I think we have to get comfortable collecting data and making sense of data. In your work with oceans, in order to know if an ocean is being overfished or if coral is being bleached faster than it was twenty years ago, or if there's suddenly some outbreak of marine disease or there's suddenly some new impact of ocean mining, or you want to be able to monitor whether some shipping lane is upsetting a bunch of whales, all of that requires data, right? Everything has to be monitored and tracked and you have to learn patterns in that data in order to be able to predict what the likely consequences are going to be of an acceleration in those trends. In that sense, the nerdy answer is that more people need to be data scientists in order to make use of that information and try to do good with it.

Ayana: What are the top three things you wish everybody knew about climate or climate solutions?

Mustafa: We have to embrace the reality of profit. Too often in the climate campaigning community there's a sense that we can just wrap up anti-capitalism and fundamental systems rethinking in the climate movement. And it should be separated because it's harming the climate cause by trying to wrap in too many other fundamental political goals. And it tends to exclude people that aren't on the anti-capitalist angle, which does a disservice to the climate challenge that everybody faces.

We should be trying to create alliances that are really broad and really diverse. That should include anti-capitalist campaigners. But it should certainly also include big industry and companies and innovators that are profit-motivated.

Ayana: This is the world we live in.

Mustafa: Because this is the world we live in. But we could also end up in a very positive situation in 50 years' time where we have close to 100% renewables, amazing healthcare is available to everybody, and transportation systems are entirely renewable. We could have completely addressed climate change, captured carbon, solved electricity storage, and we could still be in a completely capitalist society. Right?

And then the other thing I would say is, the technical solutions get a bad rap. Often I hear people in NGOs and in activist organizations dismiss the technical folks too quickly. Like, "They just think there's techni-cal solutions to everything." This demonization, this simplification of who a person is and what their identity represents as just being a techni-cal person proposing that there are only technical solutions, creates an unhelpful stigma. In general, I think we need to be more forgiving of each other.

Ayana: My perspective on that, because I feel like ecosystems don't get enough love, is not that I'm against technical solutions, but that I feel the need to speak up for nature and emphasize that technical solutions are not the *only* thing we need. Because I encounter so many people who want to ignore the carbon cycle and the water cycle. And so I agree with you, from the flip side of the coin.

Mustafa: Likewise, there's an unhelpful stigmatizing of the campaigners and activists. So, the main thing I would say is that we need to ease off on the stigma and embrace the wearing of many different hats, and of mul-tiple solutions that need to coexist and need to be elevated by each other, and not rely on accusations or blame or finger-pointing.

Ayana: Better vibes, more collaboration.

Mustafa: Yes.

FOLLOW THE MONEY

It's funny how money change a situation.

—Lauryn Hill

What if we reimagine capitalism, regeneratively?

What if your tax dollars drive decarbonization?

What if companies lobby *for* climate policy?

What if the wealthy go all-in on climate solutions?

What if banks stop financing fossil fuels?

What if climate benefit is the test for every dollar spent?

What if corporations consider Earth as a shareholder?

What if . . . ?

10 Problems

- Climate change could cause $38 trillion in global economic damages every year by 2050.

- The climate impacts of burning fossil fuels cause around $500 billion in losses every year—from property damage to government spending on recovery, construction-surge inflation, and power outages.

- The gap in GDP per capita between the richest and poorest countries is 25% larger than it would be without climate change.

- The richest 10% of the world population owns 76% of the wealth, takes 52% of the income, and accounts for 48% of global carbon emissions. The poorest 50% of the world gets only 8.5% of the income and accounts for 12% of carbon emissions.

- Since the Paris Agreement was signed in 2015, 60 banks have provided $6.9 trillion in financing to fossil fuel companies, with the top four U.S. banks alone—JPMorgan Chase, Citibank, Wells Fargo, and Bank of America—providing more than $1.46 trillion.

- To reach the Paris Agreement goals, financing for climate solutions must increase by at least 590%—to $4.35 trillion annually—by 2030.

- Pension funds, which hold more than $46 trillion in assets worldwide, are among the largest institutional investors in fossil fuels. Just 14 public pension funds in the U.S. have a combined $81.6 billion invested in fossil fuels.

- Globally, only 1.6% of philanthropy funding goes to climate—in the U.S., only 0.5%. Approximately 70% of U.S. foundation and nonprofit leaders state that they have no plans to divest from fossil fuels.

- Nearly 50% of corporations do not have a net-zero pledge, and 58% of global business executives agree that their companies have overstated their sustainability commitments.

- Trade associations for fossil-fuel-related companies spend hundreds of millions of dollars a year to obstruct climate policy via advertisements, lobbying, and political donations.

10 Possibilities

+ Every $1 invested in resilient infrastructure can yield $4 in benefits.

+ Getting to net zero is a more than $12 trillion business opportunity. In 2023, $1.8 trillion was invested in the clean energy transition, a new record.

+ Due to the favorable economics of renewables, decarbonizing the energy system by 2050 could save up to $15 trillion.

+ In 2023, for the second year in a row, banks generated more revenue from environmentally friendly investing (about $3 billion) than from fossil fuel investing ($2.7 billion).

+ In 2023, renewable energy sources accounted for 22% of electricity generation in the U.S., surpassing both nuclear and coal. And, globally, for new energy production capacity added in 2023, 86% was from renewables, mostly solar.

+ Globally, the clean-energy sector employed 36.2 million people in 2023, up 6 million from 2019. In the U.S., over 3.3 million people were employed in the clean-energy sector as of 2022, over 3 times more than in fossil fuels.

+ Reforming fossil fuel subsidies and putting a price on carbon could generate $2.8 trillion globally in government revenues in 2030.

+ More than 1,600 institutions have divested more than $40 trillion from fossil fuel stocks.

+ Foundation funding for mitigating climate change has quadrupled, from $900 million in 2015 to more than $3.7 billion in 2022.

+ In 2023, more than 23,000 companies, collectively worth at least $67 trillion, disclosed their environmental data on greenhouse gas emissions, deforestation, and/or water use—a 300% increase since the Paris Agreement.

Divest and Protest

Interview with Bill McKibben

I first met writer and activist Bill McKibben in 2000, when I was a junior in college. Bill came to Harvard to teach a seminar with Professor James McCarthy, an oceanographer and lead author of the Third Assessment Report of the UN's Intergovernmental Panel on Climate Change (IPCC). The class textbook was a huge three-ring binder filled with a rough draft of that hundreds-of-pages-long report, a global consensus on climate science.

Reading through that report,[86] while I got the general point (the Earth is warming, with awful knock-on effects), the gravity of it didn't sink in. From projected sea level rise to the hundreds of millions of people who could be displaced, it all seemed abstract to me.[87] That was more than twenty years ago, when climate change seemed abstract to most people, if they thought about it at all. But it was certainly already concrete for Bill, who in 1989 published *The End of Nature*, the first popular-press book on climate change.

When Bill reminisced with me about writing that first book, he shared that in his naïveté he'd believed that once the scientific evidence was presented, world leaders would respond with decisive action to the greatest danger humanity had ever faced. "But by the UN climate negotiations in Kyoto in 1997," he says now, "it was clear that the fossil fuel industry had mobilized to fight against this scientific reality and its implications, and they were having real success. So by that point I was getting pretty scared."

Bill has been turning reasonable fear into smart climate action from the beginning of this movement. He's an esteemed leader, collaborator, chronicler, and elder to myself and thousands of others. He's a phenomenally prolific journalist who has published twenty books and hundreds

---

86 Let's be honest, I was mostly skimming—I was *far* from a diligent student as an undergrad.

87 Sea level rise is now expected to displace between 250 and 400 million people globally by 2100.

of articles,[88] and now shares weekly dispatches in an influential climate newsletter.[89] He co-founded, in 2008 with some of his Middlebury College students, the now-global, grassroots climate activism group 350.org. From fights against oil and gas pipelines, to campaigns to get universities and pension funds to divest from fossil fuels, he's been helping to lead the way. Recently, Bill's work has been focusing on something perhaps even more fundamental than fossil fuel corporations: the banks that finance them.

———

Ayana: A lot of your activism is focused on money, on how we can stop funding fossil fuels. Was there a moment when you realized this wasn't so much about Exxon per se, as it was about Wall Street?

Bill: I began to really look at finance as part of the fossil fuel <u>divestment</u> campaign that Naomi Klein and I helped dream up in 2012. It's now ✱ arguably the biggest anti-corporate campaign in history—we're up to about $40 trillion in endowments and portfolios that have divested, including lots of pension funds, but also lots of educational institutions—Oxford and Cambridge and Harvard and Princeton and the University of California and on and on and on down a long list.

That got me started understanding that there are two levers in this climate fight big enough to be worth pulling. One of them's marked *politics*, and we've tugged hard on that one. And finally it's opened to one degree or another. People have done remarkable work over the last few years in making that happen. Above all, the Sunrise Movement, with Varshini Prakash and so many others, who were the most successful political organizers around climate there have ever been.

But aside from politics, the other lever big enough to make a difference is marked *money*, and we now understand how important that lever is.

---

88 Including a seminal piece in *Rolling Stone* called "Global Warming's Terrifying New Math" that explains how the amount of carbon held in the existing reserves of fossil fuel corporations and petrostates is 5 times more than we could burn and still keep global warming under 2°C—so the vast majority of those fossil fuels simply cannot be burned.

89 Subscribe! billmckibben.substack.com

Ayana: I learned from your reporting that the top 60 banks loaned out well over $4 trillion to fossil fuel companies to *expand* their extraction in the six years since countries around the world signed the Paris Climate Agreement.

Bill: That's right. And the four biggest American banks—JPMorgan Chase, Citibank, Wells Fargo, Bank of America—are responsible for more than a trillion dollars of that. They're by far the biggest lenders to the fossil fuel industry.[90] That money is used for pipelines and other things that produce extraordinary clouds of carbon. [Per the Carbon Bankroll report, "If the largest banks and asset managers in the U.S. were a country, they would be the third-largest emitting country in the world, behind China and the U.S."]

We finally have, thanks to BankFWD and some other groups, a detailed study of just how much carbon we're talking about. And the numbers are incredible. That study looked at the financial holdings of the big tech companies, Google, Apple, and so on because (a) these guys have a lot of money, and (b) they've all pledged to go net zero. Well, it turns out the money these tech companies have sitting in the bank is producing far more carbon than anything they're doing with their operations, because their deposits, in the hands of the banks, become investments in the fossil fuel industry. When that BankFWD analysis factored in these new numbers, Google's carbon emissions went up 111% overnight. Microsoft's financed emissions were even higher than Google's. And Netflix has more carbon coming from its cash than from all the servers around the world dishing out TV every night.

The numbers when they get applied to individual people are remarkable. If you have $125,000 in the bank, and a lot of people my age with their retirement savings and things do have that much in the bank, that's producing more carbon than all the flying, cooking, heating, cooling, and driving that the average American does in a year.

Ayana: Because it's potentially funding the expansion of fossil fuels.

---

90 JPMorgan Chase alone lent $434 billion to the fossil fuel industry from 2017 to 2022.

Bill: It's just turned over to companies to go frack with. Almost certainly for a city—say, New York City, because it has lots of money in the bank and pension funds—that money is producing more carbon than all the subway trains and police cars and fire trucks and school buses and whatever else New York City runs in the course of a day.[91]

Getting a handle on this is incredibly important, and it should not be an impossible task. Fighting Exxon is key, and we've done it with some success, but we're well aware that Exxon is going to fight to the last breath. Their business model depends entirely on digging stuff up and setting it on fire. And so they'll fight for the right to combust. But there's no reason for Chase or Citi to fight to the end. They make some money off this part of their deal book, but it's like 5%, 6%, 7% of their business. It's a declining part of their business because, clearly, renewable energy is going to be more important in the years ahead.

One of the reasons that analysis by BankFWD and their colleagues focused on these big tech companies, and demonstrated what a huge amount of carbon they were producing with their cash, is because we're hopeful that we can enlist those guys in this fight. It'd be really good if Tim Cook sat down with the dude who runs Chase and said, "Look, nice to see you as always. We can't meet our net-zero promises because you guys keep taking our money and handing it over to Exxon. So, you're gonna have to choose Apple or Exxon, which do you want?"

Ayana: Ooooh.

Bill: When that happens, I think they'll make the right choice. In the meantime, to help speed up that day, we have a big consumer campaign going on against these banks.

Ayana: And this is your new work with Third Act, the organization you founded, and its campaign around banking. The thing that I find so exciting about Third Act is that you are specifically aiming to engage an older audience, right? People who are in their third act of life, people who are probably retired, are in their sixties or seventies or eighties and wanting to make sure they do their part to leave behind a habitable planet.

---

91 Good news: New York City is committed to achieving net-zero greenhouse gas emissions in its public pension funds by 2040, one of the first cities in the nation to make this commitment, due to the activism of 350NYC and others.

Bill: That's right. Beautifully put. We've been going a few years now and we've got tens of thousands of volunteers across the country and working groups in many, many states and cities. One of the projects we're working on is this campaign around banking. And, happily, lots of other groups have joined in—the Sierra Club, the Stop the Money Pipeline Coalition, Hip Hop Caucus, on and on. The only thing we're asking of these banks, and this is what makes it so preposterously easy, is to stop funding the expansion of fossil fuels.

Everyone's well aware that, sadly, we're going to be using oil for a few more years because the oil industry has spent thirty years making sure that that's the case, blocking earlier action. And so they'll need a bank, just like anybody needs a bank to keep on carrying on daily business. Fine. What's not fine is financing any more expansion of this infrastructure. Well, the banks are saying they're not even going to come up with their net-zero policies, whatever they're gonna be, until 2027. I mean, come on. We gotta work a little harder than this because we're actually in an incredible emergency.

Meanwhile, the International Energy Agency, which is not exactly a radical outfit—Henry Kissinger founded it in the 1970s as a club of the big oil-consuming countries—said that if we had any hope of meeting the targets we'd set in Paris, investment in new oil and gas fields and coal mines had to stop in 2021, period. That did not faze the big banks a bit. They've continued to lend hundreds of billions of dollars a year for those purposes. It's as if they just cannot stop themselves in their pursuit of very short-term profit. So we're going to try to rein them in. And it's not an impossible task to imagine that.

Ayana: If we manage to rewire the financial system so that it's helping to create a better climate future, what would that world look like?

Bill: So, first of all, let's get clear on the massive scale that we're talking about. The world needs about $2 trillion a year in new investment, much of it in the Global South, in order to build out renewable energy at the needed pace and cope with climate impacts. That's a lot of money.

Ayana: That's a lot of money.

Bill: The good news is there is a lot of money in the world.

Ayana: Yes.

Bill: The bad news is that it's almost all in the Global North. If you want to think about it in the biggest terms, that's money amassed over the last two or three hundred years as the North got rich off the burning of fossil fuel. And remember, the entire continent of Africa has produced something like 2% or 3% of all the global carbon emissions. Their emissions are, for all intents and purposes, a rounding error in the calculations. Meanwhile, the United States alone has produced about 25% of the excess carbon in the air.

Ayana: The most of any nation in the world, all told.

Bill: By far, and probably no one will ever catch us. Not even China. The carbon I poured into the air when I got my learner's permit when I was fifteen back in the 1970s—that's all still up there trapping heat. The Global North got rich burning fossil fuels. And now we've got an immense amount of money, hundreds of trillions of dollars, a lot of it in the form of pension funds and retirement savings. The best chance of mobilizing those trillions is for it to be deployed in ways seen as a good investment. People who have a pension that they spent their lives working for want and need that money to be invested in something that'll give them a decent return; that's how they're planning to live out their retirement. Renewable energy in Africa or South America or Asia should provide a reasonable return. It's pretty cheap to put up, and so people will be able to afford to pay for that energy, and that money will provide a return.

Ayana: And if businesses in the Global South can get loans at a reasonable rate, then they can afford to install renewables much more quickly.

Bill: The catch here is "at a reasonable rate." At the moment, if you're in Nigeria or Ghana or Côte d'Ivoire, it's expensive to access money for these kinds of things from the North.

Ayana: Because people think it's a risky investment.

Bill: People think it's risky and in some ways it may actually be risky, so banks demand a lot of interest. They'll lend you money, but it's gonna cost you 15% or 20%, which is too much. That makes it unrealistic to do these projects. So, the idea that's emerging is that what we call the multilateral development banks, such as the World Bank, will figure out how to take a small amount of public money and sprinkle it into the middle of these deals to take away most of the risk, and that would then allow this vast amount of money in the Global North to be used for these vital climate projects. Now, I'm not saying this is the most enlightened possible outcome here. It really isn't. But in realistic terms, it's at least part of the answer that we're going to have to get at.

Ayana: These development banks are the structures we already have that are moving money from the Global North to the Global South—though it's often with dangerous strings attached.

Bill: That's right. What we need them to do is to use their financial acumen to de-risk those investments so that the truly vast sums in those pension funds and things can make it from North to South. Because most of the money's in private hands. And I do not foresee a moment when we just decide to act as charitably as we should, as morality requires. Justice works through politics, you know? That's a sad truth, but ✳ there you are.

Ayana: That's a hard truth to stomach.

Bill: Figuring out how to do this in a way that is as fair and helpful as possible to the Global South is probably going to be the main work of ongoing UN negotiations over the next few years, these questions of finance. In the largest sense, we have to take some of that wealth that fossil fuels created and figure out how to put it to work in the places that fossil fuel decimated. That's at least a beginning of rebalancing things on this planet.

There's other rebalancing that needs to happen. The most fraught and most important in some ways is the fact that we've literally made it impossible for people to live in parts of the Global South. And that unlivable territory will keep expanding over the decades to come.

Ayana: In terms of places where the heat is too hot for human physiology to even handle.

Bill: Too hot, too dry.

Ayana: And sea level rise.

Bill: That's right. The ocean is suddenly in your living room. On and on and on. Well, we're gonna have to figure out how to let people in those ✳ places come live in these places, because that's the only possible thing that'll work. And truthfully that'll end up helping the Global North too, because we're aging, tiring populations. There's going to be some re-balancing in terms of population, one hopes, over the next decades. The overall rebalancing away from the incredibly inequitable world we in-habit now has got to be the work of this century, both for moral reasons and for practical ones, because we cannot solve these problems other-wise.

Ayana: There's this idea of a <u>loss-and-damage fund</u> or <u>climate repara-tions</u>, where wealthy nations pay Global South nations that have emit-ted very little but are bearing the brunt of climate disasters, to enable them to mitigate and adapt to the impacts of climate change. And there's another layer to it, which is that these countries are also already bur-dened by a huge amount of debt to organizations like the IMF, the Inter-national Monetary Fund. So with every climate disaster, if they borrow more money, they have to repay that somehow. Then the choice they have to make is: Do we pay off the interest on these existing loans, or do we invest in climate adaptation and recovery for our people?

I know you're quite familiar with the work of Prime Minister Mia Amor Mottley, of Barbados, who's been an incredible pioneer in helping to figure out another way to handle that debt, a way to renegotiate those loan terms that would actually help Caribbean islands and countries around the world become more equipped to respond to the climate crisis.

Bill: Yes, so Mia Mottley has put forward what is called the Bridgetown Initiative, named after the capital of Barbados. It's an attempt to refor-mulate the global financial system that was set up at Bretton Woods in New Hampshire in the wake of World War II. And, depending on what parts of the Bridgetown Initiative are ever adopted, it would do a lot of things, including make it much easier for countries to borrow money in ways that don't end up being absurdly penalizing, that don't force people to pay back money that they need for doing the kind of climate mitiga-tion and adaptation work they have to do through no fault of their own, that set up special drawing rights that countries could draw on in the case of emergencies that wouldn't land them as deeply in the debtor's prison many of them now inhabit. These are all good ideas. To make them happen requires considerable sums of public money, and that can't happen without the acquiescence of institutions like the U.S. Congress.

Ayana: It's terrifying that the U.S. Congress plays such a big role in global climate policy.

Bill: I'll tell you, when I was over in Egypt at COP27, the results of the U.S. midterm elections came back in and everybody in Egypt was paying close attention to congressional districts in Colorado and wherever. Because people have to; it's survival for the rest of the world.

Ayana: What else needs to be done on the activism, policy, and corporate sides right now in order to unlock all of these changes?

Bill: I think they'll only be unlocked if we push hard. Literally this is the capital in capitalism. And I don't think that we've as yet, on the financial front, done as good a job of making the case as we have on, say, things like the Keystone pipeline or even divestments. Because it's one beat further away. It's a little harder for people to understand why Chase Bank is a crucial target than why Chevron is a crucial target. Our job is to get people to grasp that your nice suburban branch of Bank of America or Wells Fargo or whatever, really should have a large smokestack on its roof. Because it is, whether you know it or not, effectively emitting huge amounts of greenhouse gases into the atmosphere, probably more than any other building in your nice suburb.

Ayana: A lot of what you're describing is in essence a cultural shift. That's a third lever we can pull.

Bill: All of this should happen straightforwardly, and none of it will happen straightforwardly. It's going to take, as always, lots of pressure and campaigning. That's the story of the whole climate fight. Every time I end up in jail, I think, "This is incredibly stupid. Why do I have to go to jail to get people to pay attention to physics?"

But truthfully, I don't think that there is any scenario where we don't ✳ have to march in the streets. One of the things that makes me quite happy and hopeful is that we're seeing a large, new population of people come into this fight. I got tired of hearing people say that it was up to the next generation to deal with this. It's true that young people have indeed done most of the leading on this fight—from divestment to the Sunrise Movement and the Green New Deal to Greta Thunberg and the thousands of young activists all over the Earth. But young people lack the structural power by themselves to make these changes. So it's been gratifying to see older people at places like Third Act joining in this fight. Our conventional wisdom is that older people become more conservative as they age.[92] But we cannot let that be the case here.

And what's fun is to do this activism in conjunction with, and following, the younger people in this fight. For example, we were in Brooklyn for big demonstrations against banks in late 2022. We were marching from Chase to Citi to Bank of America. And of course the high school kids were out in the front of the march because they move quickly! But there were a bunch of us at the back of the march, with a big banner that said "Fossils Against Fossil Fuels."

Ayana: Ha! That's very good.

Bill: Someone had a T-shirt that said: "Damn right, we're cranky." If our civilizations are going to work, we need elders acting the way elders have traditionally acted in societies, not the way that we've schooled American elders to act. My least favorite bumper sticker in the whole world is that one you see on Winnebagos sometimes, the one that says, "I'm spending my kids' inheritance."

Ayana: Ugh.

Bill: I mean, that's just gross in every strange possible way, but it's also entirely true. We're spending everybody's inheritance and fast, and we need to stop and regather ourselves. And we can.

Ayana: I remember being backstage with you at the big Global Climate Strike in New York City a few years ago when Greta Thunberg was in New York. And seeing all of those young leaders onstage—Ayisha Siddiqa, Xiye Bastida,[93] Alexandria Villaseñor, the list goes on. All of these young people who had been organizing their schools, their communities, their families to get them out in the street. There were hundreds of thousands of people in downtown Manhattan protesting that day, and I remember looking between you and the teens on stage, and feeling myself in the middle and thinking: We have a three-generation-deep movement now; we might actually be able to get this done if we can all figure out our roles and make sure we're staying connected.

---

92 A 2022 study shows that Millennials are bucking this trend, instead becoming *less* conservative as they age. Gen Z may well follow suit.

93 An interview with Ayisha and Xiye appears in the next section.

Bill: Amen. We've basically understood what we needed to do since 1989. Our only real problem is time. Solar power, wind power, batteries ✳ have gotten remarkably cheaper in the last decade; they've crossed the line three or four years ago where they're now the cheapest form of energy on the planet. The engineers have done a great job. Over time, that alone will do it. Seventy-five years from now, no matter what, that's how we'll run the planet. Sheer economics will eventually force even Exxon to the floor. But if it takes us anything like seventy-five years to get there, as you know . . .

Ayana: We're screwed.

Bill: The planet we run on sun and wind would be a broken planet. So our job has to be to catalyze this reaction to make it happen faster. And ! that's hard because all fossil fuel corporations need to do is delay. And delay is easy. That's why it's gonna be a tough, tough slog in the next few years. Nobody should expect markets to solve this problem by themselves. They won't do it in anything like the time that we have. That's why we still need to keep pushing, even if we're tired.

Ayana: You recently wrote, "Put simply, at the moment the easy supply of money from the banking system to Big Oil drives the ongoing climate crisis. *Capitalism is straight up behaving like a suicide machine.*" What would it look like if we rewired this "suicide machine" to be part of the process of healing ecosystems and supporting life?

Bill: Well, let's just start with getting them out of the business of wrecking the planet. Because trajectory is everything. We got pretty good news in December 2022 when, after impeccable work by activists in the UK, HSBC, the biggest bank in Europe, announced that it was going to stop lending money for the development of new oil and gas fields.

Ayana: Yes, that was big news.

Bill: Very big news, for a couple of reasons. Let me say, it's not everything HSBC should be doing. They'll still be lending money for some kinds of fossil fuel infrastructure and things. But it's a big deal because they're the 13th-biggest lender to oil and gas companies in the world. They have $3 trillion in assets. And they've been a massive lender to the tune of more than $100 billion to the oil and gas industry. When Citi and Chase and Wells Fargo and Bank of America look around now, a lot of their cover is gone. One of the big boys, one of the players in their club, has ceded to reason, has heeded physics. HSBC did not want to give up this part of their business, but they eventually figured out that they better do it or else there was more trouble ahead for them. And that's always the job of activists, it seems, to make it more painful for corporations to keep doing the wrong thing than to do the right thing. Job one is to stop banks from lending to fossil fuel corporations. And once that bleeding is stopped, then yes, there's a different vision of what finance might look like going forward.

Hopefully it'll start to be, as with energy production, more decentralized and local. At the moment, most American assets end up in these huge money center banks, where they're used for this endless casino of speculation and funding one dismal thing after another. But instead, it's possible to imagine a financial ecosystem that's much more reliant on things like local <u>credit unions</u> or locally owned banks that lend out money for local projects to local people and keep more money closer to home. In general, that should be happening as we start to rely more and more on renewable energy, because oil and gas are in Texas and Saudi Arabia and Russia, but happily, sun and wind are everywhere.

Ayana: That's an interesting thing to point out. I hadn't quite thought about it that way before, that by the very nature of renewable technologies, they can be and should be more distributed. Which means that the financing and maintaining and job creation can be distributed more broadly as well.

Bill: Absolutely. If you rely on any resource, but in this case energy, that you can only find in a few deposits around the world, the people who control those deposits end up with too much power. Conversely, if you rely on something that's available everywhere, it helps walk us down this ladder of centralization back toward something better. Your local credit union is less likely to behave like the suicide machine that Chase or Citi are behaving like at the moment.

Ayana: You have to see those credit union customers at the grocery store. It's a whole different scenario.

Bill: And you've funded the grocery store or the food co-op or whatever it is.

Ayana: What you're beginning to describe is quite different from the status-quo financial system, with banks and corporations driven by maximizing shareholder profits and quarterly earnings.

Bill: I don't claim to be an economist or a great prophet in this area. But I do think it's clear that capitalism, as we've conceived it at the moment, isn't working for the betterment of the world or its people. Over the last forty years, this extreme, laissez-faire capitalism, exemplified by what we've done in the U.S., has produced sickening levels of inequality and perilous levels of carbon. Half of the Arctic is melted now and the Antarctic is quickly following suit, because temperature is going through the roof. And so we better do something else now.

I mean, sometimes you read on Twitter people saying, well, we can't do anything 'til capitalism is gone or whatever. And I'm never quite sure what that means because, well, I'm a Vermonter, so I tend to think that Bernie Sanders's politics are pretty good. And what Bernie talks about constantly is the example of the countries of Northern Europe, of Scandinavia. They're clearly capitalists. And they clearly have corporations. But they do not let corporations do whatever they want.

Ayana: Indeed, they do not.

Bill: These are regulated places where they make sure that inequality doesn't get absurdly out of control. And they've done a better job than the rest of us at dealing with environmental challenges, too. They don't let, to the same degree, corporations just pour their stuff in rivers, and they don't let them game the political system in quite the same way either. They don't let them pollute the political atmosphere with huge, invisible clouds of money.

Ayana: There are a lot of structures in place to maintain the status quo, so it often takes intense activism to pry open these windows of opportunity for change. Are there particular barriers to this transformation in our financial system that you'd like to see fall?

Bill: Well, yeah. The barriers are that our political system allows extraordinarily wealthy companies to use their wealth to game our political system. So, a barrier to change is something like the *Citizens United* Supreme Court decision. It says that corporate money is the equivalent of speech and that you can't inhibit it in any way, which is so very American, and very absurd. Corporations aren't like people. Corporations are like, I don't know, they're like bees—I admire incredibly their ability to go out and work around the clock to produce honey. But it wouldn't be useful to ask bees whether producing more honey was a good idea or not, in the same way that it's not useful to ask Citibank whether endlessly increasing Citibank's profits is a good idea or not.

Ayana: Ha! Are there ways in which you're seeing this future that we need already start to take shape?

Bill: There's lots of money beginning to be available for doing the right things, and that's exciting. You can almost sense how it's going to start accelerating. The Inflation Reduction Act, which we all worked to pass, puts close to $400 billion into clean energy. That does two things. It builds a lot of EV chargers and retrofits a lot of big buildings, but it also, in the process, makes politically powerful a lot of people who do solar energy or wind turbines or insulation or whatever. And at a certain point, and it's already starting to happen in some places, their political power begins to match the fossil fuel industry. This virtuous cycle begins to spin in state houses and even in the Capitol, and we begin to make more rapid progress. Political power is always going to play a role in how fast we're able to move.

Ayana: Are there specific places you're thinking of when you say that?

Bill: Look out in the Midwest. There's states that appear to be, if you look at who they vote for for president or something, bright red. But they're also states—Kansas, Iowa, places like that—that have extraordinary amounts of renewable energy now. Mostly wind power.

* **Ayana:** Texas and Iowa are leading the nation in wind energy production.[94] Let's zoom out a bit. Can you paint a picture of what that future looks like twenty years from now, if we have charged ahead with all the climate solutions we already have at our fingertips, and the shifts that you've described in our financial system?

Bill: I'm not a philosopher of where we should be going, but trajectories interest me. And in a place like Vermont or a thousand other places—

---

94 70% of the wind power in the U.S. is generated in red states, because it's windy in the middle of the country and the economics just make sense.

Iowa or Morocco or wherever you want to think about—you can imagine a world in the not-too-distant future where one of our two most important commodities, energy,[95] is produced primarily locally in the form of electricity from the sun and the wind.

Among other things, this allows much more money to stay close at home in the local economy. Eighty percent of the world's population lives in countries that are net importers of fossil fuel. Not having to pay, in essence, rent to the Saudis or Putin or Exxon is a big benefit for most people on the planet. Even most people in countries like the U.S. live in regions that don't produce fossil fuel. So you've got a lot more money close to home, circulating close to home, and hopefully some of it's being used to build up local institutions again.

There would be increases in local food production, and in the diversification of what farmers can grow close to home and sell to people nearby. Maybe we'd be able to see some of that money flowing back into local journalism as opposed to the highly centralized and profit-driven journalism we've seen. A world that's more localized is not a perfect world, by any means, because it still has all of us in it who are all contrary and odd and whatever else. But it's a world where you're more connected to the things around you.

Ayana: In part, you're describing a world where solving problems becomes more tractable because you can actually connect with people who can help, who have the influence to make those decisions.

Bill: That's definitely correct. But that's only half the beauty of it. Because it's not just that the kind of massive consumer capitalism that we've engaged in has managed to wreck the Earth. It's also, from what we can tell, managed to make us less happy than we would otherwise be.

To give you an example, we've watched farmers markets spread around the country and watched the delight with which many people greet them. Sociologists started following people, the way that sociologists do, first around the supermarket and then around the farmers market. Everyone's been to the supermarket, so they know how it works. You walk in, you visit the stations of the cross around the perimeter of the market. Perhaps you have a conversation about cash or credit with the

---

95 The other one is food!

cashier, although perhaps now you just do your self-checkout, leave without ever talking to anyone.

But when people walked around the farmers market, they were having 10 times more conversations. Not 10% more; 10 *times* more, an order of magnitude more. The economy as we've built it over these last decades has encouraged us to forget that we are socially evolved primates. It wasn't that long ago that we were sitting on the floor of the savannah picking lice out of each other's fur. It's no wonder that we like farmers markets. One hopes that a hundred years from now, the fever that we're living in will have passed. The fever that's raising the temperature of the planet, but also the fever of consumption that we've been living with, replaced by a more natural and more human set of connections to the people around us.

Economies won't vanish. We'll still need stuff and want stuff and that's good because people like to make stuff. But some of the disfigurations of our economic life have become so apparent that there's no denying them at this point.

Ayana: What would you say is the nerdiest, least sexy, or most esoteric thing that we need to do in order to turn this vision of a better future into reality?

Bill: Well, much of the work I did in 2022 was trying to elevate the humble heat pump to an exalted place.

Ayana: Heat pumps for peace!

Bill: That's right! The day after the invasion of Ukraine, I launched this nascent effort: heat pumps for peace and freedom. And people started understanding and getting behind it. Within one hundred days, the Biden administration had invoked the Defense Production Act to start putting federal money to the task of building more of these things.

Ayana: Lest anyone doubt the power of the pen.

Bill: There you go. Truthfully, at this point, it's quite mundane things like heat pumps and magnetic induction cooktops.

Ayana: And building codes that prevent gas lines going into new construction.

Bill: Absolutely. That's starting to happen. And building codes that mandate insulation.

Ayana: Insulation has gotta be the least sexy. So, how can people help? Is there a type of person or a type of skill that would really make a difference? Like a legion of programming nerds or Beyoncé?

Bill: Well, Beyoncé, of course, but that goes without saying.[96] One of the things we need are lots of people willing to do quite dull work. There are about 20,000 cities and towns in the U.S., which is actually a manageable number. That means if you had 5 people from each town, a core of 100,000 people who were knowledgeable about this stuff, and willing to go sit through all the endless planning board meetings and public utility commission meetings, and keep putting unrelenting pressure and keep providing all the information, that would make a huge, huge difference.

Ayana: Is this something Third Act is training folks to do?

Bill: Yep. We've been training people up in particular to try and work on public utility commissions in different states because they're extremely important and unbelievably dull.

Ayana: Can you describe what the heck a public utility commission (PUC) does? Why are they so important?

Bill: Well, the public utility commission's role should be to represent the public in its dealings with utilities in setting rates, in making policy about what kind of power plants they're going to build, and so on. That's what they were originally set up to do and that's what they should be doing. But of course, almost universally they've been captured by the utilities that they theoretically regulate. And one of the biggest reasons for that is no one pays attention. No one goes to their meetings. No one covers them in the newspaper.

Basically the utilities have figured out that they can rule PUCs at will. And there's often a quick revolving door back and forth between them and the utilities they're supposed to regulate. In some cases—Arizona was a particularly egregious example—the utilities were using ratepayer money to fund candidates for the public utility commissions that theoretically regulated them.

---

96 Beyoncé, if you're reading this, most live music fans think climate change is an important issue and that artists should use their platforms to talk about it. Call me if you need talking points!

Ayana: Well, that sounds problematic.

Bill: It's corrupt as hell. So it's super important for there to be citizens
involved. Another way of saying all of this is we need some real citizens.
Citizenship has sort of gone out of style in our country. And partly that
dates back to this Reagan-era idea that markets were going to solve all
problems. And if markets were gonna solve all problems, why would we
need to care about civic institutions or being citizens or whatever else?

Ayana: Calling all citizens. Any last thoughts to share?

Bill: The thing the climate movement has managed to do is shift people's
sense of what's normal and natural and obvious. The sum total of all
that work has shifted the zeitgeist. We can tell it from the polling and we
can tell it from the fact that the U.S. Congress finally passed a piece of
climate legislation, the Inflation Reduction Act, 34 years and 45 days
after climate scientist Jim Hansen first testified before the Senate, in
1988, that climate change was real and dangerous. All of our politics
are translated through the medium of physics into the world that we
have to inhabit. And now we're in a moment when these movements
that we've built need to shift at least a little from exhortation and demonstration to execution and deployment as well.

Ayana: Implementation.

Bill: Whether or not the world can do it in the time we have is an open
question. It's why the waste of the last thirty years, thanks to the fossil
fuel industry, is so incredibly tragic and evil. But that water is under the
bridge. Our job now is to do the best with where we are.

Today's economy is divisive and degenerative by default.
Tomorrow's economy must be distributive and regenerative
by design.

—Kate Raworth

Corporations, Do Better

With K. Corley Kenna

Corporations are hugely responsible for the mess our planet is in, and we need their help turning things around. Governments can't do it alone. Individual action is not enough. We need businesses to direct their expertise and resources toward transformational climate action—aggressively reducing emissions, quickly divesting from fossil fuels, actively protecting nature, and supporting the communities hardest hit by this crisis. And, it turns out, climate solutions are good business.

*   Of more than 500 global companies, 69% report higher-than-expected financial returns on climate initiatives. McKinsey estimates that getting to net zero is a more than $12 trillion (*trillion!*) opportunity. Meanwhile, more customers are demanding to know whether products are worth their social and environmental costs. While a bunch of large companies have *plans* to both get off fossil fuels and protect nature, there is woefully inadequate *action* toward reaching their much-hyped targets. More than 50%
!   of corporations (of 2,000 surveyed) do not even have a net-zero pledge.

Luckily, businesses thrive on challenge, on continuous improvement and competitive innovation. But, clearly, we can't just wait for them to voluntarily and speedily do the right (and often profitable) thing. Here are some key steps for all companies and industries to take to get serious about solutions to our planetary crisis, followed by steps governments and citizens can take to ensure the private sector does its part.

Ten things every corporation can do:

1. **Stop the greenwashing.** The vast majority of large corporations have quantifiably done almost nothing to reduce their carbon pollution. But you wouldn't know that from their glossy marketing campaigns. They are simultaneously downplaying the problem and convincing you they're addressing it, dangerously sapping our collective clarity and urgency.[97] That must end.

---

97 Shell, Exxon, and BP are *retracting* their climate promises, expanding drilling and ex-

2. **Find new ways to make a buck.** Businesses need to re-think not just *how* they produce, but *what* they produce. This should include shifting to regenerative materials (those that don't just take from the planet but also give back), creating new revenue streams around resale and repairs, and turning waste into products (such as clothes made from ocean plastic pollution or homes made from straw bales). It's also key to consider what we should *stop* producing entirely—yes, single-use, hard-to-recycle, fossil-fuel-based plastic products are at the top of that list. Durability and circularity are sustainable. Rabid consumption and consumerism are not.

3. **Commit to the long term.** The obsession with quarterly earnings must end. A focus on maximizing short-term profits is often in direct contradiction to the long-term needs of our planet and communities. Boards of directors and shareholders have a powerful role to play in this strategic shift to plan for and adapt to climate change. The most resilient businesses take the long view. Even fossil fuel companies are building their offshore oil rigs to account for sea level rise—as they cause it.

4. **Make a real plan to account for and reduce emissions. Execute the plan. Be transparent.** Businesses need to assess exactly how, across their supply chain, they contribute to the climate crisis, and then figure out how to transform to fix that. Companies need to understand, and make clear to their customers, where in their operations their waste and carbon emissions come from—and set specific targets for reducing them. Currently, only 38% of companies that report their emissions include all their scope 3 emissions.[98]

---

traction, reducing investments in renewable energy, and raking in record profits. In 2020 and 2021, 80% of Chevron's advertisements mentioned sustainability, yet only 1.8% of their expenses went to any projects other than fossil fuels.

98 "Scope 3 emissions refer to all indirect upstream and downstream emissions that occur in a company's value chain, excluding indirect emissions associated with power generation (scope 2). They cover the extraction and transformation of raw materials,

Yes, corporations should be on the hook for their scope 3 emissions.

5. **Actually decarbonize and protect nature; don't just buy offsets.** Many corporations rely far too heavily on carbon offsets, deferring decisions about <u>deep decarbonization</u> far into the future. We simply can't plant enough trees to avoid hard choices. Plus, there is increasing evidence that many of the offsets purchased have come from projects that didn't in fact reduce emissions—like forests that were not at risk of development or, increasingly, are going up in flames.

6. **Lobby *for* climate solutions, not against them.** Many CEOs, boards, and investors already know that addressing climate change is both the right thing for the planet and key for their bottom lines—including protecting their assets and supply chains from climate change's floods and fires. Yet they often sit silently while their trade associations and affiliated PACs fund initiatives *against* climate action. In a 2022 analysis of ten major oil and gas companies, all of them were found to be lobbying *against* their stated climate goals, and many have pro-oil lobbying budgets far greater than what they've pledged toward their climate commitments. Instead of hypocritically preventing progress, corporations must direct their lobbyists to advocate *for* robust climate policy aligned with their stated climate commitments.

7. **Collaborate before you compete.** So much progress can be made at the "pre-competitive" stage. Companies that manufacture in the same factories or in the same countries can collaborate on ramping up renewables, greening production facilities and the grid, and advocating for policies that support a just transition. In a single factory in Vietnam, for

---

manufacturing, logistics, distribution, use and end-of-life of products."—Science Based Targets initiative

instance, you might find five or more brands. The scale at which change must happen demands this kind of sectoral pre-competitive cooperation.

8. **Use scale and market influence to accelerate change.** What if most companies, not just a select few, commit to supporting the development of regenerative organic agriculture, zero-emissions shipping, and low-carbon building materials? These solutions exist today. Scaling them and creating buyer's clubs could ensure a guaranteed demand that would help these green businesses grow. Wealthy companies can make a virtue out of their ability to do big things at scale, creating markets for solutions.

9. **Talk about the process.** Businesses tend to only communicate about the environment when they have a big announcement or milestone to tout. Instead of only talking narrowly about successes or "greenhushing" entirely, it's important to talk about the process and progress, to tell the messy truth, trials and errors and all, to lead by example and bring the rest of their industry along with them.

10. **Give back.** In addition to how companies operate, there's also the question of what they do with their profits. The outdoor apparel company Patagonia,[99] for example, has paid an "Earth tax" since 1985, which later evolved into 1% for the Planet, an initiative through which nearly 6,000 businesses in more than 100 countries have committed to donate 1% of annual sales to environmental organizations. That should be the bare minimum. In 2022, Patagonia changed its ownership model to make Earth its only shareholder: From now on, all money not reinvested back into the company goes to climate solutions and protecting nature. It's still a for-profit company, and still working to minimize its environmental footprint, but we are in a

---

99 I serve on the board of directors and Corley is VP of Communications and Policy.

planetary crisis and the founder and owners decided they were rich <u>enough</u>[100] and put their money where their values are.

Any company can start charging ahead on these fronts right now. They can. They should. But most aren't. So we need to get smarter and more serious about holding corporations accountable—that requires governments, and it requires you. Here are five ideas that *only* governments can implement:

1. **Stop subsidizing fossil fuels.** Stop fueling the crisis with taxpayer dollars. In 2022, handouts for fossil fuel corporations hit $1.3 trillion globally—the same year Big Oil pulled in a record $4 trillion in profits. How about we <u>subsidize</u> the good stuff instead of the bad stuff? Good stuff can include: first-mover incentives, training for climate jobs, community solar, the American Climate Corps . . . the list is pretty much endless.

2. **Mandate climate reporting to establish rigorous accountability.** The EU is leading the world by creating a standardized system of mandatory reporting, and large companies operating in California will soon be required to disclose their emissions. In the U.S., to date, we have largely relied on companies to voluntarily report their climate emissions, but efforts are underway to change this. Such mandates go a long way to not only hold companies accountable to their promises, but also create more consistent and transparent mechanisms for progress. Reporting requirements will flip the script from "trust us," to rigorous accountability, to measurable progress.

3. **Set tough baselines for pollution.** Congress established corporate average fuel economy standards for vehicles in 1975 as a result of the 1973 oil crisis. The aim was to make cars

---

100 Would that more wealthy executives understood the concept of "enough."

more fuel-efficient, and it worked incredibly well. Automobile manufacturers were forced to innovate, and they did: Cars went from averaging roughly 13 miles per gallon then to 26 miles per gallon today. What if Congress tried (again) to set similar baselines for greenhouse gas emitters in all sectors, forcing faster decarbonization? What if the EPA was fully resourced and consistently and aggressively pursued its mandate to protect human health and the environment and regulate greenhouse gas emissions?

4. **Make companies clean up their messes.** Right now, companies we work for and buy from are getting away with polluting, leaving communities and local, state, and federal governments holding the bag. Instead, polluters should fully bear the costs of these "externalities," and governments should ensure that our laws to protect clean air and water and protect us from toxins are much better enforced. If need be, take them to court.

5. **Use both carrots and sticks—incentives and loans, plus taxes and fines.** Tax incentives, grants, and federal loans will supercharge an energy transition. The funding in the Inflation Reduction Act was a great start. In the state of Georgia, billion-dollar investments in green industries have led to so much job creation that Republicans and Democrats are fighting over who should get the credit. Carrots are important. But so are sticks to prod companies to do the right thing. We need governments to hold companies financially accountable for their greenhouse gas pollution, like through a <u>carbon tax</u>—while ensuring that expense isn't simply passed on to consumers and communities. In the 1980s and '90s, cap-and-trade programs were used effectively to address acid rain and ozone depletion. Such incentives could work just as well today.

Make no mistake, getting these corporations and governments to take these actions will not be easy. It will require consistent and persistent pressure and advocacy; it will require citizens and employees to hold them accountable. So here are four things each of us can do to influence corporate behavior:

1. **Be skeptical. Don't fall for greenwashing. Demand specificity and transparency.** We have to move from demanding climate commitments to demanding detailed climate *plans*. Companies should be setting short- and long-term goals, adjusting them each year for what they learned and where they were inadequate. Unless your favorite company is sharing where most of its emissions come from and how it is working to eliminate them, consider their climate message—no matter how glossy or grand—a smokescreen.

2. **Be a climate citizen.**[101] We must stop electing government leaders whose campaigns are paid for by the corporations refusing to get serious about climate. As a voter, you hold the power to elect leaders—at every level of government, from school board to city council to president—who will pass stronger climate laws.[102] Make this your electoral litmus test! Make clear that climate is your top issue not only when you head to the ballot box, but all year round. Don't be shy about contacting your elected officials to advocate for specific climate policies. And don't be fooled into thinking regulations are inherently bad; well-crafted ones can keep us safe and advance climate solutions.

3. **Make change from the inside.** As a shareholder or employee, you have a powerful voice to affect company practices. Demand that your corporations sever ties with the lobbyists who are preventing progress, and instead establish ties with advocates for climate policies that are actually

---

101 "Climate citizen" is a term we learned from Kate Knuth, who shared her vision for what it means in an essay in *All We Can Save*.

102 Per the Environmental Voter Project, 8 million registered voters across the U.S. who have environment/climate as their number one voting issue did not vote in the 2020 election. To help inform your choice of candidates, the League of Conservation Voters tracks the environmental voting records of members of Congress: scorecard.lcv.org

in line with the stated corporate climate commitments. From wherever you sit, push forward on the ten items in the corporate list above.

4. **Vote with your dollars.** Choose companies that stand for quality—offering products that are durable, versatile, and low-impact—and are committed to really giving back to communities and the planet. And avoid those that don't. One way to do this is by supporting brands that are certified benefit corporations (B Corps)—companies that are legally required to put care for people and the planet, along with profit, at the center of their business strategy.

For decades, environmental justice communities have been elucidating how businesses largely caused the four intertwined crises we face— climate, biodiversity, injustice, and democracy—and made a windfall doing it. That can and must change. And let's be real. CEOs and corporate boards know what to do, or at least where to start. It's mostly greed and a lack of imagination that's holding them back. In the words of Vincent Stanley, Patagonia's director of philosophy, "Dystopia is not inevitable. But it will take all the intelligence and navigational skill we can muster to collectively rebuild the house while living in it."

It should no longer be a question of *whether* a company will eliminate greenhouse gas emissions, but how quickly and how *justly*. Corporations must be accountable for solutions at scale—not just talk a big game, or do less harm, but do *dramatically* more good.

Money is made up. The planet is real. And yet it's the made-up thing we bow down to.

—Katharine K. Wilkinson

Since Billionaires Exist

Interview with Régine Clément

Someone has to go first. Someone has to take the leap and bet on an emerging technology, a new company, or a new way of doing business. When it comes to climate investing, it's governments and wealthy individual investors that dive in first. And wealthy investors, unlike banks or governments, are not constrained by large corporate structures and protocols—they are flexible and can afford to take big bets for possible big returns. Now, should a small group of extremely wealthy people have outsized influence in determining which climate solutions succeed and fail, and on what terms? My answer is no. But we don't get to hit "pause" on the climate crisis while we overhaul our economic systems. We've got to walk and chew gum, to go where the money is and activate it.

Wealth disparity around the world is unconscionably large and expanding. The richest 10% of the world population owns 76% of the wealth, takes 52% of income, and accounts for 48% of global carbon emissions, while the poorest 50% of the world has only 2% of the wealth and 8.5% of income, and accounts for 12% of emissions. (The proportions are similar in the U.S.) We need to continue working to address inequality while also getting torrents of money behind climate solutions. In other words, regardless of whether we think it should even be possible for billionaires to exist, they do. And while that's the case, how can they help accelerate the transition away from fossil fuels?

There is *a lot* of money to be made by investing in climate solutions.[103] And there are a lot of very wealthy people who have what is termed a "family office," a team to manage and invest their assets. If money + strategy = power, many of them have it in spades. So, how can we insert climate into those strategies? How can they unleash their capital in useful ways? Enter investor education and impact investing.

Back in 2012, Régine Clément was serving as the trade commissioner

---

103 For example, from 2011 to 2020, investments in renewable power generated total returns 7 times higher than fossil fuels (422.7% vs. 59%). McKinsey estimates that getting to net zero is a more than $12 trillion opportunity.

and head of energy and environment at the Canadian Consulate General in New York City. She was running a clean tech accelerator program and noticed that some of the participating entrepreneurs, those funded by family offices, were able to continue to raise capital in the challenging years following the 2008 financial crisis. As Régine recalls it, "Silicon Valley venture investors had left the space. The markets were in disarray. And completely contrary to everybody else's behavior, some of these families[104] were sticking with the companies, investing more money, rolling up their sleeves and doing the work that was needed to make sure that these companies survived the downturn and could continue to grow when the market stabilized."

This got Régine thinking that climate investors might benefit from a sort of accelerator program, just like the entrepreneurs she was supporting. Then she heard about CREO, which was at the time a network of about twenty family offices focused on venture capital investing in solar energy, energy efficiency, and the California carbon market. She sleuthed out an intro to this discrete network, became inaugural CEO of the nonprofit CREO Syndicate, and has since grown it to more than 200 family offices in more than thirty countries, on track to collectively invest $100 billion of capital into climate and sustainability solutions by 2025.

I was introduced to Régine in 2019, when CREO was building an ocean economy program. I became curious about the possibilities of this self-described "community of investors dedicated to solving the climate crisis." Niche—and essential.

———————

Ayana: What's the idea behind CREO Syndicate?

Régine: Our mission is to catalyze high-impact capital into climate solutions and the decarbonization transition. And our work toward this is in five areas of impact: greenhouse gas mitigation, carbon removal, biodiversity outcomes, circular systems (as a way to manage waste), and pollution control. We invest in everything from sustainable forestry and

———————

104 If you, too, find the way "families" is used throughout this interview to be unrelatably disconcerting (my family is certainly not doing this stuff), then just insert "ultra-high-nèt-worth" in your head before each use of the word.

land management, to regenerative agriculture and sustainable aquaculture, to lab-made proteins, to plant-based packaging, to more efficient manufacturing processes and buildings that use less energy and water, to solutions that address food waste, to zero-emission fuels for shipping and aviation, to electric cars and charging stations, to heat pumps for homes. We are starting to look at climate adaptation as well.

Ayana: That's a lot!

Régine: We launched as a network of trailblazing family offices that had invested in Clean Tech 1.0 in the late 1990s and early 2000s. They were operating on the assumption that the future would look very different, that climate change would become a material risk and opportunity. The rest of the market wasn't really seeing this yet.

Ayana: Great return on your investment, keeping the entire planet from collapsing.

Régine: Indeed. But that's an interesting point, right? How do you actually integrate climate projections into risk-return models? There are a lot of risks and opportunities in climate investing. It's a very complex transition we're tackling.

Ayana: What kinds of risks? The normal risk that the companies you're investing in will fail?

Régine: Companies need more than capital to scale; they also need a market. One of the golden rules in venture capital is that you want to ✳ have an addressable market size of at least a billion dollars for the economics of it to work. But in the climate space, sometimes these markets don't yet exist. So you have to build the market while you're building the technology. That's hard and increases the risk of failure.

But I feel strongly that not only is there an opportunity to invest for high positive impact around climate goals, but that you can actually make returns, too. So this isn't philanthropy. In fact, many of these families started from philanthropy and realized that there weren't enough dollars there to actually support the transition. Government and philanthropy won't be enough. Policy won't be enough. We need to activate financial markets.

Ayana: Yeah, we're talking about a huge amount of money that needs to be mobilized.

Régine: In 2023 the world hit $1.8 trillion in global <u>investment</u> going toward the climate transition, into clean energy and clean mobility. That is a big deal—a lot of work went into that. But we are not yet winning, we are not bending the CO_2 curve. Let's not forget that $1.1 trillion went into oil and gas that year.

And it's estimated that it will take at least $119 trillion to do this transition by 2050. We need more capital going toward scaling solar, onshore wind, offshore wind, electric vehicles, and toward fifty or sixty other technologies that are at different stages of the commercialization curve. The challenge is, they all need different types of financing.

Solar took several years, if not a decade, to get to conventional investors. Some of our members partnered with the Department of Energy (DOE) under the Obama administration to finance some of the first large-scale solar projects in the states. Conventional, more restricted capital couldn't come in until these investment opportunities were de-risked. Those families took that higher risk to develop a nascent industry, for positive impact and potentially higher returns.

Ayana: There's a whole chain of things that needs to happen.

Régine: Yes, and then over time, banks will say, "Oh yes, of course, I've seen this already a hundred times and therefore I know how to finance this." That is essential for us to get to the trillions we need into the transition. More than ten years later, now that solar has been scaled, we see conventional, less-risky capital coming in and investing, but with much lower returns. You need both types of investor for the formula to work.

With the Inflation Reduction Act, $360 billion went to the DOE loan program that's now run by Jigar Shah, who is trying to figure out how government investment can best help de-risk the next set of technologies we need to decarbonize our economy.[105] And now the CREO community is trying to repeat what it did with solar with advanced nuclear, green hydrogen, sustainable aviation fuels, and carbon capture and utilization. And we are starting to think about how we adapt to climate change while mitigating greenhouse gases—like, how are we going to manage rising temperatures requiring more air-conditioning units, when we are trying to decarbonize the energy sector?

---

105 Jigar is interviewed in the following chapter.

Ayana: So there's this group of family offices trying to figure out how, by being at the leading edge, they can open the door for a lot of other climate funding.

Régine: I like to call them trailblazers. They are essentially willing to make bigger bets with a longer-term view, while supporting these companies in lots of non-financial ways to have a positive impact on addressing climate change.

Since these investments are not standardized, a lot of work needs to be done to assess and manage risks. This is why the community is important, because it's hard, on your own, to access all of the insights, data, partners, potential customers, and all of the soft and hard infrastructure you need for your investments to be successful.

The way the overall financial industry works doesn't align well with this intensity of work. Most mainstream investment funds have a high volume of capital to manage—they have to go fast and it has to be easy. And the set of regulatory rules under which they operate prevent them from taking the types of risk our members can take.

Ayana: So CREO members are more risk tolerant?

Régine: Our higher risk tolerance is important, but perhaps more important is our ability to be flexible with capital. Whereas, if a private equity fund sees an interesting opportunity, they generally have a ten-year span to invest and exit to make the returns they promised their limited partners. And they can only invest equity capital, not debt, credit, or other types of capital. They're much more constrained in the ways you can invest.

Ayana: As you see it, we need to harness the capitalist system as part of climate work. You've written, "Put simply, we are a society overly driven by capital and wealth." That raises the question: How do we work within the system so that we can get where we need to go?

Régine: I think about that probably every day. As with any system, capitalism can be shaped by government guidelines and rules. The idea of green capitalism, for example, takes the capitalist system and adds a set of rules and regulations that price in negative environmental externalities (like greenhouse gas emissions). For example, adding disclosure rules for publicly traded companies, so that investors can make better decisions based on the risks associated with CO_2 emissions.

And then, if you achieve green capitalism, is it sustainable over time? Only if we achieve a regenerative system. A capitalist system inherently requires both constant growth and scarcity of resources. That right there is a problem.

Ayana: There is a particular absurdity in having both perpetual growth and scarcity as fundamental principles.

Régine: And yet that exists in capitalism. But we've made a lot of progress. The financial industry is shifting. Could we get where we need to go with a capitalist system that has the right rules and regulations in place? I think so.

Ayana: I kind of think that might be our only option at this point.

Régine: Probably, and it will be hard. Creating rules and regulations that can properly account for the complexity of the socioeconomic system, the science, and the constraints of the financial industry is not for the faint of heart.

Ayana: There are a lot of variables.

Régine: A lot. Simultaneously, we need to work on bringing new moral norms to capitalism. People define the rules of the system, and so it is on us to define the rules for a new capitalism. The concept is not even new; we see the use of moral norms in investing. For example, the Islamic approach to investing with Sharia principles supports the idea that, yes, you invest to provide returns, but you do it in a way that doesn't create harm in other spheres. And we see the use of exclusions for arms dealing and tobacco as a common practice in Western societies. The creation of impact investing, and trying to find ways to tackle climate and social outcomes in an integrated way, is an expansion of these approaches.

Ayana: You've said previously that "the challenge of climate change is perhaps best defined as our challenge to end destructive capitalism." You've talked about capitalism as *amoral*, and said there are lots of different versions of it we could have, lots of different guardrails or moral stances we could imbue it with, but that fundamentally it's a tool. I love the question you posed in *All We Can Save*: "How can we use the mechanics of capitalism as it currently exists to transform it?" How would you answer that question?

Régine: Oh.

Ayana: You asked it! I'm just trying to get you to answer it.

Régine: Ha! Yes, I'll remember that next time.

So, we are incentivized to seek economic growth. That is considered progress. Economic growth has been primarily supported through the extractive industries of oil and gas, and of minerals that support energy, materials, and fertilizers. It has also generally fueled a consumption culture. Now, I'll say not everybody agrees with that growth premise. There's more literature coming out about that.

Ayana: About de-growth and the downsides of constant growth as the goal.

Régine: Yes, but if you can rethink growth in a regenerative sense, through a circular system, if you look at other criteria such as well-being as a measure of success, instead of only GDP, if you bring a new way of interpreting the basic concepts that guide the capitalist system, then you could create a system that values economic, social, and environmental progress and helps us manage our resources in a more sustainable and just way.

Ayana: Putting this into action, CREO itself isn't investing; you're a non-profit fostering a network of informed investors.

Régine: Yes, we deliver research and educational programs, and facilitate investments, in partnership with our members. For example, if members are interested in energy storage, we can leverage expert knowledge and the experience of our community of investors, we can develop market maps, including the scope of investable opportunities, and provide key questions to ask to assess the opportunities.

Ayana: Like, how can we get more batteries?

Régine: Yes. Or what kind of battery technologies are best suited for long-duration storage?

Our team is out in the marketplace talking to entrepreneurs and fund managers and seeing what solutions they need to raise capital for. And we'll do a quality check on these opportunities, make sure companies have the right teams, the right knowledge, the right relationships; that they have a professionalized back office to be able to deliver on these strategies. We are a partner on the knowledge-building side (educational programming), and a partner for deal origination (without taking any fees).

Ayana: Toward CREO's goal of $100 billion, you've so far shepherded $5 billion or so out the door in these investments?

Régine: In fact, the numbers are much bigger than we thought. We haven't updated our website.

Ayana: Doing the work can take precedence over website updates. That's okay.

Régine: Exactly, that's always my excuse. So, in 2021, we surveyed our members for the first time about the investments they're making. And fifty-three members reported back that in 2021 alone they invested or reallocated over $14 billion of capital into climate and sustainability.

The other number that was interesting is that all-time investment in this space by these fifty-three families was over $34 billion. What does that mean when you extrapolate to all 200 or so of our member families? We're going to do another survey this year and try to find out.

Ayana: You should offer everyone a $10 gift card if they fill out the survey. So you get better response rates.

Régine: Haha. All right, we'll try that. For now, we believe we are at least at $50 billion deployed all-time, halfway toward our goal.

Ayana: I think you need to get to a trillion. If we're talking gigaton-level carbon solutions, we should be talking in trillions. No?

Régine: We should absolutely be aiming for trillions. If you look at the marketplace, families represent about $10 trillion globally out of about $500 trillion in capital markets. When we say the transition requires $100 trillion, this is not necessarily $100 trillion of *new* capital. It can also include redirected capital. This is where a family office would, for example, say to a fund manager: "Our family has had a twenty-five-year or even a multi-generational relationship with you, and now we have new goals: We want to support the decarbonization transition through our investments. To continue to work with you, we need you to be moving toward a climate-aligned strategy, including asking the companies you invest in to have a clear climate transition plan. We will give you a year, two years, three years to figure that out. We can help you develop specific metrics that we can track together. And if you fail to align with our goals, we won't invest in your next fund."

Ayana: This could be super influential. Asset managers and investment banks don't want their investors saying, "Get your shit together on climate or I'm not giving you my money."

Régine: It's especially influential if these families can work together, and leverage network effects beyond the investment dollars, to accelerate changes in the system. The plumbing realignment takes time. Time is really the challenge here.

Ayana: It's encouraging to know that this amount of money is moving and there are all these ripple effects happening.

When you said that the challenge is to end destructive capitalism, I assume you're not trying to burn it all down, so what do you imagine this transformed capitalism would look like?

Régine: This is a big question.

Ayana: *The* big question.

Régine: I don't believe we're going to move away from asset ownership. I just don't see a world where that would actually happen. Therefore, we need to establish the values, rules, regulations, and guardrails that ensure that asset ownership does not get out of hand in the context of people, planet, wildlife. Do I believe that the entire world can move toward a full values-based approach to capitalism? It's possible, but I suspect that's not something that changes within one generation.

Ayana: Not fast enough for addressing the climate crisis.

Régine: Exactly. We need to decarbonize as fast as possible. We also need to establish both a values approach and a rules-and-regulations approach. And if we look at where we were ten years ago, vis-à-vis today, we have made tremendous progress. We have gone faster than I ever would've expected on the one hand. On the other hand, our ecological systems and our climate are not going to wait. So the challenge continues to grow and accelerate.

Ayana: What are the major barriers to accelerating the solutions? What's holding us back on this transformation around capitalism and investing?

Régine: So we recently did a quick analysis of the friction points that are preventing capital from flowing more readily, because I was asked by someone who is special to me, "Why aren't we scaling faster? We have the technologies, we have the solutions, what's happening?"

! One of my most important conclusions is that it's a lack of <u>leadership</u>. It's not just shying away from investment risk. People who could be leaders are staying on the sidelines because of fear of not getting it right and being attacked. People don't trust one another. People are angry. There are all kinds of accusations. And when people are feeling attacked and feeling shamed, it's not a good place to do good work.

Ayana: No, it is not. Also, it does seem people are less willing to take risks on climate stuff than they are on tech or other sorts of investing. That worries me. We need not just more leaders, but also, across the board, more courage.

★ **Régine:** Absolutely. We do need more courage. We all do. And the network effect of being part of a community makes a difference. So does not being afraid to talk about the failures. Investors fail outside of climate, too.

Ayana: What would you say is the nerdiest, most esoteric thing we need to do to get things rolling? Like not cool or sexy, but totally necessary.

Régine: Well, I want to be deliberate in saying there's no one thing. Really, there isn't. But, personally, I would love to see more dialogue across opposing views in this space. The decarbonization transition must include efforts by as many stakeholders as possible to go as fast as possible. I would like to see us engage more deeply and productively with the fossil fuel and chemical industries to accelerate the transition in an optimal way—their industrial expertise and labor force, for example, is useful (if not critical). I believe that trying to work with them, while continuing to exercise pressure, will lead to a better outcome.

We need to bring stakeholders who have traditionally not worked together very well to the table. And that means building relationships and trust, which sounds basic, but is absolutely needed. And there is a true climate injustice element to all of this that I don't want to minimize.

We need to get much more creative and stop saying, "That's not possible. No, we've never done it this way. That won't work." We need to instead bring a mindset of, "Okay, let's consider what would it take to bring that change."

The truth is we don't need more [economic] growth to improve people's lives. We can accomplish our social goals right now, without any growth at all, simply by sharing what we already have more fairly, and by investing in generous public goods. It turns out justice is the antidote to the growth imperative—and key to solving the climate crisis.

—Jason Hickel

Your Tax Dollars at Work

Interview with Jigar Shah

The U.S. government has a long history of investing in emerging technology and startups, whether that's airplanes, LED lights, the internet, or electric cars. You probably know that the government funds R&D for renewable energy. But you may not know that the government also makes loans to clean-tech startups, providing the financing to get proven technology out of the lab and into the marketplace. If this is surprising to you, ditto. But also, yes please, because we need it.

The Department of Energy (DOE) Loan Programs Office was created via the Energy Policy Act of 2005. It was put into that act by the late Senator Pete Domenici of New Mexico, who thought the DOE could play a critical role in helping commercialize energy technology, nuclear in particular. It wasn't until four years later, in 2009, with the American Recovery and Reinvestment Act stimulus bill under President Barack Obama, that the Loan Programs Office got a significant budget to work with. The remit includes fossil and renewable energies, and vehicle and critical mineral companies.

The person who now runs this office is Jigar Shah. Jigar and I met in 2011, when I was fresh out of graduate school and excited to be doing wonky stuff in Washington, D.C. I was a wide-eyed Sea Grant policy fellow at the National Oceanic and Atmospheric Administration. He was already seasoned, and working as CEO of the Carbon War Room, which aimed "to accelerate the adoption of business solutions to advance the low-carbon economy." And he had already co-founded and helmed SunEdison, whose investments helped solar energy grow from something niche into a billion-dollar industry. Before landing at DOE, he was also founding president of Generate Capital, expanding beyond solar to more broadly finance energy and infrastructure projects, and published *Creating Climate Wealth: Unlocking the Impact Economy*. Needless to say, I felt out of my league when I first met him as a policy neophyte.[106] I was a little fish and had a lot of ques-

---

106 At the time, I actually emailed a climate friend to say how excited I was that Jigar had accepted my Facebook friend request!

tions. He graciously humored them. And a dozen years later, I'm still impressed and back with more (and more informed) questions.

———————

Ayana: I was surprised when I heard that you had taken a job in the administration, because you seem like such an entrepreneurial rabble-rouser to me.

Jigar: I was also surprised.

Ayana: But with hindsight there's definitely a through-line to your career that connects entrepreneurship with government interventions.

Jigar: There is. I'm a mechanical engineer by training. When I decided to go out into the world to make a difference, commercialization was a big deal for me.

Ayana: What do you mean by commercialization?

Jigar: Solar panels worked in 1992 when I entered college, and they worked in 1996 when I exited college. The technology, the actual panel itself, hasn't changed all that much since then. It's gotten better, more efficient, but there's a solar installation at Georgetown University from 1984 that still operates today. Yet it wasn't being viewed as a mainstream approach to producing electricity. And so I set off to figure out, *Well, why is that? Why do some technologies stop at the lab?* What I found is that there needs to be a rabble-rouser who actually wants to commercialize it. But you also need financing to support it. It's not for the faint of heart.

Ayana: And that's where the government can step in. This DOE office you are now running is the one that, for example, gave the big government loan that saved Tesla from an early demise.

Jigar: Yeah, and we're proud of that. I think that loan changed the face of personal car transportation.

Ayana: Tell us about the Loan Programs Office—I had to look it up when I saw that you had been appointed. It's a powerful office, but one that is not well known or understood.

Jigar: The Loan Programs Office evaluates extraordinary projects conceived by private sector entrepreneurs and innovators who can't find funding. We review their projects, and give them a fair hearing. All of our loans have to have a reasonable prospect of repayment, so our underwriting standards are the same as a commercial bank. We say "No" just as much as they would. But the difference is that we're supposed to be the first lender, and not scared of new technology. And that's easy for us because we have ten thousand engineers and scientists associated with DOE, many of whom hold the original patents to the technology that's coming in for a loan. In contrast, the commercial banking sector, whether by nature or regulation, is not supposed to do weird stuff. They're not supposed to go first.

Ayana: Weird like risky and unknown?

Jigar: Mostly unknown. I don't think most of this stuff is risky, as the Loan Programs Office has proven through its fifteen-year history of making loans.

Ayana: Because people are paying the loans back.

Jigar: People are paying them back. But the projects we fund are certainly unknown.

Ayana: You are working with a lot of money. Let's put this in concrete terms. What scale of financing are we talking about?

Jigar: When I first entered the office, we had about $44 billion of loan authority that we still hadn't used.

Ayana: Why wasn't it being used? People couldn't figure out what to fund, or was the program not fully running?

Jigar: I mean, we are the office that was made famous by Solyndra.

Ayana: Ah, that's right. Solyndra was the failed solar startup that caused huge drama during the Obama administration when they couldn't pay back a loan.

Jigar: Oh, yeah. I think that took up two pages in Obama's memoir. So, the office was dormant. Then current DOE Secretary Granholm during her confirmation hearing said she and President Biden aimed to revive it. They convinced me to join, and I went out to entrepreneurs to see how much demand there was for this office, because it could have been zero, right? And we got seventy-seven applications seeking roughly $79 billion.

The Bipartisan Infrastructure Law expanded our authority tremendously in 2021. Then the Inflation Reduction Act in 2022 gave us more resources. And these are loans, not grants, so we make money for the * federal government.

The IRA gave us an additional roughly $100 billion for our existing program, so that's the Innovative Energy Loan Guarantee Program (that's solar, wind, nuclear, biofuels, geothermal), the Advanced Technology Vehicle Manufacturing Program, and the Tribal Energy Loan Program. And then we were given this new program called 1706, which is the Energy Infrastructure Reinvestment Program, that could be up to $250 billion of additional resources.

Ayana: Okay. That's $350 billion. You've got *a lot* of money you're working with now.

Jigar: We have a lot of money. And as you know, dealing with climate change requires a lot of money.

Ayana: This is a multi-trillion-dollar-a-year problem, climate change.

Jigar: Yeah. When you think about what it would take for the U.S. alone to hit the president's 2035 decarbonization goals, it's probably at least $10 trillion. And then more for the 2050 targets of decarbonizing the whole economy.

Ayana: 2035 is primarily 100% renewable electricity goal.

Jigar: Exactly. But also we gotta get started on all the rest of this stuff.

Ayana: You can't in 2049 be like, "Oh, right, transportation, agriculture, manufacturing, and buildings."

Jigar: And the reason why people are so excited about solar and wind today is because it's cost-effective. And it wouldn't have been cost-effective if we hadn't provided billions of dollars of loans in 2011 for the first 500 megawatt-plus solar projects, which Wall Street didn't want to touch. Or the first four wind farms, which Wall Street didn't want to touch. Or the first three geothermal projects that Wall Street didn't want to touch.

Ayana: Wall Street, get your shit together.

Jigar: Well, I agree. But Wall Street is very conservative. And they're highly regulated to be conservative. They're saying, do we have a ten-year track record that shows it's going to work? So we need to get started on industrial decarbonization, and hydrogen, and carbon sequestration and storage, and nuclear, and EV charging networks, and all sorts of things within the Loan Programs Office now, so that they can be scaled up to country-wide scale by 2035 and 2050.

Ayana: As you're describing it, your job is to seed these industries of the future and help those technologies develop in the same way that solar and wind have proven themselves. There are a lot of climate solutions in this earlier stage that could use a jump-start.

Jigar: Yes, and in the rearview mirror, people think that things went really fast with new technologies, but during the process, they went really slow. Think about Tesla: When we provided that loan, people were not excited about driving in an electric vehicle. And even when the Model S came out, people were like, "Really? We're gonna drive in that thing? It's $90,000." And then they went public in 2010 and every year people thought they were going bankrupt.

It wasn't until probably 2018, when the Model 3 came out, after they'd gone through manufacturing hell, that people were like, "Huh, I could see myself in an electric vehicle." Today everyone takes for granted that Ford, GM, Volkswagen, Mercedes-Benz, Audi, BMW, and everybody else has an electric vehicle. But in 2009, when we first provided that loan, it was not at all clear that <u>electric vehicles</u> were going to make it.

Ayana: You mentioned hydrogen as something that needs to be scaled. Can you explain how that's part of our energy system?

Jigar: Hydrogen is not an energy source, it's an energy carrier. We have to spend energy to make hydrogen, and then that energy gets stored in the potential of that hydrogen that we've created. The reason we do that is because we have a modern lifestyle that needs hydrogen; whether it's for fuels that need to be desulfurized so that we can use them in jet engines, or whether it's in ammonia for making fertilizer, or for chemicals that we use for everyday processes, you need an H molecule to be able to make them. We use 10 million tons of hydrogen a year in the United States.

Ayana: That 10 million tons is a feedstock for fuel.

Jigar: That's right.

Ayana: Is there a company or project your office is funding that you think could change how fast society can transition off of fossil fuels?

Jigar: There's a project called Advanced Clean Energy Storage (ACES) in Delta, Utah, where an 1,800 megawatt coal plant is getting shut down. The facility is on top of salt caverns, so they're installing electrolyzers to convert renewable energy into hydrogen, which they will store in the caverns.

Each salt cavern will store 150 gigawatt-hours of hydrogen. To put that in perspective, the total amount of power stored in the utility-scale batteries we expect to be on the grid by 2030 will be 100 gigawatt-hours in the entire United States. Just one of these caverns will store 150% of everything that we think will be on the grid by 2030 for utility-scale batteries, lithium-ion batteries.

Ayana: That's wild.

Jigar: Right? It got so crazy after we put out this loan that someone on Wall Street started buying up all the salt caverns in the West, and has created a salt cavern real estate investment trust that people can invest into.

And who would've thought that battery storage is something we would support in the grid? That only came about in 2010, 2011. Now everyone is scaling it up like crazy. And now we're saying, actually we can store enough power to run the entire West for a day in these salt caverns.

Ayana: Wow, this is huge. Until now, our electricity grid has been based on fossil fuel and nuclear power plants supplying energy in real time. But now, with batteries, we can store electricity for when we need it— potentially on this major scale.

Jigar: It's massive in scope. And it's tiny in footprint. This one location in Delta, Utah, could basically power all of Los Angeles for a week with the salt cavern capacity they have.

Ayana: Because this hydrogen is a form of electricity storage.

Jigar: Yeah. And if you decide that you never need to turn it back into electricity, then you could make any number of green chemicals or other green products with this hydrogen.

We make a lot of excess solar and wind energy in the spring and the fall because people don't really use air conditioning or heating in the spring and the fall, but a lot of solar and wind energy is still being pro-duced. You could turn all that extra electricity into hydrogen and store it for the season. It doesn't leak from a salt cavern. You could store it there for six months if you wanted to.

Ayana: I'm sure there are eighteen other examples you could give that would equally blow my mind.

Jigar: So many.

Ayana: Talk to us about your vision for the future, if we get it right on energy. What will our energy system look like in 2035?

Jigar: We start by saying that entrepreneurs are the lifeblood of Amer-ica. We invented everything in the world that you think is cool.

Ayana: You're sounding super capitalist for a government policy person.

Jigar: You know, you could take the capitalist out of capitalism, but you can't take the capitalist out of me. But look, for forty years, we decided that we were going to let all this technology we invented go to China, to Europe, to India, to other places and be commercialized there. And today we're saying, no, we're going to take the technologies we invent and commercialize them here.

Do I know exactly how we're going to decarbonize the planet? No. My job is to make sure that nuclear power is cost-effective, and a tool in the tool belt. And that carbon sequestration storage is cost-effective, and hy-drogen is cost-effective. And virtual power plants, solar, wind, geother-mal, and low-impact hydro. Whatever it is that we have the imagination to invent, it's my job to make sure that it's not left on the cutting room floor because these entrepreneurs couldn't get debt to finance their first project.

Ayana: This is that "bridge to bankability" I've heard you talk about.

Jigar: Yes, then there's gotta be a robust conversation between the public sector, the NGO sector, the private sector, and others around which of these tools that we've helped to commercialize are going to be scaled to trillion-dollar scale and not just billion-dollar scale. And which ones we're going to allow to transform our lives because of justice considerations, workforce considerations. But what I am in charge of is making sure that the tool is ready for prime time when someone decides they want to use it.

Ayana: Do we have all the tech we need to get to 100% clean electricity? Are we close?

Jigar: We certainly have all the tech we need to meet the president's goal of 50% carbon reduction by 2030. That we definitely have all the tech for.

Ayana: That's great news.

Jigar: Whether we have all the tech we will ever need, I doubt that's true. There's lots of emerging tech like fusion and other things that people want to see come to reality, and I'm here for that. The research-and-development side of DOE should continue to innovate, and should use biomimicry, taking lessons from nature in all we do. There are lots of ✳ other problems we have, like governance problems, but I'm confident we won't have a technology problem to solve climate change.

Ayana: So the tech that you're talking about for reaching this 2030 goal, that's primarily wind, solar, and storage at this point?

Jigar: Yeah, and also hydrogen. We produce 10 million tons of hydrogen a year in the United States and all of it's gray, produced from natural gas.[107] But there are already announced projects that could produce up to 10 million tons of clean, green hydrogen by 2030, mostly made using an electrolyzer to convert clean electricity into hydrogen—that technology was invented well before World War II. We have the ability to green all of our hydrogen that we use in our country. And we have ways of reducing the carbon footprint of other industrial processes (like making steel and cement) using hydrogen as well, instead of coal.

---

107 The term "natural gas" is used to refer to the gaseous fossil fuel primarily made of methane. "Natural" implies that it is good for you/nature, which it is not; depending on how it's transported (types of ships, leaks in pipelines, etc.), emissions from gas can be twice as much as coal. So I generally opt for the more descriptive terms "methane gas" and "fossil gas."

Ayana: The goal is to have no emissions from our electricity by 2035. And maybe by 2050 we are not using fossil fuels anymore at all?

Jigar: It's unlikely that we're going to end fossil fuels by 2050, for a variety of reasons. One is because in many countries around the world—one of which I was born in, India—many people use one one-hundredth of the amount of fossil fuels or emissions that an average American uses. But the people there still aspire to a "modern" lifestyle. And I don't know that they're going to get there just through solar, wind, and nuclear power. They might need some fossil fuels. And so we need to be careful about making sure we recognize that the human condition needs access to abundant energy. That is how we feed everybody. That is how we give people elevators. That's how we allow people to have modern transportation solutions. But I do think that we have the technology today to do all of that in a way that significantly reduces carbon dioxide emissions to the point where we can come into balance with our ecosystem.

Ayana: Let's talk about nuclear energy. I'm always hearing some version of "advanced nuclear is right around the corner." When you mention helping to finance nuclear through the Loan Programs Office, what kind of nuclear are you talking about?

Jigar: First off, we're only talking about fission right now, not fusion. Fission involves breaking apart elements to create power as opposed to fusing them together. And that kind of <u>nuclear power</u> is already with us. France is largely powered by nuclear plants. In the U.S., about 20% of all of our electricity comes from nuclear power. We turned on the Vogtle nuclear plant in 2023 in Georgia, which is awesome. It's not like we don't know about nuclear power or haven't done it in this country. But in general, it has proven very difficult to do cost-effectively, especially when it comes to new plants.

Ayana: And the Vogtle plant is an example of that, right? Didn't it take like thirty years to build?

Jigar: Well, good news is it was only thirteen years.

Ayana: I knew there was a three.

Jigar: And it was over budget by more than double. Not necessarily something we wanna replicate. But the worst is over, and now they have real plans and real technology that is final, final, final. And now Poland has decided that they're going to build the next six reactors using the same design as Vogtle, and they think they'll be able to do it for about half the cost on a per-reactor basis.

Ayana: And is this a significantly different nuclear technology than what people normally think of, than the kind that makes me concerned?

Jigar: In general, what's different about the technology today versus the technology forty years ago is that the technology forty years ago relied on active safety systems—moving water around, doing things that actively cooled the reactor. Now everything is passive. Even if people totally bungled the way they're operating the reactor, which doesn't happen, but even if they did, the reactor is default-set to passively shut down in a way that's safe.

Ayana: That is a big difference.

Jigar: Yeah, so that's the first thing. The second thing is that the original nuclear plants are so large that everything was custom-made. Like the valves are bigger than me, six feet or seven feet around. Now we've moved to small modular reactors.

Ayana: What is a small modular reactor?

Jigar: These are more like 300 megawatts instead of 1,000 megawatts.

Ayana: And physically, how small?

Jigar: Physically they can be less than the size of a football field.

Ayana: And 300 megawatts, is that a lot?

Jigar: It's enough to run a small town. And all the valves they need are already made for the natural gas sector, so they're like one-tenth the cost. The reason why you go to small modular reactors is you can use the existing supply chain for parts.

Ayana: Where would these reactors be located?

Jigar: You would co-locate them at an industrial facility. Dow is saying that instead of burning natural gas to power their industrial facilities, they would co-locate a nuclear plant there. They would use some portion of the 300 megawatts of power it produced, but also use the waste heat that comes off of the nuclear reactor for all of the industrial processes that require heat, so they won't have to burn any natural gas.

I think you'll see a lot of these nuclear reactors co-located at big industrial facilities. And the industrial facilities always also have great security complexes around them already because they've got dangerous chemicals or things there.

Ayana: Given that it takes a long time to build nuclear—that the permitting and construction process can take more than a decade—if the goal is to decarbonize our electricity sector within the next ten years, do you think nuclear is going to be a big part of that, or is it more for the next phase?

Jigar: Let me say it this way: I don't think it's possible to get to our 2035 goals without new nuclear. And I'm a huge fan of solar and wind—that's been my career. When I started in the solar industry, solar was 0.01% of all the power in the country. We're now approaching 4% of our total grid coming from solar, and on track to hitting 20%. I am extraordinarily proud of the industry. But we're not going to get solar to 80% of the grid and we don't want to.

＊ Exchanging one addiction, which is coal, for another addiction doesn't make any sense to me. You want a diversified technology mix. You want 20% to come from solar, 20% from wind, 20% from hydro, 20% from geothermal, 20% from nuclear. You want a mix.

There are also characteristics of each. Nuclear is 24 by 7 power that runs 100% of the time, where solar is more like 25% and wind may operate 50% of the time. And so you need to optimize the transmission grid, optimize battery storage, optimize all these things. When you put all of these things into a SimCity model—now I'm dating myself from video games . . .

Ayana: I loved SimCity. Bring it on.

Jigar: Well, you put together all these different pieces and you're like, "Oh, what if I move this, and what if I put this in?" et cetera. The lowest-cost, highest-resiliency and -reliability structure results in three or four scenarios. And in all of those scenarios, you need new nuclear.

Ayana: What percent of the mix do you think nuclear needs to be?

Jigar: Right now we're at almost 20% of U.S. electricity generated by nuclear. Modeling shows it's gotta go up to about 30%. And then, a lot of these existing reactors are going to retire by 2050—some of them came online in the 1960s. Running for eighty years is awesome, but after that you kinda have to shut down. So you have to replace all of that capacity by 2050 as well.[108]

Ayana: I've been seeing a lot of sparkly press about <u>fusion energy</u>, which you mentioned the Loan Programs Office is not funding. What's your take on the state of that technology right now?

Jigar: DOE is funding fusion on the R&D side of things, but not on the commercialization side. There were a whole bunch of awesome designs in the 1980s, and a plan to build a pilot nuclear fusion power plant. But it never happened. Today we can take those 1980s designs, stick 'em in a supercomputer and ask: Will they work?

Turns out, yes, they do work. We should do this. And people raised a few billion to build a pilot plant. But so far, fusion has barely been able to produce more energy than it takes to start the reaction, which is a very low bar for "commercial viability"—I'd consider that more technical viability. But I'm still super excited about them hitting that threshold. I'm giddy about it.

Next, fusion scientists have gotta figure out how to produce 50% more electricity than it took to start the reaction and then 100% more. And as they move through that, at some point someone is gonna say, "I'd like to put one of these reactors at my military base," or wherever. And then somebody will design a project. I don't know when that's going to happen. In this decade, I think.

Ayana: So this is not going to help us hit our 2030 decarbonization goals.

Jigar: I don't think so. But remember, there's no such thing as a deadline.

Ayana: We're gonna need a lot of clean energy for the rest of human history.

Jigar: Yeah. We'll need fusion to work by 2100. I mean, the human experiment continues, right? We're going to need more energy to do all the things that we want to do. Hell, one day we might even decide to go to Mars.

---

108 315 of the retired or operating coal plants in the U.S. could potentially be retrofitted with advanced reactors in a coal-to-nuclear transition, generating 263 gigawatts of electricity.

Ayana: Let's not put that at the top of the priorities list. What role do you see for carbon dioxide removal technologies as a way to get some of that stuff out of the atmosphere that's already there?

* **Jigar:** There are three major carbon removal processes that we have now. I'll go through them briefly because people confuse them. First, there's the method for sequestering the pure stream of CO_2 exhaust produced by manufacturing ammonia, ethanol, and natural gas. It's ridiculously easy to capture the CO_2 stream, stick it into a pipeline, and then put it into a Class VI well.

Ayana: What's a Class VI well?

Jigar: That is the EPA's designation for what it takes to safely put CO_2 under the ground into a rock formation that won't allow it to escape. And we've been doing it for over ten years.

Ayana: We're locking it away.

Jigar: Yeah. Illinois has the largest carbon sequestration, from its ethanol plants. And they've been capturing about a million tons a year that would otherwise be emitted into the atmosphere with their manufacturing, and putting it into a pipeline and sticking it into a Class VI well and trapping it safely for the rest of time. Capturing CO_2 in this manner is remarkably cheap.

The second approach is carbon removal from natural gas and coal power plant emissions. At natural gas power plants, the exhaust is like 7% or 8% CO_2, and for coal about 14%. Finding that CO_2 in the exhaust and concentrating it into a 100% pure stream and then putting it into a pipeline, and then sticking it in the well, is a lot more expensive. And that's where the DOE is doing R&D. That carbon removal technology may never be used for natural gas and coal power plants because we have all these other cleaner technologies to produce power, but that R&D is useful for chemical plants or other plants who have CO_2 as a fraction of their exhaust.

That first approach, to capture pure streams of CO_2 and put it under the ground, probably costs $29 a ton. At the coal plants and natural gas plants, the costs are probably closer to $100 a ton.

And then the third approach is direct air capture. That's taking CO_2 out of the air, capturing it, concentrating it, and then putting it under the ground. Doing that cost like $900 a ton a few years ago. Some people think with big scale-ups, they can get to $350 a ton soon, which is great. And then the secretary of energy has a goal of getting that down to $100 a ton by 2030.

Ayana: So we've got carbon capture and storage to figure out. And also battery storage. All these battery technologies rely on minerals like lithium and cobalt and manganese and graphite. There are a lot of environmental impacts and human rights abuses associated with minerals mining, but there are, of course, also massive environmental impacts and human rights abuses associated with mining fossil fuels—for their entire history.[109] This is not a financing question per se, but what's your thinking on how this new energy economy, which is reliant on these minerals, doesn't repeat the sins of the old extractive industries?

Jigar: Well, it actually does come into the loan context because we don't allow them to repeat the sins, otherwise we won't give them a loan. We're forcing folks who are looking for money from us to work with an NGO that has an ever-evolving best practices framework. That includes standards like using electric equipment to reduce the pollution in the area, and compensating local workers fairly. So the mine in Mozambique that produces graphite that we then process in Vidalia, Louisiana, has a union and they employ local workers from Mozambique.

I think there is a right way to do this. Human rights abuses have mostly been documented in the cobalt supply chain from the Democratic Republic of the Congo. And we've reduced cobalt use to the point where ✱ we can get almost all of the cobalt that we need for the electric-vehicle revolution from recycling.

Ayana: Oh, interesting. Yes, ideally at some point we could just keep re-using what we've already mined.

Jigar: And that will be true. We've funded Redwood Materials and ✱ Li-Cycle, both of which are recycling EV batteries at a 95%-plus level. And with research and development, they could probably get to 99%. But we have a billion cars in the world. We've only replaced a certain number of them, so we're going to have to mine more resources.

---

109 Mining for minerals used in clean energy may need to be scaled up from 7 million tons in 2020 to 28 million tons in 2040, but that is 535 times less than the 15 billion tons of mining and extraction required for a fossil-fuel economy.

Ayana: Hopefully more retrofitting and converting existing vehicles too. That would mean a lot less mining.

Jigar: Yeah, I'm pretty positive on all this, but also we all have to be cognizant that, like you mentioned, the sheer devastation that the fossil fuel industry has done in many of these resource-rich countries is so high. But we have a chance to do it right this time around. The fact that there's a chance that we could do it wrong shouldn't be a reason we don't do it. It should be a reason why we ensure that we do it correctly this time.

Ayana: Overall you sound so optimistic. Perhaps part of that is because the technology is so exciting and it's evolving so quickly. But do you, in your heart of hearts, believe we're going to hit our decarbonization goals?

Jigar: The thing I remind people of is that we were also highly unlikely to win World War II. We took over all of our automotive plants to build airplanes and tanks. If the ecosystem starts to falter, the president of the United States could, along with our allies, say, "We're going to take a World War II–like footing, and we're going to seize all of our industrial capacity and force them to make the things that we need to dramatically reduce carbon emissions."

Look at Europe. In one year, they reduced their use of Russian gas by 80%. Everyone said it was impossible, that there's not enough heat pumps in the world. There's not enough workers to install insulation. There's not enough electric vehicles. There's no way the Germans can build an LNG import terminal in seven months. They did.

We are capable of a lot. But it does take a political moment. And you and I both know that we will reach that political moment. I'm from Gujarat in India, and it now regularly hits 120°F [48.9°C] in the summer there.

Ayana: Which is basically unlivable for the human body.

Jigar: Oh, definitely. Even in the shade, it's 100°F [37.8°C]. And if you can't afford air conditioning, you can't live there. I'm not blind to the manifestations of climate change and the wars being caused by climate change and the conflicts and the tribalism, and the return to authoritarian governments. It all disturbs me, but my best and highest use is to make sure these entrepreneurs and technologies are ready for prime time when the world calls for them to be deployed at trillion-dollar scale, when the political moment hits.

Ayana: Apart from politics, what do you see as the barriers to this transformation? Like if we could just get X, Y, and Z out of the way, we could charge ahead to this decarbonized, clean energy future?

Jigar: Well, right now everybody has their request. Some folks want to make sure that everything we do is done with <u>union labor</u>. Others want to make sure that everything has a direct benefit to justice communities. Others want to make sure everything is sourced from American manufacturers.[110] I'm happy to deal with all of these asks. But at some point we have to move faster. At that point, I'm hoping that some of these things will have become an integral part of the fabric, so people aren't jettisoning those concerns, because we know how to do these things and still move fast. And I'm proud of all of the best practices that we're implementing along these lines with each one of our projects, because a lot of people thought it was impossible.

Ayana: That's reassuring to hear.

Jigar: So let's get problem-solving. Stop complaining, start solving problems.

Ayana: That is the vibe. What three things do you wish everybody understood about energy climate solutions?

Jigar: One is that we have all of the technologies we need to pursue decarbonization at scale now. We will get additional technologies, and we need those, but we already have all the ones that we need to move super fast now.

Two, we can't solve all the problems at the federal-government level. Most of the problems we have are local problems. Only 4% of rooftops in this country are filled with solar panels. Australia has 30%. The reason we don't is because local towns and cities make it impossible to put solar on your roof. And we don't recycle like Sweden does, for example, because waste management is regulated at the local level. We can't solve

---

110 73% of U.S. voters support scaling the production of American-made clean-energy technologies (84% of Democrats and 67% of Republicans).

that from the federal-government level, but you can solve it as an activist in your community.

Ayana: Rabble-rousers in city council meetings. Let's go.

Jigar: Then the third thing is we need a lot more people to choose the trades over college. We are at least a million people short on the trades. That's a huge thing we need to fix.

Ayana: Wow. And with "the trades," we're talking about electricians and plumbers and builders and pipefitters, et cetera.

Jigar: Yeah. And it doesn't have to be gendered. Almost everything in this country was built by women during World War II because the men were off fighting. Everybody is eligible for this.

For a long time we allowed for the message to be that everyone has to go to college or else you're never going to be something. And that can't be the message today. It has to be that some people want to work with their hands and you can still make a six-figure income working with your hands. We should stop judging people for wanting to be an electrician or a pipefitter or a truck driver.

Ayana: I'm so enamored with people who have actual skills. Swipe right on electricians.

CULTURE IS THE CONTEXT

My mission in life is not merely to survive, but to thrive; and to do so with some passion, some compassion, some humor, and some style.

—Maya Angelou

What if new storytelling matches our new reality?

What if films acknowledge climate as the context?

What if the news is 50% climate instead of 1%?

What if journalists focus on solutions?

What if disposability is out and durability is in?

What if we collaborate across generations?

What if we view climate action as a moral imperative?

What if . . . ?

10 Problems

- Corporate TV news (ABC, CBS, NBC, and Fox) spent only 17 hours covering climate change in the entirety of 2023, across 435 news segments—less than 1% of overall news programming.

- Only 2.8% of the 37,453 scripted films and TV episodes released in the U.S. between 2016 and 2020 mentioned climate terms like solar energy, fracking, sea level, or renewable energy, and less than 0.6% specifically mention "climate change."

- The fashion industry is responsible for about 10% of global carbon emissions and 20% of industrial water pollution. Textiles alone use nearly 10% of the plastic produced each year.

- Americans dispose of about 13 million tons of clothing and footwear each year, of which only 13% is recycled. The average fast fashion garment is worn as little as 7 times before it is discarded.

- Roughly one-third of all food produced in the world is wasted, accounting for 8% of global greenhouse gas emissions.

- Roughly 75% of U.S. public-school science teachers cover climate change, but only devote 1 or 2 hours to it per year on average.

- 45% of the U.S. population does not participate in any outdoor recreation. Americans spend nearly 7 hours per day in front of screens.

- In the U.S., the wealthiest 10% are responsible for 40% of the greenhouse gas emissions, with the top 1% responsible for around 16% of the country's emissions.

- Aviation contributes over 2% of global carbon emissions, with the most frequent fliers responsible for 50% of those carbon emissions. Air travel is expected to triple by 2050.

- For 1 in 4 child-free adults, concerns about a dangerous climate-changed future are a major reason for not having children.

10 Possibilities

+ In 2023, 22% of climate coverage on U.S. broadcast TV news features solutions.

+ Only 25% of U.S. viewers say they hear about the climate crisis in scripted TV and film. Nearly 50% of viewers are open to seeing more.

+ If the average number of times a garment is worn before being discarded doubled, greenhouse gas emissions from the textiles industry would be 44% lower.

+ Approximately 27% of the U.S. population, and growing, has access to food waste composting programs.

+ An EV can use energy 4 times more efficiently than a gasoline vehicle. Globally, 14 million EVs were sold in 2023 (up from 120,000 in 2012), and the share of EVs on the road is now over 14%.

+ 60% of urban trips are shorter than 5 kilometers, and 25% are less than 1 kilometer. Using a bicycle instead of a car for short trips could reduce travel-related emissions by 75%.

+ 75% of registered U.S. voters agree that schools should teach children about climate change.

+ For political conservatives and for fathers in the U.S., conversations about climate with their children doubled their concern.

+ Young Republican voters (18 to 39 years old) are more environmentally concerned than older Republicans, and are more likely to favor policies to reduce the effects of climate change.

+ 62% of U.S. adults say they "feel a personal sense of responsibility to help reduce global warming," but 51% say they don't know where to start.

I Dream of Climate Rom-coms

Interview with Franklin Leonard and Adam McKay

I love science and the scientific method and how we can test hypotheses and learn things about the world. I love facts. But, if anything, our climate scenario has been one brutal lesson after another about the selfish, deadly quest for money and power, facts be damned. I'm looking at you, fossil fuel companies whose own scientists warned of the climate crisis back in the 1970s. Also you, corporations bulldozing forests and spraying poisons. We see you, banks that finance it all and corrupt politicians who make it all possible. We're on to you advertising agencies and PR firms greenwashing the hell out of it.

Clearly, we need to reassess how we're communicating about climate. Scientific papers, news reports, and documentary films aren't getting the job done, accurate and thoughtful though they may be. Facts (alone) won't save us (sigh): It's a battle for hearts, not just minds, and hearts are opened and tugged by stories.

How do we weave compelling narratives about this highest-stakes drama playing out on the planet? Here are a few hints from polling across twenty-three countries that was led by the nonprofit Potential Energy: (1) people don't like bans (e.g., of gas stoves)—instead focus on upgrading to clean technologies; (2) "fear versus hope is the wrong debate" because *love for the next generation* is far and away the most popular reason for climate action; and (3) the narrative "later is too late" outperformed all alternative messages, including creating jobs, improving health, fighting injustice, and preventing extreme weather. This is helpful orientation as we consider: What stories, whose stories do we tell? How do we shift the zeitgeist to make climate pollution untenable and climate solutions the status quo?

Though I am absurdly out of touch with popular culture, I appreciate that culture is the context for everything. It shapes policy, ranks priorities, guides political will. I also appreciate that a constant barrage of bad climate news is unbearable. So my North Star for climate communication is to take climate seriously, but not take *ourselves* too seriously. In other words, how about some jokes? As filmmaker Adam McKay and I put it in an op-ed we co-authored:[111]

> When people laugh together it gives them perspective, relief and, most of all, a semblance of community. Research shows that humor can lower our defenses and make hard truths easier to hear. . . . This is not to say that everything should be jokes and memes. There's no one right way, one proper tone, for communicating about climate. We need all manner of compelling creations to reach all types of folks . . . funny, dire, hopeful or all of the above.

Adam's illustrious career started with improv comedy and then years as head writer for *Saturday Night Live* before he made movies including *Anchorman*, *Vice*, and *The Big Short*, and co-created *Funny or Die*. I met Adam at a screening for his film *Don't Look Up*, which is essentially a climate movie posing as a satirical comet-disaster movie. I remember sitting tensely in the audience, relating all too well to the disregarded scientist characters who were trying to warn the government and public. I was in a stinky-nervous-sweat panic, wanting to yell at the screen, "Listen to the scientists!" (Spoiler: They did not listen. That did not end well.)

Powerful, inspiring films begin with an excellent script. Franklin Leonard, founder and CEO of The Black List, knows that better than anyone. Franklin and I go way back to freshman year of college. He graduated and went to Hollywood, where he worked at various big-deal production companies, like Universal Pictures. He read a lot of scripts, and watched a lot of movies. Turns out a lot of the best scripts weren't being made into movies, so Franklin created The Black List to surface incredible screenplays (regardless of a writer's industry connections), and help get a broader spectrum of films produced by a broader spectrum of writers. It's working. From 2007 to 2023, the films that made their annual shortlist have won a total of 54 Academy Awards from 267 nominations, including 4 Best Pictures and 11 screenwriting Oscars—

---

111 I loved co-writing this with Adam because I could drop in a comment that said "Insert joke here" and he would do just that. See: "Why Our Secret Weapon Against the Climate Crisis Could Be Humour," in *The Guardian*.

and generated 90% more revenue at the box office than films not on their list. Most exciting is that their meritocratic screenplay evaluation process and support for writers has created many opportunities for social impact in storytelling. On the topic of climate, The Black List has teamed up with environmental nonprofit NRDC. As Franklin describes it,

> Our partnership with NRDC is explicitly about identifying writers who are either already writing something in the direction of climate storytelling or who may write their next thing in that space, and giving them the financial support and also the relationship support—whether that's people in the industry or scientists—to make sure that whatever it is they're trying to write, they can write as well as possible. It's like, 'We got you; we'll connect you to the people you need.'

It goes both ways: Hollywood needs expert advice, and experts need Hollywood storytellers. And what a powerful synergy that can be. Scripted TV shows and films are a popular source of information on social issues for over 75% of viewers, yet only 25% say they hear about the climate crisis on the screen. We can change the climate conversation, with a little help from our friends.

————

Ayana: Given that climate change is the biggest crisis humanity has ever faced, and there are endless possible storylines—from apocalypse to possibility, to dramas and thrillers and romance—why are climate movies so scarce?

Adam: The big thing to remember is that <u>Hollywood</u> reflects the world at large and what people are interested in, which tends to be stories we've been telling over and over again for thousands of years. In the immediate sense, people respond to stories through a kind of muscle memory—it's emotional and inherited and cross-generational. And with climate, there's one central problem: It's never happened before. It's hard to overstate how much we have never encountered anything like the breakdown of the entire livable climate.

Our media—especially corporate media and media that's owned by billionaires and media that takes fossil fuel advertising dollars—has completely failed to express the urgency. Whether that's on purpose or a subconscious conflict of interest, who knows? The biggest problem I encounter when talking to people about climate is that most people just haven't heard the scientific information, so they tend to think, "It's gotta be hysteria." Like, "Yeah, it's bad, but no way it's *that* bad because nothing's that bad."

Ayana: Franklin, does Adam's perspective resonate with you?

! **Franklin:** It resonates greatly. As a society, we have not grappled with the reality in which we live. We have not grappled with it as individual human beings, myself included, nor as an industry—Hollywood is part of that failure.

Film, television, these are businesses that are making art, but that are also making a product that they're trying to sell to people. But they tend to be shortsighted and not even terribly data-driven about what their audiences want. They're just making what they *think* the audience wants.

Ayana: Like the misconception that films with Black casts won't make that much money, and then there's *Black Panther* breaking box-office records.[112]

Franklin: Right, what people think will sell is based on conventional wisdom that is all conventional and no wisdom. I would argue that resistance to telling great climate stories would probably fall in that category. Part of the solution involves telling climate stories that don't feel like taking your vitamins, but feel like they are just good, entertaining comedies or action movies.

Ayana: As Dr. Kate Marvel has said, where is *Fast and Furious: Electric?* Or *Public Transit?* Or *Fast and Furious: Amtrak!*

Franklin: Exactly. Where is the *Fast and Furious* movie that lives in the world of non-combustible public transit? I don't know what that movie is, but there's definitely an action movie to be made that's set in a world where as a society we've changed how we move from place to place. What does a heist movie look like in that environment?

Ayana: Ooh. My favorite genre.

---

112 Hollywood could gain $30 billion a year if it adopted more racially inclusive practices, per a McKinsey study.

Franklin: You can put all these futures into the backdrop and make the movies entertaining. When I think about climate storytelling, a lot of the work that I admire is in the *Black Panther* franchise.

Ayana: Yes, vibranium and maglev trains and green buildings and trolleys and that whole world they created is such a compelling vision of climate futurism without urban sprawl.[113]

Franklin: Exactly. Marvel Studios did not say, okay, we're going to make a climate movie, let's kick off *Black Panther*. No, they decided: *We want to make* Black Panther, *we think there's a ton of money to be made doing it, but implicit in making that movie is all of these other things.* I think it's inevitable that climate storytelling becomes more important, but because of the nature of how Hollywood is organized, the industry itself is never going to be the driving force, which is why we have to be vigilant about coming at it on so many other fronts.

*

Ayana: The movies don't have to be about climate. But it seems to me that if we don't acknowledge that climate is the context within which every story on Earth is now unfolding, then we're sort of missing it, right? Because this is the world we live in. We are all experiencing disasters *now*, which are increasingly less natural. So it seems strange to me that Hollywood has essentially been ignoring this major global phenomenon.

Adam: There's an accidental collision of two breakdowns going on. In fact, maybe without one you wouldn't have the other. First, it so happens that institutions across the world have been hit with a thwacking, giant, rogue wave of money and corruption, legalized corruption in a lot of cases. These institutions, whether you're talking Hollywood or the media, they've all been strapped to a different purpose than the collective good. They're now entirely driven by stock prices, shareholder value, quarterly profits. Franklin's right; Hollywood doesn't think in terms of "What are the most constructive movies we could make?" There's not one single meeting happening in Los Angeles right now where anyone is saying, "It's great to make these movies with car chases that make a lot of money, but what's something we could do to really help people?"

!

---

113 Kendra Pierre-Louis, whose interview follows this one, wrote about this in "Wakanda Doesn't Have Suburbs," her essay in the *All We Can Save* anthology.

!   Second, and more dangerous, is the fact that the media, the way we communicate collectively, has been entirely privatized and is entirely driven by clicks at this exact time that we need collective honest communication. That really is troublesome. Because no one's going to do anything just because it's the right thing to do, or because we need it, especially with a story that is not yet a part of our collective imagination. That's the tricky place we're in. The only bridge that gets you there is science.

Ayana: Is this why *Don't Look Up* was nominally about an asteroid and not climate?

Adam: Yeah, but also the movie is silly. It's pretty funny that it's a trope— one we've seen in a ton of disaster movies—that the comet is going to hit the planet and everyone rallies to deal with it. Well, nowadays, would they rally? So, comedically, a comet was a little snappier, and it happens to be a pretty good analogy.

Ayana: And when we met, outside a basement screening room in some hotel in New York City, I remember immediately giving you a hard time, like, "Great, so you made this movie, but at the end of it people are going to want to be a part of the solutions. What's your plan for harnessing that?"

Adam: Yep.

Ayana: And I remember thinking, he's probably like, "I just made a whole movie with all these famous people and everyone's gonna watch it. Isn't that enough?!" But instead of dismissing me, a stranger, you took my prodding to heart, brought in collaborators, and doubled down on the film's impact campaign.

Adam: Did I ever tell you, by the way, Netflix commissioned a fascinating study? It was cool because what they found was, first off, that people just seeing the movie, the percent of them who said they were willing to take
*   action to address climate change was like 3%. But when people also interacted with the impact-campaign content, that nearly tripled, to 8%. So that conversation we had after that screening had a big impact.

Ayana: Well, that's amazing to hear. Glad I strong-armed you into giving other people homework.

Adam: Yeah, keep doing it because they estimate something like 200 million people saw that movie. So those statistics are a big deal.

Ayana: Okay, one thing I'm hearing from both of you is we have to figure out how to make people a lot of money while making movies about climate. It seems that should be a pretty easy nut to crack, given that there's an infinite number of stories on/around/related to/with subplots about the climate crisis and all the solutions that are possible all over the planet. There's so much rich fodder, not just for primary storylines, but also for side intrigues.

Franklin: Yeah, the reality is that making something great is very hard. If you imagine for a moment all of the storytellers all around the world that were going to write something or create something, the vast majority of those films are not going to be things that anyone is going to invest significant money in or try to distribute globally. Not every filmmaker is Adam McKay. Great storytellers may want to tell climate stories, but they may also have other thematic artistic storytelling preoccupations.

Ayana: But can we have a <u>rom-com</u> meet-cute where two neighbors fall   ✦ in love starting community composting?

Adam: A hundred percent.

Franklin: A hundred percent.

Ayana: Or, she's at a city council meeting lobbying for bike lanes and his public comment was so sexy.

Franklin: Absolutely. And it should be on all fronts too. It can't just be the giant Netflix movie that is going to be a contender for the Oscars. Where is the Hallmark Christmas movie climate story? By doing that, you might be making something that connects with a lot more people, that may change their behavior or get them to look further into these issues.

Ayana: Like, "We got him skis for Christmas, but there's no snow, and little Johnny is crying. What do we do?!"[114]

Franklin: The first step is always a great story, well told. And it's important that we acknowledge that doing that is extremely difficult.

---

114 On December 25, 2023, only 17.6% of the lower 48 U.S. states had snow cover, the lowest in at least two decades. RIP white Christmas movies being relatably realistic.

Adam: You know, given that a lot of our commercial media is steered almost entirely toward making profits, there's another road to go. Audiences, especially younger audiences, can spot BS and are sick of it. And that is a giant opportunity—it exists with theater, with film, with novels, books, podcasts, across the board, maybe even a whole new type of medium that we can't even imagine right now—to just go around the money because the money is gridlocked. The money has one message, "Keep the status quo," and that's not how life works.

I think there's a big change coming in the next few years where you're going to see an explosion of independent filmmaking, the likes of which we've never seen before, stuff that is raw and true and original. And the same with theater. The same with civil disobedience connected to theater. Franklin has cultivated this alternative road more than just about anyone. But I think, Franklin, your road's about to become a highway—I mean bullet trains and electric cars, of course.

Franklin: It's hard enough to be an artist in this world, and the last thing I want is to dictate to artists what they should be writing about or how they should be doing what they do. But I want to make sure the people who *are* doing amazing climate work have the infrastructure and resources to continue telling their stories.

Adam, I think you share this concern, that even if you did have an amazing script on this front, the folks who you would theoretically take it to to get it made are resistant, are definitely not predisposed to being receptive to it. And I don't know how we solve that problem on the scale we need to, as quickly as we have to solve it. There are projects that are going to go into development now that will still be in development seven years from now.

Adam: Without exaggeration, I just wrote an email to a big, giant director passionately pleading with them to take a look at this climate project we have, which is entertaining, moving, a little bit funny. Yes, it's about climate, but I think the storytelling is sound. And surprise, surprise, the problem I keep running into is people don't understand the urgency.

What we need is a team of five scientists who are friendly and cool to talk to this director, to go meet with studio execs and express this urgency, because it's the missing ingredient. I want to set up a hotline where we could have creative people call scientists and ask, "What's going on?"

Ayana: Great idea. When we do see filmmaking about climate, it's often about disaster, so I'm curious: Is there a way to tell stories about *solutions* that's exciting and palatable? Because that's where I wanna go. Everyone knows the future we are trying to avoid—*The Day After Tomorrow* was actually effective in raising awareness about the future we *don't* want. But what other future could we have instead is a topic that we're not seeing enough media about.

Adam: Well, this is the result of decades of billions of dollars being thrown into our media ecosystem from giant oil companies telling us that we need to be afraid of the solutions, telling us we're going to be forced to follow new laws, that we won't be able to drive our wonderful, cool cars. Meanwhile, the solutions are *awesome*. The solutions are just better. It's borderline utopian. Energy could be free. That's for starters. Pollution will be mostly gone. The world will be reborn.

Probably the future with architecture is going to integrate the natural world. It's more than Frank Lloyd Wright designing a house with a waterfall going through it. What we're seeing now is a new type of design for homes, for buildings, where you have moss on the walls, because guess what? Moss works as insulation, it improves the oxygen in the room, and there's some of it that's bioluminescent so it can serve as a beautiful night-light. And trees that grow through buildings, buildings covered in vines and plants, common spaces with ponds.

And the solar panels are getting smaller and smaller—probably what they'll look like in maybe twenty years are little triangle-shaped gleaming gems, hitting the sun. They're going to look beautiful. We're going to have to paint most buildings with a new paint that's highly reflective. "Oh my God, I don't want my house to be white." Uh, have you ever been to the islands in Greece or looked at a picture of them? They're stunning. Not bad to look at. We could have roads paved, instead of with asphalt or whatever, with solar panels, which is already starting to happen.

The solutions are fantastic and the solutions appeal to all of our instincts. If you're someone who maybe leans more right-wing and you don't trust big institutions and you want the right to make your own choices, there's energy independence, unplugging from these giant utilities, not having to be strapped to the grid. And if you're on the other end of this spectrum and you're progressive and you believe in communal life, that's going to be part of this solution as well. We're going to be sharing resources more and more, we're going to have a world with more common spaces, more parks. You're going to be getting rid of gas stations. It's hilarious how fantastic the solution is. And yet there's a huge amount of propaganda and misleading information that's been put out there when it's just a glorious new valley with beams of sunshine going through it.

The solutions are as enjoyable, wonderful, and life-improving as anything you can imagine. And necessarily as a result of that, we're not going to work fifty, sixty hours, seventy hours a week anymore. The workday will shorten. They're probably going to have to have universal basic income of some sort, because something like 3.5 million people are employed driving trucks around the country. That kind of stuff is going to go away. You're probably going to have bullet trains that can long-haul cargo. But yes, my headline is: The Solutions Are Awesome.

Ayana: Franklin, I wonder if that's actually part of the problem, because in dystopian movies we can see the classic hero's journey. And in the more utopian scenarios, there's not necessarily the same tension, like, "Okay, that's nice, but boring—where's the drama?"

Franklin: Stories are sequences of conflicts—position, counter position, conflict resolution, and so on and so forth at various scales, both intimate and in terms of the broader story. Could you create a big action movie in a utopia? Sure. But that utopia has to cease to be a utopia the second there is a conflict. It's not that it's boring, it's that you don't have the fundamental principles that propel a narrative, in my experience.

Ayana: What about stories of progress? Because there's a question of, like, how we get from here to there, which is not going to be easy. The story is not that we're done; the story is constant overcoming until we get close.

Franklin: Stories about how we could get there can definitely work, again, as long as they're entertaining and look like the kind of thing that an audience is going to want to see. Just to go back to the *Black Panther* example. As the backdrop, it doesn't even have to be utopian. It's just, "Look at these things that could change our experience in the world."

✱ I believe that film and television are both mirrors and windows. They tell us the way the world is, but are also instructive about what's possible and what could be, and can serve as inspiration. I believe in a world where there are a bunch of kids who are watching movies and seeing a scientific innovation that doesn't exist, a product that doesn't exist, and they may say, "One day I'm going to invent that."

Or directionally, they're going to be trying to create a world that looks a little bit more like Wakanda or a little bit more like some other, more utopian vision of the future. And providing examples of that, even highly speculative, theoretical ones, creates the conditions by which the generation that experiences them in the movie theater goes out into the world and tries to make those things real.

If you tell climate stories well, if you present a vision for the future that is compelling and cool and has all of the cool gadgets and things that we would all love to see in the world, people are going to pay money to see it. To this day people are obsessed with the Marty McFly *Back to the Future* shoes, right? That technology exists because someone saw it in the movie and said, wait, we can make those.

Ayana: Compelling and cool is definitely where we need to be positioning climate solutions.

Franklin: Or look at the "Scully Effect." Something like two-thirds of women in STEM during a certain period in the last twenty years cite Dana Scully from *X-Files* as an inspiration for them.

Ayana: Oh, wow.

Franklin: It's fundamentally unquantifiable, but it's also undeniable that the aggregate effect of millions of young women watching Dana Scully was more of them choosing to go into STEM.

Ayana: Or Shuri in *Black Panther*.

Franklin: Exactly. And what do these women accomplish? How do they do things differently than their peers did before they entered the space? What inspiration do they serve to other young men and women who then go into those spaces as well? It's a dramatic ripple effect. You want ＊ to create as many ripples as possible in order to create the endgame.

Ayana: We've talked around it, but I wonder if there's something fundamental we've missed, which is, Why is it even important to tell stories about climate change at all?

Adam: Franklin said it beautifully, mentioning the Scully Effect. And Einstein has written beautifully about this, how he viewed physics as an act,
★ first and foremost, of imagination. In dealing with the most catastrophic threat in human history and for life on this planet, we are going to need imaginative new types of stories to bridge that gap between what we've known and what we now need to know.

Ayana: Franklin, why do we need to tell stories about climate change?

Franklin: Because it's the most important issue confronting human existence, and the job of storytellers is to explore exactly those things. I don't think we can have a healthy storytelling culture if it is not engaged in exploring the realities of the world in which we live.

Adam: That is a major point. You see it in the history of Hollywood in the 1960s, where the world had changed but Hollywood kept trying to jam the same old movies in. They were doing these cheeseball musicals and old traditional John Wayne Westerns and because of it the studio system collapsed. Famously, that's what led to the '70s film explosion renaissance. *Easy Rider* in the late '60s kicked us off and we saw brilliant writer-directors. New types of storytelling emerged because that old framework wasn't working. We're in exactly that same spot right now.

Look at the ratings for broadcast news. Look at the readership for newspapers. Look at the ratings for traditional television. They're telling a story that is vestigial. It's a story from 1997.

Ayana: Golden age of hip-hop, but we'd like to evolve in some other ways.

Adam: Oh, the hip-hop was good back then. No question about it. But the idea that there's two political parties having a battle of the ideas?
★ The idea that an environmental issue is just an issue? We're sold these narratives that no longer apply to the realities ahead of us. Actually, our storytelling needs to catch up with reality. And this is a profitable, career-enhancing endeavor. I mean, look at those filmmakers in the late '60s and '70s and what legends they became because they paid attention to where the world was at.

Ayana: The big question that we're here to grapple with is "What if we get it right?" What would the world look and feel like if we were actually, properly, with the required urgency, addressing the climate crisis in our culture? How would the process of producing or financing these cultural products change?

Adam: I'm not sure that financing would necessarily change. The idea that dealing with reality tends to be profitable is a pretty sound one. But the level of communication, the way we talk to each other, would change radically. Right now, network news is soft, kind of glancing news. What the news would look like instead is that when you turn it on, there'd be a bug in the corner with the parts-per-million CO_2 count. And also a bug for where we're at for warming.

There'd be no fossil fuel advertisements—that would be the equivalent of having advertisements for how much fun Russian roulette is for kids. The fossil fuel ads are going to look insane in about four or five years. Like, how did they ever do that? There's going to be no ads for gas-powered cars. The whole way you're going to engage with the idea of news is going to change. The sad thing is it's probably going to change because of crushing catastrophic reality. Whereas if they all took a beat, and took in reality, most of these people doing this broadcast news and a lot of print news would realize the way they're doing it is insane. It's the result of a deranged reality. So, yeah, it's going to change the whole emotional timbre. I'm starting to see it happen a little bit in conversations with people. They're starting to take in the enormity of it. And 90% of future news is going to be climate or climate-related. The same way news in London before the Nazis started sending in warplanes was entirely about preparation for the oncoming invasion.

Franklin: I think that's right, and when you change the media context ✶ within which writers are working and within which studio executives are making decisions about what to make and what to invest in, that changes everything. If you imagine a world where we are constantly seeing these climate images and this is the focus of the conversation, it is inevitable that the things people write, the things people aspire to make, will involve those ideas.

Ayana: There's an analysis by Media Matters about how few minutes of climate coverage there is on mainstream TV news. It's around 1% of ! overall news programming. So that rings true to me, Franklin, that if we had a level of news saturation that reflected the seriousness of the challenge, people would be dealing with it differently.

What advice would you give to screenwriters, filmmakers, storytellers who want to tell climate stories that can break through?

Franklin: Make it entertaining. Make sure that the person reading it wants to turn the page to find out what happens next, and make sure that when they're done, they're a little bit sad that they can't spend more time in that world with those characters. If you do that, you're on your way. There's no guarantees, but if the reader doesn't feel like that, there's not going to be much opportunity there.

Adam: I would say don't rely on the old stories. Don't rely on what you were taught in screenwriting class. Use that as a guideline, a base, but don't rely on it. Look toward the actual reality that we're living in and where we're headed. And, Ayana, you said this earlier, there are hundreds of thousands, millions of stories in this new reality. They will pour out of the realization of where we're going and how we're living right now. You're doing something right for the world by writing stories about the collapse of the livable climate, about corruption, and our inability to grapple with this new world we're living in.

People are going to feel the truth behind what you're doing. It's going to crackle, it's going to be scary, it's going to be funny. It's going to be touching in ways that stories from the past can't be because they don't fit the world we're living in. So it's extremely exciting telling these stories about this new world that we're in, even though, yes, it's scary, too. And it should never feel like medicine. It should never feel like you're doing it because it's the right thing to do. These stories should be head-to-toe alive and funny.

Ayana: I want the climate solution rom-com subgenre so bad.

Adam: Amen.

Ayana: The stories we tell matter, including the love stories.

Franklin: My banner quote on Twitter is from Alan Rickman: "The more we're governed by idiots and have no control over our destinies, the more we need to tell stories to each other about who we are, why we are, where we come from, and what might be possible." That has been the primary organizing principle of my work for the last decade, as it seems like we're more and more governed by idiots and have no control over our destinies.

Ayana: So, leaning into possibility becomes ever more critical. Not hope or optimism or pessimism, not utopia or dystopia, but what is *possible?* How can we help each other see that, find our roles in it, lean into it, be a part of solutions? I find it frustrating to see climate stories often presented as this binary, as we either succeed or fail. Because it does matter if we get it 80% right versus if we get it 10% right. Hundreds of millions of lives hang in the balance.

Adam: The world is not a straight line. We're in a new age. We need to see a new type of storytelling to match this new reality.

Houston, we have a problem.

—Apollo 13

The Planet Is the Headline

Interview with Kendra Pierre-Louis

The real-life story of the climate crisis is more intense, wild, and over-the-top than any Hollywood thriller or heist movie. Our reality would get laughed out of the writers' room as too on-the-nose. But does this story-of-all-stories get wall-to-wall news coverage? Nope. Despite a horrifying parade of extreme weather events fit for an apocalypse film, less than 1% of corporate-news airtime in 2023 was about climate—up from 0.4% in 2020! Despite climate being the greatest threat to life on Earth in 65 million years, since the asteroid that took out the dinosaurs, it got a combined 23 hours of coverage from major networks in the U.S. over the course of that *entire* year. As of 2023, 38% of Americans report that they only hear about climate in the media a few times a year, max.[115]

As Franklin and Adam emphasized, getting climate reporting right is key to shifting what's happening in Hollywood, and to changing culture more broadly. Until my conversation with them, I hadn't been planning to include a chapter on the news. But as we spoke, two things quickly became clear: We needed to dig into journalism, and that conversation needed to be with Kendra.

Kendra "Gloom Is My Beat" Pierre-Louis, as she states her name on Twitter, is a science journalist who has been focused on climate for about a decade now. She's worked in newsrooms from *Popular Science* to *The New York Times* and is now reporting at *Bloomberg*. We met at what was billed as a serious pro-science event that turned out to be more like a wacky pro-atheism conversion event. We both had the same "Huh? What?" reaction, caught each other's eyes, and proceeded to whisper snarky commentary to each other in the back of the room. Basically, the start of any good friendship.

Kendra consumes a *shocking* amount of media and is one of the most well-read people I know. She's a human encyclopedia prone to tangents,

---

115 This paragraph originally appeared in the op-ed I co-authored with Adam McKay. I tried to find another way to put this, but this is still exactly what I want to say.

which is much more hilarious than it is frustrating—even during meetings when we worked together on the *How to Save a Planet* podcast, where I was thrilled to recruit her in as senior reporter.

On the whole, Kendra is disappointed in journalism (same), but sees clearly how it could morph into an accelerant for addressing the climate crisis, especially with more focus on solutions (also, same). Here's her vision for that transformation, with a few of the meanders edited out, so you don't lose the path.

Ayana: If you were in charge of a newsroom—*The New York Times*, CNN, NPR, local news—what would you change about how climate is covered?

Kendra: In order to understand a little bit about what I would change, it first helps to understand two big things about climate and the news. First is that unless we're in the middle of a climate-linked disaster, climate change is still relatively slow-moving and invisible. And that makes it difficult for news outlets to cover it. Second, news generally has a very problem-focused lens: "If it bleeds, it leads." And that lens doesn't lend itself to solutions. Or if it does, the solutions it covers are simplistic: Here's how to recycle, here's how to buy an EV. As journalists, we're not trained to cover systems change. The idea motivating journalism is, "If we tell you there's a problem we've done our job, and it's up to society to figure out what the solution is."

Ayana: That's helpful context.

Kendra: If I were in charge of newsrooms, I would make sure our journalism was helping people understand how to be participants in a democracy. That's a niche that local news used to fill. They would tell you when your community board meetings were happening, they would tell you when your school-board elections were happening, they would even tell you how to run for school board. And I think that type of journalism is as important as the deep investigative pieces that highlight the problems.

Ayana: On climate specifically, what would you change about how the news covers it?

Kendra: One thing that is crucial for journalism generally, but especially acute in climate journalism, is to shift away from our narrow focus on problems. We need to zoom out and ask questions around solutions. The person who's reading or listening to or watching your news story is not just a consumer, they're not just a passive observer of the news you're giving them, they're a participant. The media should be answer- ✱ ing the fundamental question: What do people need to know in order to be better-engaged members of society? How do we make it easier for people to participate at all levels of governance?

Ayana: What would you change about the types of things we cover or how we cover them?

Kendra: First of all, I would get climate out of the climate ghetto. It's getting a little bit better, but, in general, in a newspaper, you turn to the climate section. Or on TV news there might be a climate segment. In reality, climate touches all aspects of society. And reporters on all desks need to ✱ understand where they should be inserting climate into their stories.

Ayana: In the business section, in the real estate section, in the technology section, in the travel section, we should be seeing climate as the context. If you were running a newsroom, how would you make that happen?

Kendra: Some of it comes down to training or retraining. A lot of people don't understand climate change. If I ran a newsroom, I would have training sessions for my staff members to understand the implications of climate for their beat so that they understand how they overlap.

If a newspaper has an auto section, for example, I would shift it to a transportation section because we know that we need people to get out of cars, so why are we so narrow? Why is there so much coverage of EVs as personal vehicles instead of other things that fall under the electric vehicle umbrella, like electric buses or high-speed rail, or trolleys or bicycles?[116] If you have a transportation section, it functions differently.

Another change would be to the way we cover politics generally, not just as it relates to climate. Political coverage is very much about the horse race of elections. There's this expectation that your role as a citi-

---

116 Or sailboats, electric boats, and hydrofoil ferries!

zen in this country is to vote every two or every four years and that's where your role in democracy begins and ends. That's not true. There are all of these ways for you to participate in democracy, from planning board meetings to running for office yourself. It doesn't have to be big state- or federal-level positions.[117] But all of that gets swept to the side in the coverage. You don't read about civic engagement in the newspaper. Maybe you google it, but all of that is left to the individual to learn on their own, someplace else.

Ayana: Is there a particular example you can share about journalism and civics?

Kendra: Okay, so here's an example from my own life. I did not understand until I was twenty-seven that anyone can effectively write a bill. There's a process to getting it before a legislator and getting it introduced. It's not uncommon for bills to start not in Congress but with a think tank or a lobbying group.[118] Many bills can get started at the citizen and at the grassroots level. But because most people don't know that, that gives disproportionate power to the people in lobbying groups or in institutions or oil companies who understand how this system works. We learn about bills that are in progress, but we never learn how that bill came to be.

Ayana: Yeah, I didn't really understand that until I was in my thirties.

✱ **Kendra:** One of my pet peeves is that as an industry, the way we report on politics treats it like a spectator sport. But politics needs to be participatory, and in order for that to happen, people have to know how they can participate. We might report on a bill, but we won't tell people that there's a public comment period. We talk about the process of the bill, but we don't necessarily let people know that you can write a bill. There are ways of getting engaged, but there's this expectation that people just know. And I think most people don't know.

For the vast majority of us, who probably never got civics in school, politics is a black box. And then the way in which the news reports it re-

---

117 Shoutout to Lead Locally, supporting climate candidates in local elections across the U.S. See: leadlocally.org

118 Case in point, how the Green New Deal came to be, which is explained in the upcoming interview with Rhiana Gunn-Wright.

inforces that black box. People need a Civics 101. We never learn how to ✱
engage with our members of Congress. We never learned that we, as the
public, are allowed to weigh in on bills. There are opportunities for public
comment on not just bills but also regulations.

And I think, oftentimes, stories focus too much on "Well, this party
likes this bill, or this party doesn't like this bill, or this will be a win for this
issue for these groups." What gets obscured in pursuit of this horse race
is what the bill actually says, what the implications are, and whether or
not those implications are good or bad. The words that I hate the most in !
stories are "activist groups want." With climate, most of us want a livable planet. Most of us want air we can breathe. Most of us want drink-
able water.

Ayana: That part is important because sometimes the trick is that the
title of the bill sounds really good. It sounds like something you want, but
you don't actually know what's in it, or that it's actually written by a fossil fuel lobbyist, or what alternatives might be better, or who benefits and
who might get screwed.

Kendra: Or even what the long-term implications are.

Ayana: Or whether there's a more cost-effective or more equitable option. We need more comparative analyses of climate legislation instead
of, "This is how many Republicans are against it. This is how much it
costs." Weigh the options for me. There was way more press about how
much money the Inflation Reduction Act would cost than about what
was actually in the bill. Why does the news cover politics like this right
now?

Kendra: I have a few theories. Part of it is that our brains are wired for
story, and the horse race is an easy narrative for people to fall into. And
using that story framework is easier because you don't have to dig into
the substance of the report. When you're talking about legislation that's
in process, a reporter might have to read hundreds or thousands of pages
over and over again to figure out what the substantive changes are. And
that's a pain. I've been slowly working my way through the Inflation Reduction Act and it makes my eyes bleed, especially since so much of it is
in reference to weird tax codes and you have to go between documents.
It's not easy.

Also, people are trying to offer a take as quickly as possible. It's much easier with a bill to be able to call up a source who's on the Republican side and a source who's on the Democrat side, and get their position and read the summary to get an idea of what the bill says and then write that take.

! **Ayana:** We're doing he said/she said reporting on climate as opposed to asking, "Is this a good policy proposal?"

Kendra: And that gets you into the problem of false equivalency, which is getting a voice on each side, he said/she said, to make it seem like you're neutral. An extreme example of this is what we saw in the early days of climate reporting, where one side said climate change was happening and the other side said it wasn't. And even now, when the science behind climate change is on even firmer ground, you still see that pattern of reporting. We don't spend a lot of time in the news debating whether or not gravity is real, right? We don't give a lot of column inches to flat-earthers.

Ayana: What would you say is the current quality of climate reporting overall in the U.S.? If you were handing out report cards, what would our grade be?

Kendra: Oh God. I would give us C+ or B−.

Ayana: Okay, I'm actually relieved to hear it's a passing grade. But there's clearly a ton of room for improvement, as you've laid out. Are we doing better than last semester? Worse? Holding steady?

Kendra: If you compare climate coverage now to, say, twenty years ago, we're doing better because we have lost a lot of the false equivalency. So that's great. You're beginning to see climate change pop up more and more in newscasts outside of "a very special climate episode." But we're not talking about it with the frequency and the urgency that it needs. More publications are pivoting to solutions, but the solutions are often narrowly focused on personal consumption and not on the systems we're embedded in. We're really short on actionable information people can take to their legislators or even take to their kids' schools to push for the bigger changes we need to implement.

In many ways, the way we report on it helps people better understand the long-term risks we're facing, but leads almost to a sense of paralysis. People feel stuck, they don't feel like there's anything they can do.

Ayana: Yeah. You've brought up a few times this need to focus on solutions. "Solutions journalism" is a term these days for this category of reporting. When we were collaborating on *How to Save a Planet* this was something we were trying to figure out—the "how to" part, giving people specific, meaningful ways to get engaged. And that was your first deep foray into this type of journalism. It seems you're all-in now. What was that arc for you?

Kendra: I'm gonna be narcissistic for a second.

Ayana: Okay. I appreciate the warning.

Kendra: On a personal level, when I joined *How to Save a Planet*, I was a little bit burnt out on reporting traditional climate stories. I felt like I was writing stories that people had to brace themselves to read. They knew when they clicked on my stories, that I was about to give them some stuff that was gonna depress them. I think I did a decent job at injecting levity in stories when the topic didn't involve people dying. But even so, I was writing stuff that I knew was going to bum people out and I knew it was going to bum me out. And I wasn't offering an exit. I felt like the work was creating an emotion in readers but I wasn't offering them anything to do with that emotion.

Sometimes with news, you have to do that. You have to write stories that detail problems to make people aware. For instance, I did a story in December 2021 about the implications of rising sea levels on groundwater. It was an important story to do because so many people didn't even realize it was an issue. And in the year since, there have been more stories on the topic. It definitely struck a chord in people. Now there's growing awareness of it as an issue. It's being factored into planning more.

Ayana: It certainly factored into my mother's understanding of where we should move. I sent her an option near the coast and got an all caps reply referencing that piece you wrote, with "WHAT KENDRA SAID" as her veto.

Kendra: Haha. But at the same time, how many times can you read a story about wildfires? How many times can you read a story about hurricanes? They're going to keep happening in the same pattern because the underlying factors that put people at risk aren't changing. And we're not empowering people with the information to change that. The information people need is about the macro level, like what actions we need to take on climate change and reduce greenhouse gas emissions. But it's also on a local level, like how do you get your planning board to stop green-lighting buildings in areas that you know are going to flood? We're not giving people the tools they need to advocate for change.

Ayana: The shift toward a solutions focus is something I'm seeing in science. When I started graduate school, there was this huge debate about how far scientists could go with their research and communication before they'd get considered activists. The idea was that once you're considered an activist, people assume you can't do science objectively anymore—as if caring about, say, whether coral reefs keep existing would make you less reliable in studying them. But with biodiversity and ecosystems deteriorating, scientists are becoming more outspoken, and engaging on policy, as they realize there's a cost to being passive observers. Are you seeing a parallel rise in solutions journalism in major newspapers, or is that still a small niche?

Kendra: There is more focus on solutions journalism, but it's often couched in the frame of personal responsibility. I think because there's still such a fear among journalists of being labeled an activist, there's not as much reporting on the big systemic issues. A lot of it is still pretty focused on, here's the best EV for you, here's how to weatherize your house.

Ayana: When you came over to podcasting, even our team hadn't fully embraced a solutions-journalism framework. I was arguing so strongly for it. Honestly, it seems reckless to lead people so far and then just end the episode with, "Okay, well, that's the problem. The climate apocalypse is nigh. Good luck everybody. Have a nice weekend." So as we developed the show, the team agreed we needed to include calls to action where we would tell people who had been excited about the week's topic, whether that was finance or farming or renewables, what they could actually do to help advance climate solutions. How did it feel to be given this opportunity or directive to have a call to action as part of your journalism?

Kendra: It was interesting because we wanted calls to action, but we didn't want too many personal-responsibility calls to action. We didn't want to do an episode on recycling and plastics and tell you to bring a tote bag everywhere you go. That was not the way we were thinking about solutions. The fundamental shift for me was to look at policy and legislation.

Often there's a climate-linked problem, whether it's fossil gas or whatever, and the default expectation is to treat it at the end point. Like, now we've polluted this river, how do we get the pollution out of the river? Or how do we treat the pollution before it goes into the river? Instead of asking, well, why are you generating this pollution in the first place?

Working on *How to Save a Planet* forced me to think more systemic, more granular, and backwards. It shifted how I even defined the problem in many cases. After you left co-hosting, we had an episode in September 2022 (one of our last episodes before we were canceled, RIP) about the ways in which people had taken those calls to action and individualized them. We had one guy who ran for his public utility board and he is perhaps one of the only people to ever openly campaign for the position, which is kind of hilarious.

Ayana: Ha! That's great.

Kendra: And we had a group of college students in California block an Exxon project, and they did it using information from an episode that one of our co-workers, Rachel Waldholz,[119] produced about climate finance. They were able to make a different argument, not just an ecological argument but a financial argument, about why their municipality should block the reopening of this Exxon offshore drilling platform.

Ayana: This is phenomenal.

Kendra: One of the things that I loved about the podcast was that we were giving people information and they were going out and using it in ways that we hadn't conceived. We were giving people information about how social change happens. And then they were taking that to leverage change in their own communities. And for some reason, sharing that information is considered activism in many traditional newsrooms. And that doesn't make sense to me. We give people information about lots of other things. Why is this any different?

---

119 Rachel also helped me prepare for the interviews in this book. When we prepped for this one, she said, "Billionaires should fund local news. That would solve so much."

Ayana: I love the way you put that. I certainly wish more journalists would talk about the implications of what they were reporting on, and about the levers for change. As you mentioned, a lot of our calls to action involved advocating for policy change in some way. Often the suggestion was "Call your representatives." For example, the Kigali Amendment of the Montreal Protocol treaty commits nations to phasing out refrigerants like hydrofluorocarbons that are potent greenhouse gases. So we suggested people call their senators and voice their support for ratifying that.

People are busy and they don't want to waste their time by filling out a petition or something if they don't see how it matters. But calling your representative really matters because very few people actually do it. The staff write down your message, and they tally the requests of constituents since they want to get re-elected.

Kendra: Yeah. And then for one episode we made you actually call your representatives.

Ayana: I got so nervous! You know I do not like talking to strangers on the phone, Kendra. I was so anxious that I was sweating, but I called Senator Schumer's office to express my support for what was then called the American Jobs Plan.[120] I'm glad I did it.

Kendra: It's so funny because in the real world, you are so much more calm and cool talking with random people and I'm the awkward turtle. But because I'm a journalist, the idea of picking up the phone and calling the town hall, or calling random strangers, is no big deal for me.

Ayana: Well, clearly we're a good team. So, if you could go back a decade to when you first started covering climate, what advice would today-Kendra give ten-years-ago-Kendra?

Kendra: A few things. Think about the solutions as standalone stories. For example, I think <u>producer responsibility</u> bills can stand alone as a story.

Ayana: What is "producer responsibility"?

---

120 The Build Back Better bill morphed into the American Jobs Plan, and an iteration of that was eventually passed with the moniker Inflation Reduction Act. More on that in the next section, "Changing the Rules."

Kendra: Here's a good example: We know that we're all choking on mountains of plastic waste. And we also know that plastic isn't, for the most part, really recyclable. And that most of the plastic that enters our lives is not plastic that we asked for. You order something off the internet, it comes in a box that's filled with plastic packaging. You buy a mouse for your computer and it's wrapped in plastic packaging. You have little control over this. And so producer responsibility legislation essentially says, "Hey, it's not on the consumer to reduce their plastic use."

Ayana: "Hey, corporation, deal with your packaging."

Kendra: Yeah. So much of this plastic consumption is involuntary on the part of the consumer, and the municipality ends up having to pay for the disposing of it through taxes. Producer responsibility legislation would mean that either the producer has to deal with that waste financially— they have to pay a certain tax—or that certain materials are simply banned. Either way, the entity that is making that packaging or putting ✳ that plastic on the shelves is the one that has to deal with the waste associated with it, because it's the producer who's creating the problem. It's a tactic for bringing the problem back to the source.

Ayana: So, for example: "Beverage company, you're in charge of all the plastic bottles you make. It's not my problem, it's your problem. You figure out how to deal with those."

Kendra: And some governments are extending this to clothes. If you're making fast fashion that only lasts eight to ten wears, its disposal is your problem, not our problem.

Ayana: What other advice do you have for a-decade-ago-Kendra?

Kendra: Show don't tell. If nobody's dying in your story, it's okay to make a joke. Lead with humanity, because when we're talking about climate change, in many ways, at the risk of being species-ist, we're talking about saving humanity.

Ayana: Species-ist?

Kendra: Yes. I'm gonna be real. I don't care if the mosquitoes all die. In fact, I wish they would.

Ayana: Haha. What are the stakes for the news getting it right?

Kendra: The survival of humanity on the planet.

Ayana: Okay, help us understand that leap.

! **Kendra:** There's a ton of data that when you lose a newsroom in a community, government corruption goes up, there are all sorts of downwind effects. We've gotten used to so much information all of the time that we can forget that it usually begins with a journalism outlet. And in so many ways our perspective of the world and our understanding of the world is shaped by the news and how the news reports on certain topics.

I would also tell young me that the frames you use to write your journalism matter almost more than the facts. Because how many times have we read a story where all of the facts in it are correct, but the frame is so off that it leads you to the wrong perception?

Ayana: What do you mean by "frames"? Can you give us an example?

Kendra: A bad climate frame is the way news outlets report on climate denialism and make it seem like many more people reject the science of climate change than actually do. The reality in the United States is that the overwhelming majority of people accept the science. The denialist camp is generally 10% to 14% of the population, so that means most people are either fully or partially on board.

Meanwhile I've been giving public talks for four years now, and the most persistent question I get is "What do we do about the climate deni-
* alists?" My answer is we don't have to do anything about them. They're a small minority and that is not where we need to be focusing our attention. But the frame for so long has been about the denialists that people don't believe they're such a small percentage of the population.

Ayana: So the problematic frame is, "Climate-science deniers, so many, such a big problem."

Kendra: And that leads to several things. It means people who are in congressional districts where their legislative official is a denialist don't feel empowered to call him and say, "Hey, change this," because they think they're the only ones in their congressional district who feel that way, and that's not true. Ninety-eight percent of counties in the United States have a majority of people who accept the science of climate change. That means they actually have numbers on their side. But they don't think they do, so they don't have those conversations.

Ayana: What would the good frame be?

Kendra: It might be the one time to go to that diner in a red state and, as a reporter, talk to people and find that they all believe climate change is happening.

Ayana: Ooh, I like that.

Kendra: And then they might tell you the issue is they just don't believe in government interference. Or they don't want taxes.

Ayana: The other thing you've mentioned as a bad frame is "Change your individual and household behaviors in order to solve this," as opposed to, "Here's the policy problem that limits the choices even available to us at the household level."

Kendra: What's interesting is how many stories there are about how I need to reduce my personal carbon footprint and how few stories I've read about how to get a bike lane in my neighborhood.

Ayana: Totally. On another topic, how should journalists be handling disinformation and greenwashing? Because it seems relentless, popping up everywhere. Some articles feel awfully close to corporate press releases.

Kendra: Especially when they're talking about powerful companies—fossil fuel companies and car companies, companies that have a lot of money and a lot of weight—news outlets can be overly deferential. One reason for this is that these companies use a tactic that in sports is called "working the ref." A player or coach will yell at the referee. When the ref calls a foul on their team, they yell that the foul was a bad call and the ref is biased against them. They are relentless. And eventually it works: The refs will start to shift their behavior. The more the player or coach screams bias at the refs, the more the refs start to shift their behavior in hopes that the player or coach will stop yelling at them.

Some of the worst players in the climate space are engaging in this game with the media, working the ref. Many individuals who work in the media come from backgrounds of relative power and relative prestige, and they're not used to being yelled at in this way by other powerful and prestigious parties. It makes them uneasy. And so they acquiesce to these powerful actors.

To your question about how to deal with disinformation, the answer is that certain news outlets need to woman up. They need to get a little bit tougher. They need to recognize that sometimes the people and entities they're reporting on are going to try to defame and discredit them because their reporting is, in fact, accurate. News outlets need to have their reporters' backs and stop watering down stories.

Ayana: How do you think newsrooms should be approaching the coverage of climate activism and the climate movement?

Kendra: I'm going to back into this question. I did a story in 2020, during the height of the George Floyd protests, because I was watching the actual protests on the internet and then I was reading the coverage of those protests, and there was a massive disconnect between the two. In the videos that I was watching, it was clear in many, many cases that it was law enforcement that was the aggressor. But in the stories that I was reading, it was unclear who the aggressor was. It was like, "Violence broke out," which is vague, right?

Ayana: Ah, that dangerous passive voice.

Kendra: I ended up digging into the journalism literature and learning about this idea that's been around since at least the 1980s called the "protest paradigm." Essentially, the idea is that, overall, news as an industry is very conservative and very regressive around the idea of protest. So in many cases, when they report on a protest, they will report on the negative consequences of that protest, like the way it stalled traffic. For example, in 2022, the workers at Kellogg's went on strike. And so much of the news coverage was focused on what brands of cereal you wouldn't have access to and not actually why people were going on strike.

! That kind of reporting is typical. Journalists will focus on the protest tactics and they'll focus on the imposition to the public, but they won't focus on the motivations behind the strike or disruption. And only more extreme protests get coverage, especially for left-wing protests. Because of working the ref, journalists are less critical in this way of right-wing protests. Reporters should focus on the issues and not just the imposition. Because the role of protest, in part, is imposition, right?

Ayana: Yes, the disruption is to get attention. What are they trying to get attention for? And is there a solution or a change that people are advocating for?

Kendra: And does the disruption balance out the thing you're protesting? Because if I'm shutting down the New York City subway because I want a sandwich, the weight of the protest is disproportionate. But we often don't get that discussion of whether the action is proportionate to the level of the harm being protested.

Ayana: So there's a lot the news media could do better. If we get it right, if we do all the things that you suggest, how would that better climate journalism make for a better world?

Kendra: Better journalism will lead to people having more realistic yet ✱ more climate-inclusive or more climate-beneficial ideas of what the future can be. Better climate journalism will lead people to be more effective at pushing for the changes they need in their community to better live within our planetary limits.

In order to reduce emissions, it's not going to be a one-for-one swap. We're not just going to get rid of all of our gas-guzzling cars and replace them with electric vehicles. That isn't sustainable. We're not just going to electrify our single-family homes. We're going to live differently. We're going to live in communities that are a little bit denser, more duplexes and triplexes and quadplexes. We are going to live in communities that have fewer cars and they're going to have better, faster, more effective mass transit. They're going to have more bike lanes. The air would be much cleaner. Our communities would be livelier. The noises would be different. We wouldn't be hearing cars as much and we would be hearing people, which is a big difference.

Ayana: Like children playing and people laughing?

Kendra: Right. And if all of a sudden you're not driving everywhere, then the weird strip mall doesn't make as much sense because it's a drive-to destination. Things would become a little bit more human-scaled. More markets, but smaller. Corner stores your kid can actually run to for an errand, so you don't have to do it all the time.

Ayana: Oh, I totally remember being sent to the bodega with a list from my mom.

Kendra: My sister lives in a subdivision that has no stores. It doesn't even have a post office. There's nowhere to walk to.

Ayana: The future is more pedestrian-friendly.

Kendra: We're going to have to rethink how we do zoning—less zoning ✱ for detached single-family housing and a lot more inclusive zoning, which allows for many kinds of homes to coexist. So fewer suburbs, definitely fewer exurbs, more communities, more villages, more neighborhoods. When you ask people if they like driving to work every day, an overwhelming majority say they have no choice, and about half wish they had other options. Even people who love cars think of cars as a "sometimes thing." But we don't give people other alternatives. So many of our decisions are predetermined by corporate actors.

In places like New York, where flooding is going to be a growing issue, there's going to be more vegetation, more bioswales, more green spaces, because you're going to need something to soak that all up. Something that I've been thinking about a lot is that the thing that can determine whether or not your neighborhood floods is whether or not the street gutters are cleared out. And I have a feeling that in the future there are going to be more community groups for stuff like that.

Ayana: The Gutter Brigade. Didn't see that one coming. And all this would come about because journalism and media are showing us what the solutions are and how we can help to make those things happen?

Kendra: Yes, by giving people the information that they need to agitate for the changes they want in their communities.

Ayana: I don't particularly consider you an optimist, but your response about how fixing climate journalism would change the world sounds actually optimistic. Like, if we got climate journalism right, we would solve transit and housing and corporate responsibility. That is quite a strong view about the power of journalism. Why do you believe that magnitude of change would be the result?

Kendra: Well, if you look at other aspects of journalism, you can see how powerful the effect can be. Look at the way we report on crime, for example. The way we report on crime is what gives us mass incarceration. Crime reporting is one of the worst forms of reporting on the planet, in my opinion, because it reinforces horrible misconceptions. Crime reporting maintains a culture of fear, and perpetuates the idea that crime is going up even when crime is flat or down. That, in turn, leads to more prisons and other overreactions.

I see what happens when we get it wrong, and I also understand that sometimes, when we get it right, there are huge ripple effects. I did a story when I worked at the *Times*, which I thought of then as a nothingburger story. It was about how more and more people are moving into the wildland-urban interface, the WUI.[121]

Ayana: People whose homes are right on the edge of forests, for example.

Kendra: Forests or grasslands, areas that are far more likely to burn in case of a wildfire. And I got an email about a year later from a California state rep telling me they'd read my story, and because of it they'd introduced legislation trying to limit how much new housing is being built in the WUI.

That came about from a small article that, if you had asked me, "Is this story going to create social change?" I would've said, "No, that's not possible." I know what happens when we get it right and when we get it wrong, and the ripples that have happened in both cases. So for me it's not optimism.

Ayana: There are a lot of different elements to getting it right: journalism, media, culture, activism, business, government, policy, technology, democracy.

Kendra: And we need to get it right in education. There isn't enough climate in K–12 education.[122] I did a story for *The Atlantic*, about how earth science is one of the most maligned sciences, especially in high school. The fact that I got an earth science education at all is a hiccup related to the fact that I was educated in New York State, and New York State treats earth science like a real science, but most states don't.

Ayana: What would it look like to get it right on climate education?

Kendra: Climate change would touch everything. I still remember in elementary school learning about the greenhouse effect. There's a way of incorporating it into biology, into earth science, into chemistry, into physics. Obviously there should be some sort of standalone module on climate change generally, but it's something that can be integrated across the curricula. I don't think it should be limited to the sciences. We should be learning about it in history class. We should be learning about it in English. There's a growing realm of climate fiction. It should be cov-

---

121 Delightfully pronounced "woo-ee."

122 New Jersey was the first state to require climate to be covered in all subjects, in all grades. New York, Connecticut, and California have since followed suit.

ered in psychology. Climate grief is a real thing; we should be talking about it.

Ayana: What would you say is the nerdiest, least sexy, most esoteric thing we need to do to make this vision of climate journalism a reality?

Kendra: We need to find a way to pay for journalism. There's been a mass erosion of local journalism because it can't pay for itself. And then on the national level, much of that journalism is either paid for by tech billionaires, or it's through a model that's pretty extractive.

Ayana: Extractive in what sense?

Kendra: Extractive in the sense that local news is often a feeder for national news. But national news gets to profit off of it. So, oftentimes, a local news outlet might break a story, but they're not necessarily the ones getting the website traffic and profiting off of it. Generally people subscribe to one, maybe two papers. So every subscription to a national media outlet represents someone who's not going to subscribe to their local paper anymore. That's just the reality of it. But at the same time, in many cases, national media is, at least in part, dependent on local media.

Ayana: Who's your favorite climate journalist?

Kendra: I think everyone should be reading Lisa Song at *ProPublica*. And I have a soft spot for *Inside Climate News*, I think they do interesting work. And *Grist* has really stepped up its game, moving away from being a climate blog to a hard-hitting climate journalism outlet.

Ayana: So, Kendra, you're familiar with this way to end an interview: calls to action. How can people get involved in making this vision of a better future for climate journalism a reality?

★ **Kendra:** Subscribe to publications, financially support publications that you respect and believe in. The squeaky wheel gets the grease. When they publish things that you don't agree with or that you think are problematic, call them out on it. Better yet, call them out on social media. Don't pay for outlets you don't support and don't share links from outlets you don't support, or do so judiciously. And if you want to become a journalist, be the journalist you want to see in the world.

Dear Future Ones

Jacqueline Woodson

Dear Future Ones who do not understand
an old woman puzzling over January crocuses
and the delicate shoots of tulips poking through winter leaves

who shakes her head now thinking
Twenty years on this planet and my daughter can count
snowstorms lived through
on a single hand. Oh Dear

Future Ones, I apologize

for the magical thinking
that led me to believe this world as I knew it
would always be here and bred you into with my mind on
other freedoms cuz
how many dreams of freedom can a single body hold
now old, you see me back bent and frowning
grateful for iceless sidewalks all winter long
and terrified. Oh Dear Future
Ones. You do not know a world

that isn't already burning, a time when flowers bloomed in seasons
we understood.

Kids These Days

Interview with Xiye Bastida and Ayisha Siddiqa

In September of 2019, around 300,000 people took to the streets of Lower Manhattan for the Global Climate Strike. Roughly sixty teenagers had organized it, supported by 350.org and other climate nonprofits. The rally was just so joyful. As I listened to a DJ spinning Afrobeats between the speeches, I thought: These kids throw good parties. They've made climate activism cool.[123] Maybe we *can* win.

I was honored to be perhaps the oldest person invited to address this massive and gloriously diverse crowd—there is *nothing* more gratifying than the approval of teenagers. Standing backstage with climate elder Bill McKibben, looking up at these young leaders on stage, I realized the climate movement is now three generations strong. And I found myself solidly in the middle. It was a powerful realization of our collective potential, of the strength in numbers we could have, of the shift in public opinion this could ignite, of the insights and strategies intergenerational collaboration could unlock.

On that day, there were youth-led climate strikes in hundreds of cities all over the world. It was the largest climate protest in world history, with an estimated 4-million-plus participants.

Young climate activists were breaking through, drawing a new magnitude of attention to the urgency of the climate crisis, and deliberately framing climate change as a justice issue, one that jeopardized their futures. The conversation that follows is with two of the youth organizers I met at the strike that day, Ayisha Siddiqa and Xiye Bastida.

Ayisha, from Pakistan, and Xiye, from Mexico, both ground themselves in their Indigenous heritage and generational connection to land. Ayisha is now a climate advisor to the UN Secretary General, a fellow at New York University's Climate Litigation Accelerator, and a leader of the Future Generations Tribunal, working to ensure that youth are shaping

---

123 Thus began my often-ironic use of the term "kids these days."

how we address our environmental crises.[124] When we spoke, she had just returned from the World Economic Forum meeting in Davos, Switzerland. She was there on a mission to talk about climate litigation as a tool to "fill in the compliance gap of climate policy." When I asked her whether this litigation would be against corporations or governments or both, she said, "Yes," with a smirk.

Xiye, a college student at the University of Pennsylvania, is likewise organizing her generation and traveling the world to push political and corporate leaders to use their power to address the climate crisis. Amidst it all she is somehow maintaining a 4.0 grade average: "I know I don't have to do well in school to be part of the climate movement, but I just love it," she says. "All my classes are climate-related. I love bringing all of the things that I learn into my activism."[125] For the last few years, her activism has been centered out of Re-Earth Initiative, the climate justice youth organization she co-founded and leads.

I so appreciate the heart they both bring to their work, the humanity they highlight, and how they help others realize their own. The two of them also have a deep friendship, which is sweet to witness. In a joint post on Instagram they offered:

> This is your reminder that the only thing more humbling and powerful than fighting for Mother Earth is finding those warriors who care just as deeply as you. Those are the people to keep close ♥ ♥

———

Ayana: When we met at the Global Climate Strike, I remember being so impressed and literally looking up to you from backstage. What was that day like for you?

Xiye: Ayisha and I talk a lot about that day. It's a moment that has defined our story so much because we're not just activists, we're organizers. That's a big distinction because you're not going to get a movement without <u>organizing</u>. There's a lot of work that goes into organizing. We

———

124 She is also the author of the first poem in this book.

125 Since this interview, she graduated. Class of 2024!

worked two months straight every single day with sixty other youth and with adult coalitions to put that together.

Ayisha: I was older than some of the folks who were organizing. I was in college. And my role was essentially getting college students to attend. After school every day we would meet in the basement of the Society for Ethical Culture on the Upper West Side to plan the strike. I live in Coney Island, so I'd leave class around five, go meet up with everyone for an hour or two in our different working groups, and I'd get home around nine. We were coming from so many different parts of the city.

It was also some of the most fun I've had. We had a person dressed up as a dinosaur handing out the pamphlets. We even brought a huge boombox to the middle of Bryant Park and we were gonna put it on and start singing, but then the police were like, "Don't do that." We'd go from subway car to subway car and hand out the strike pamphlets and we'd sing our song on the train like panhandlers.

Ayana: What's the song?

Ayisha and Xiye [singing]: "We're gonna strike cause the waters are rising. We're gonna strike cause people are dying. We're gonna strike for life and everything we love. We're gonna strike for you. Will you strike for us?"

Ayisha: And the response we got as students asking people to walk out of their jobs, walk out of their classrooms, was so positive. People wanted to, and they thanked us. That was really cool.

Xiye: Youth from all over the world were doing this. When we went to sleep the night before, strikes had already started in Australia, New Zealand, the Philippines. . . . I went to sleep with that excitement like, "Oh my gosh, I can't wait to wake up to see what we can achieve and how many people are going to show up." It was that moment when you know that you can't do anything about it anymore, but you gave it your all.

Ayana: And when the strike actually happened, what was that day like for you?

Xiye: When I arrived, about an hour before the strike was officially starting, the venue was already packed. Not knowing everybody who shows up is what a movement should be. Too often in the climate movement you walk into a room and just see the same faces.

My favorite part was walking down Wall Street hearing our chants reverberate off the buildings and making the ground vibrate. Everything from that day is kind of a blur. It was surreal. I was seventeen. Ayisha was nineteen. Now I'm twenty and I'm like, how did a seventeen-year-old do that? It's absolutely wild to think back on it.

Ayisha: To know that you are in the midst of history as a teenager is empowering. It is the moment you realize you can change things. You really can. Because it takes a lot to convince somebody to get out of their house, or to leave their classroom, for the purpose of protecting the environment, but they want change. I saw people I went to elementary school with that I hadn't spoken to in years. I saw my professors show up. ♥

Xiye: You know who I saw? All the popular girls from school. I was like, "Oh my gosh, yes!"

Ayana: Haha, that's definitely a win. You made it cool to strike for the climate.

Xiye: In advance, we went to City Hall multiple times to get the city government to allow students to strike, to miss school without penalties. We would say: It's not a protest of school, it's a protest for life. The fact that the city let us all miss school with no repercussions was one of the big reasons why a lot of people showed up.

Ayisha: Yeah. The mayor was in the strike. You name it, they were there. We had gotten a permit for 5,000 people in Battery Park, not 300,000.

The strike was strategically timed for September 20, 2019, which was the Friday before the UN General Assembly meeting really got going. And what came out of the General Assembly was in such stark contrast to what had just happened in the streets. It wasn't just New York that had protested, students all over the world did. When in history have we seen this before? Kids walking out, leaving their school chairs empty all over the world.

Ayana: Millions of young people, synchronized.

Ayisha: And the truly weak result from the General Assembly[126] was just so patronizing that I took it upon myself, along with my friends, to try to change the intergovernmental system.

---

126 None of the top emitting countries committed to stronger climate targets.

Ayana: Just change the United Nations. No big deal.

Ayisha: I started thinking of a new way to organize.

Xiye: We also chose the 20th because the UN organized a Youth Climate Summit on Saturday the 21st. There were about 700 youths who came from all over the world—almost every youth involved in climate at the global level was there. Every single COP that I go to, I've been to four now, I meet people who said, "Oh, I was there in New York on September 21, 2019." And I never met them then because the UN didn't organize a way for us to meet each other. I always think, What would've happened if we actually had meaningful collaboration amongst youth starting then?

And they chose Saturday for all of the youth to come because nobody is in the UN on a Saturday. The world leaders are not there on a Saturday. Only about four or five youths were able to go throughout the whole week. I was one of them. I saw Greta give her speech. I saw Trump walk past me. And what struck me is that all of the presidents applauded when Greta spoke. It's the same thing that happens when politicians show up during strikes. We always think: What are you striking? We're striking you.

It's like politicians don't realize that they have the agency to change things, like they forget that they are decision-makers. And seeing that so obviously in front of you is like, "What are you crying about? You can do something. You can sign something. You can change the way that your country operates." They just don't do it.

Ayisha: Exactly. The people who we were asking for change were attending the strike but still not realizing the urgency or the magnitude of what we were saying. That's so unfortunate that there's a dilution of protest as just a theatrical thing. That you can show up, you can take pictures, and then you can go home. You've done it. You've held the banner. And I don't know if that dilution of the message or the protest was purposeful, but it was rather unfortunate. Because their participation didn't translate into anything afterwards. People went home. And we were still left with the crisis at hand.

Ayana: Xiye, you said that's a day you think back on a lot. Why is that?

Xiye: Because sixty teenagers organized a 300,000-person strike for climate. We also think about it a lot because we haven't been able to do it again. It was the height of our ability to bring people together. Now we aspire to make the movement that big again. COVID and other things slowed down our public momentum, so we have focused on building our internal momentum, our internal ability to be a movement.

But most of all, I think about it a lot because I want to stop thinking about it. I don't want that moment to define me. I don't want the best to be in the past. I see people being able to organize themselves in a week in Germany to stop expansion of a coal mine. They organize themselves so quickly. They get on a train, they show up, they support each other. And we can't do that in the United States because we don't have the infrastructure to do that, the networks to do that. That's what makes me not want to dwell on what we were able to achieve in the past, but think and strategize about how we can replicate that strength now.

Ayana: What do you think makes it hard to keep that momentum going or to replicate a big strike like that?

Xiye: Every time I pick up a history book I see that people went on strike for thirty days, for two years, that the revolution lasted ten years. Historically, from an idea being proposed to it being implemented takes years. And a lot of us in the youth movement are reinventing the wheel by having to go through these experiences ourselves rather than learning from past movements.

The climate justice fight is not just about the CO_2 in the atmosphere; it's also the continuation of other movements. And most of these justice movements were led by young people who had nothing to lose. As we grow older, we have more and more to lose, so we need to always remind ourselves what's most important.

One strike doesn't change everything. I want to get to that point again when we are showing up in the streets, but we are also diversifying our tactics. Some activists and organizers decided to go to law school or into the health sector, or work in some other kind of company. And that's good. We need every single sector to be infiltrated by the people who protested on the streets. We need people power flooding the streets and intellectual power flooding the halls of power. That's the balance we've been trying to get to.   ★

Ayana: It seems to me that phase one of the youth climate movement began around 2018, and that we're now in a different phase. It's gone from school climate strikes to litigation threats at Davos. Gen Z kids who have grown up as climate activists are now graduating from college, able to be in different rooms, have different conversations, have different access to power, build new strategies and connections. What do you think phase one of the youth climate movement achieved?

Ayisha: I think the strikes were actually a sequel rather than phase one. The vernacular for the youth climate justice movement in this country developed because young people were protesting the Dakota Access Pipeline at Standing Rock. And if we are talking specifically from the angle of nonviolent civil disobedience, as Xiye always makes clear, we are taking the baton and furthering what existed long before we were even born.

But what has it accomplished? So much. We are seeing major cultural change in our lifetimes, and sometimes it takes stepping out of all of that's happening around you to realize it. From academia to the United Nations, there have been efforts to make offices for youth activists, advocates, and organizers to change policy. There is inclusion of climate justice in curricula across the highest academic institutions in this country.

Ayana: Although climate justice work and organizing has clearly been around for a while, the youth climate movement was key in making the term "climate justice" mainstream, and making that the goal, instead of just addressing climate change.

Ayisha: And it's taken all of these different nodes like the youth climate movement, academic work, and the work of Black and Brown communities to make that change.

Ayana: Absolutely. I remember so clearly a common chant at your strikes: "What do we want? Climate justice! When do we want it? Now!" And also: "Systems change, not climate change!" The first time I heard teenagers shout that, I was like, "Oh, wow. They *really* get it."

Xiye: We've had to fight for those things to be the norm. The climate movement is so diverse, and people's experiences are so different, that we had to convince a lot of the activists in Europe, where the climate-strike movement started, to use these terms. There was a huge realization among them—and I applaud them for having open minds—about the things other people are experiencing. And that opened space for climate justice. For that to be *the* theme of the global youth climate justice movement is so, so special.

We have the ability as youth climate activists to make changes to the public discourse in fast and striking ways. And I think adults have a lot to learn from that. The climate movement now is not so much about national parks, ecology, or the Clean Water Act as it is about justice, human rights, and loss and damage.

Ayisha: Our movement truly is an international movement. And that has not happened many times in history, which also means that it cannot be, and isn't, one paint and one shade fits all. The geopolitical situations in countries, the geopolitical situations, the climate situations are all different. And we've had to navigate that. When we attend events, we try our absolute best to make sure that the southernmost part of the world to the most northern part of the world are represented.

Ayana: From my outside perspective, it seems like you have a great community of young people who are organizing and leading and pushing on climate justice. Xiye, you're in college. Ayisha, you're at your first full-time job. You're going through life stuff as well—probably falling in and out of love too, which can be consuming. So I wonder if the community is strong enough to guide you all through this.

Xiye: We've learned the importance of community in all of these spaces, ✱ because it is a tactic of the system to individualize you. The system wants you to be an individual. And if we don't remember that we have that huge community behind us and in front of us and all around us, we are going to be sucked into the system in a way that we don't want to. And we cannot let it happen.

Ayisha: A lot that has happened around me has felt as if it was supposed to be like this. That I was supposed to meet so many different guardians of the Earth from different parts of the world with the same mission. In so many different cosmologies—of Native people, Indigenous people, rural people, tribal people like myself—the Earth is a living being. And she is conscious. And she has consciously made opportunities for her guardians to interact.

It makes a huge difference when there's community in the room. Xiye is my community and so are my friends. That doesn't happen overnight. You need to invest in people and in these networks and relationships.

Ayana: What do you think it would look like to strengthen the community of young people working on climate? What do you need?

Ayisha: I'm seldom asked that question. So first I want to express gratitude to you for asking. And to answer, we have to step back a little, look at it from the aerial view. You have folks who are teenagers, who have taken this responsibility of entering some of the most difficult spaces in our culture. It is an emotional burden. That makes <u>intergenerational collaboration</u> that much more important, because if I'm asking advice from a peer who is on the same step on their path as me, they will only be able to tell me what they've seen in the past. They won't be able to tell me what's coming. That's why we need elders to take our hands and help us.

That's non-material. Materially, climate activists, youth land defenders, and Indigenous land defenders in particular, are very ill-resourced. Oftentimes you are invited to an international event but you don't have a bed to sleep in, or you don't know where you're going to eat. We have to stress about all of that, plus what we're going to say, plus keeping in contact with the community, plus being whole.

What we need is a local and international system of family, a system of kinship, people we can call on and say, "I need a bed to sleep in tonight. Can you offer it to me?" And it is provided. "Hey, I'm in this situation. Can somebody help?" Imagine how things would change if those resources, those basic necessities, were available.

Ayana: First of all, call me *anytime*. I'm here for all of it. Second, what you're describing is certainly possible. It's not that those things are too hard or too expensive. But it does require having, as you used the word before, infrastructure around you to help. And that does take intergenerational collaboration.

Xiye: Plus, for me, if I am given resources, it feels like I'm profiting off of the movement that's calling out the destruction of Mother Earth. And that's uncomfortable because how come we get attention from calling out the truth? How come we get platforms for describing what we are seeing right in front of us? How come we get opportunities and get into college and get career points for being in touch with the destruction that Mother Earth is undergoing?

When all these things happen, we feel guilty taking resources for the work that we do. It's taken so long for me to accept the hotel room or to ask for one—and it's usually my mom who does it anyway because I still can't really do it, because I feel like they're already giving me an opportunity. But when it comes to big conferences where 30,000 people attend, the conference doesn't care about the 30,000 people and their stories or demands, it cares about the 100 world leaders.

Ayisha: At these international events and the corridors of power and fame that we are allowed to enter, there are first-class citizens and second-class citizens, even there. If you come in representing community and advocacy, nobody's going to move for you in the halls, literally. There should be a social analysis of how people physically move and interact with others when they're in these conferences and what happens interpersonally.

Ayana: Both of you have, in various ways, become faces of the youth climate movement. And I can imagine how bizarre that could be. If people want to take a glamorous photo of you or profile you in a fancy magazine, sure, you want the exposure for your ideas. But it's weird to have that associated with your face or your outfit or whatever.

Xiye: Yeah.

Ayisha: It is incredibly unfortunate that we are expected to be these kinds of prophets or saviors. But in order for you to be accepted as a savior, you cannot have nice things. You have to make do with the bare minimum. And if in that process you are sacrificed, oh, well.

Ayana: Do you feel like you're expected to sacrifice yourselves for the movement, for climate?

Xiye: It's a calling, it's a purpose. It deeply feels like there is no other thing that I would do in my life or that I would need to do or want to do. I saw this meme the other day. It said, "Why do you care so much about the environment? Because I live here." That's it.

We often get asked in interviews, "Why do you care?" And I think people are curious about that because they want to feel the same way. They have forgotten that they have that within them too, that we all have that connection, that it's an inherent human need to protect your home.

Ayana: It seems so basic, like, of course we should want to preserve life on Earth.

Xiye: But I don't think most people have put themselves in a place where they can connect to nature. Because think of putting your feet on the ground. Think of putting your face in the ocean. Think of smelling the salty water. Think of bathing in the river, or climbing trees. All of us have those abilities to connect and feel. And we are fragmenting that connection by living in a world where that is not the norm anymore.

Ayana: Oftentimes we think about what the climate movement is fighting *against*. We're against fossil fuel extraction, against more pipelines, against injustice, against exposing people to pollution or toxins. But I'm curious to hear from both of you, how would you describe what you're fighting *for*?

Ayisha: I'm not fighting to save the world. I don't think I can. I'm fighting for my little corner of the world. And my world is my family, the community I come from, the few places that I love. I'm fighting for the Chenab River and the fields that I was raised in in Pakistan. And also for my home in the U.S., and the Atlantic Ocean.

I'm fighting for the ability to see my friends fifty years from now and to know that, I'm quoting Xiye here, the people I will love one hundred years from now will have access to the fundamental things that make us human, which is clean air, water, soil.

Ayana: Xiye, what are you fighting for?

Xiye: I'm fighting for the wetland in my hometown in Mexico to be clean, for the river that my dad used to bathe in to be clean. I'm fighting for having answers when my kids ask me, "What did you do?" I'm fighting for our generation to have a legacy we're proud of. I'm fighting to fix the severed connection of humans and Mother Earth. And I'm fighting to feel like the world is not on my shoulders because we all share that responsibility to care together.

Ayisha: Recently, we were both at COP28, and there we were fighting for a fossil fuel phase-out. Getting fossil fuel language into the agreement for the first time was important because once it's in writing they can't go back. It's a small lifeline that people and communities still have.[127] We'll see.

---

127 The COP28 agreement states the imperative of "transitioning away from fossil fuels in energy systems, in a just, orderly and equitable manner, accelerating action in this critical decade, so as to achieve net zero by 2050 in keeping with the science."

Ayana: If we think about the long list of climate solutions that we have available to us, in twenty years, when you're around my age, what does the world look like if we get it right?

Ayisha: This overwhelming anxiety would be lifted off my shoulders. There will be more happiness. We will have the opportunity to do something else with our time. For me, I'll have the opportunity to dive deeply into poetry and art, have time and space to create.

Xiye: This question reminds me of something my energy professor said: "Every coal plant, every fracking facility has a lifetime of twenty to fifty years. They will all be retired in your lifetime. When you're my age, the whole energy system could be different." And that gave me so much hope and a concrete way of seeing how we're going to change the world. I want to see the full energy grid transition to renewable. That is the world I want to live in, one where we stop self-sabotaging and take on these opportunities to create a better world.

Like Ayisha said, if we get it right, I would be able to do what I love. Since I was five years old, all I wanted to do was be a veterinarian and a mom. And I told my mom, I can't be a veterinarian because the world needs me. I can't be happy unless I'm doing enough for the world. But I've always had the idea that when I'm sixty, I'll go to veterinary school or live on a farm where I can take care of animals and learn their anatomy and do all the things I thought were cool when I was a kid.

Ayana: Does it feel like a dream deferred, missing out on these things you thought you'd be doing in your teens and twenties?

Xiye: Yes, it feels like a dream on hold. I wanted to go to UC Davis, the best veterinary school in the world. I was going to learn Latin to learn all of the taxonomy of all the animals. I had my life planned out. But in a world that is not stable, you don't have the luxury to just do what makes you happy, but you have the luxury to put happiness into the creation of this new world. And that is my mission now.

Ayisha: I also had a plan for myself and it didn't turn out that way. I'm coming to accept that this is my role in service and I'm okay with it.

Ayana: What was your plan?

Ayisha: I was studying neuroscience with the intention of going into medicine, and I've studied English with a concentration in poetry. But this is what has happened. It's a huge responsibility, but it's an incredible blessing. And I want to make the absolute most out of it before my time's over.

Ayana: Whenever I think "Oh, this is so difficult," I also think how good it feels to know that I'm doing my part, doing my best to be useful.

♥ **Ayisha:** Not only does it feel good to be useful, it's the rent that we pay to live on Earth. Mother Earth is not a hyperbole or a rhetorical analogy—Earth gives birth to us. Earth is constantly giving birth. And for all that we take, we must offer back. That is reciprocity.

Ayana: There's a poem by Marge Piercy called "To Be of Use" that I turn to often. The last two lines are "The pitcher cries for water to carry / and a person for work that is real."[128] I come back to that feeling for myself when I want to delete all my social media accounts, run away into the woods, and have nobody know my name.

Let's talk a bit more about this better climate future we could have. You've both spent a lot of your formative years in New York City. Can you imagine what New York City looks like if we get it right?

Xiye: I want it all to be green. I want no cars. I want to hear birds. I want to hear animals. I want to be able to smell that we are on an island by the ocean. I want to be able to see the stars. I want good, insulated buildings, and that people don't overuse AC—that is such a pet peeve of mine. I want New York City to be beautiful, green, colorful, and not noisy.

Ayisha: What I would imagine New York City to look like, because I live in a community that is a result of historic redlining and segregation, is that Black and Brown kids way out in Brooklyn have access to the same nutritional food as those who live in the middle of the city. I imagine an energy transition. Twenty or thirty years from now, New York City is a place that is livable and sustainable. Our metropolitan and transit systems are much more efficient.

The city is home to so much diversity and culture, but also money. Money is constantly prioritized. And if you don't make enough to be able to live in the city, you're pushed out. I imagine that the people here

---

128 The full poem appears in the "Transformation" section.

now—families and communities—would still be able to have homes. And access to clean water.

Ayana: Both of you also have deep roots in other places. Xiye in Mexico, Ayisha in Pakistan. What do you imagine those places would be like if we get it right?

Xiye: Well, if we get it right, my town would be stabilized culturally and ecologically. We would be wetland people again, fishing and farming. We wouldn't have to worry about the rain coming or not coming, the yields of crops changing so much. People would have stability, reliability on nutritious food and fresh water. People would have the chance to reconnect with what my parents call "biocultural heritage," and to build upon a world that we can trust. Because right now we cannot trust the future.   ! We cannot trust that you will have crops to harvest. We cannot trust that we will have water to drink, and that fish can swim again in every river. And that state of anxiety of not knowing the basics of human sustenance in the future is what breaks people and severs the community's capacity to imagine and act. So if we get it right, my town would be the paradise that it once was.

Ayana: I love that future vision. Ayisha, what are you thinking?

Ayisha: If we get it right, we have tomatoes, we have wheat. There's less domestic violence. Girls and women get to go to school. If we get it right, there's less hunger, there's less illness, less people dying for things that can be treated. Less malaria, less cancer, less cholera, and less bloodshed.

Ayana: For those who can't connect those dots immediately to climate, can you weave that together for us?

Ayisha: Although Pakistan has historically been one of the top ten nations impacted by climate change, it became a topic of discourse only in 2022 because there was flooding that covered one-third of the country and displaced more than 30 million people.[129] What was initially a cli-   ! mate disaster then turned into a healthcare crisis, which turned into a

---

129 From 2012 to 2021, around 21.6 million people globally were displaced each year, within their countries, due to climate-related disasters. This number is rising dramatically, with 32.6 million people displaced in 2022 alone.

domestic-violence crisis, which has then led to more pain for women and children right now. And now we are experiencing inflation and food insecurity like never before. So a pound of tomatoes that would be 20 rupees is now 200 rupees.

And when your money is worthless, even if you're a middle-class person, you cannot buy the basic necessities. We're having a grain-shortage crisis. And in my village we keep a huge tank of grain that we turn into flour, and it's been empty. After the flooding, our cotton crops, grain crops, they were all destroyed. So when the climate crisis hits parts of the world that have historically been turbulent, that have had political instability, social instability, women and gender issues, it exacerbates them.[130]

Ayana: Part of what both of you are saying is that if we get it right on climate, you get to just live and be a part of your culture and explore your interests and be with your people and enjoy the world around you.

✱ **Xiye:** Climate justice is joy because the climate crisis is violence. So if we get it right, we get joy. We get the ability to do the things that fulfill you rather than fight against the things that perpetuate the crisis, which are all violent. Floods are violent. Extraction is violent. Animal agriculture is violent. Everything that created the climate crisis, going back to colonization, all of these things have been cycles of violence. And we have to break those cycles. That is what we're here to do. You cannot restore what has been broken, you cannot heal, without joy and without that aspiration for joy. That is why we do all our work. Because you want your kids to be happy, you want your grandkids to be happy.

Ayana: How are we going to get there? What next steps do we need to take right now? Reveal to me all your grand secret plans.

Ayisha: Oh boy. I can't.

Ayana: Give us a hint.

Ayisha: Well, we need lawyers. Our movement needs all the material things that I was mentioning, and the resources and expertise of people who have gotten their struggles to the other end. This time around, we need to do internal work equal to the external work.

---

130 For example, in Kenya, domestic violence increased 60% in areas that experienced extreme weather, and in South Asia, each 1°C (1.8°F) temperature increase was associated with a 4.5% rise in the prevalence of intimate partner violence against women.

Xiye: I would say that with the previous phase we got people's attention. The whole world's attention. All of the industries' attention. People used to ask me a lot, Why do you strike? And I'd say, "I strike so that I get invited into the halls of power. Because if you're not noticed, nobody will know that you have something to say." So the phase now has been the conferences, the speaking, the interviews, the press, the writing.

Ayana: What would you say is the nerdiest, most esoteric, least sexy thing that we need to do in order to start to turn this vision of a better climate future into reality? Like, not a cool thing, but a necessary, boring thing we need to check off?

Xiye: My professor said, "People who go to meetings change the world." ✱ So we just need to go to a lot of meetings!

Ayana: Haha. Not the answer that we want to hear, but there's truth in that for sure. Ayisha, what comes to mind for you?

Ayisha: I think we need more economists involved. Just for developing countries to meet their NDCs,[131] they will need $1 trillion in external funding. That $1 trillion is available—the world spends like $1.6 trillion annually on the military.

Xiye: We also need to build structure in our organizations.

Ayana: You need more COO types and accountants?

Xiye: Yes, we need to be able to accept resources. We need to be able to have structures and systems that make money work, that make resources work, that connect people successfully.

Ayana: This is very possible. When we think about who are the leaders we need, a lot of the time people look to the youth climate movement and say, "The young people will save us." Which is ridiculous. I know you feel the absurdity of that, because this needs to be an intergenerational movement.

Xiye: There are so few adults, mentors in the climate movement, you being one of them, Ayana, who actually listen to us, and who see our perspective as important and helpful and useful.

---

131 NDCs are nationally determined contributions. Those are the commitments countries make under the UN Paris Agreement for reducing greenhouse gas emissions. More on this in the following interview, with Kelly Sims Gallagher.

Ayana: My sense, broadly, is that adults are reluctant to listen because they feel guilty, because they know you're right, that we should be pursuing climate justice with all of our hearts and energy. Young people have unique and powerful moral clarity and older folks often don't want to upend their lives by fully acknowledging the dire truth of it all. But setting that aside, how would you describe the leaders we need for this moment?

Ayisha: We need brave leaders. We need leaders who understand that everything is at stake and are operating like it's an emergency. Our current leaders are really nonchalant.

Ayana: It does sometimes feel like nonchalance will be the death of us. What a way to go out. With everyone being too chill.

Ayisha: Yes. And that's the critique that's oftentimes used against women, right? Too emotional, da da da. Nurture and care and emotion is actually what keeps things alive. That's what Mother Earth teaches us. And we need more of that. And that's also probably an answer to why this movement is overwhelmingly non-male.[132]

Who will save us? What if nature saves us? What if that's the question we should ask? And what nature teaches us is gentleness, nurture, intentionality, time, and care. Those are not insignificant things. Those are what bring forth life.

Ayana: Xiye, who are the leaders that we need?

Xiye: I think we need women leaders because of all of the qualities that we bring in the care of life, beyond being rational and critical. All of these things that make the world a square, and focus us on money and power, are the things that have derailed us from our true purpose, which is to build community, to build connections.

We need leaders who can remember. We need leaders who are not just looking forward to the next innovation, but leaders who understand the past. We need leaders who are courageous. We need leaders who can love.

And we need leaders who are also ready to step aside for the next leaders to come in. So often we forget the next generation. I know I cannot

---

132 68% of the Global Climate Strike organizers were female.

believe there are already sixteen-year-olds coming up in this movement behind me. We have to know how to move on from position to position, from area of need to area of need as dynamic leaders, grounded leaders.

Ayana: Absolutely. What are the top three things you wish everyone knew about climate, climate justice, climate solutions?

Xiye: Number one, everything is an opportunity for growth and change and transition. Number two, you are part of the solution. And number three, the answers are not in the intangible. Everything that we need to change the world already exists.

Ayisha: First is that our wars for resources are immediately related to the climate crisis, from the jet fuel that is used for the planes that drop the bombs, to the conflicts we're seeing in Afghanistan and Syria and Gaza.

Two, you are nature. And in defending nature, you are also investing in your tomorrow. And you are worth investing in. The tomorrow and the future that we're all working toward, it belongs to you, too, and you deserve it and you're worth it. You are worth all the love.

Three, it is a superpower if you are feeling the anxiety and all of the emotional burden of the environmental crisis. That is not an illness. It is testament that you are human, that you are part of the world, and that you are actually sane.

There Is Nothing Naive
About Moral Clarity[133]

What if we approach the climate crisis with the moral clarity of children? Some things are simply right and some things are simply wrong.

It is right to steward life and justice on this magnificent planet. It is right to quickly transition to renewable energy. It is right to protect and restore habitats and species. It is right to hold corporations accountable. It is right to ensure a just transition, leaving no one behind. And it is right to enact strong government policies that will accelerate all of this.

On the other hand, it is wrong to make this magnificent planet unlivable. It is wrong for the corporations that got us into this mess to continue to profit while they set the world on fire. It is wrong to drive one million species extinct by changing the climate, destroying habitats, and dousing the planet with pesticides. It is wrong to create hundreds of millions of climate migrants and then close our borders when they seek shelter on our shores. It is wrong to force the most marginalized and vulnerable to bear the heaviest climate impacts—from surging seas, to raging fires, to pummeling storms, to enduring droughts, to frequent floods, to unbearable heat.

Some people might tell you that seeing stark right and wrong is naive. It is tempting to succumb to endless compromise as the norm. Resist.
* And let's be clear: Moral clarity is not the same thing as naïveté.

It is naive to expect that governments and corporations will do the right thing without our insistence, or that someone else will handle it. It is naive to focus only on what we can do as individuals, instead of what we can do together, in community. It is naive to assume that the needs of poor communities, communities of color, and Indigenous communities will be taken care of, unless we ensure that they are prioritized. It is naive to think we can "solve" or "stop" climate change. It is also naive to give up.

---

133 This is an excerpt from the speech I gave at the Global Climate Strike in New York City in September of 2019.

In 1967, Dr. Martin Luther King Jr. gave a speech against the Vietnam War, and those remarks could not be more apt for this climate context and this moment:

> We are now faced with the fact that tomorrow is today. We are confronted with the fierce urgency of now. In this unfolding conundrum of life and history there is such a thing as being too late. Procrastination is still the thief of time. . . . Over the bleached bones and jumbled residue of numerous civilizations are written the pathetic words: "Too late." There is an invisible book of life that faithfully records our vigilance or our neglect. . . . Now let us begin. Now let us rededicate ourselves to the long and bitter—but beautiful—struggle for a new world. ♥

What if we enter the rest of our lives with as much simplicity and ✱ moral clarity as we can muster? What if we get *that* right?

CHANGING THE RULES

What we pay attention to grows, so I'm
thinking about how we grow what we are all
imagining and creating into something large
enough and solid enough that it becomes a
tipping point.

—adrienne maree brown

What if developing countries leapfrog past fossil fuels?

What if we pay climate reparations?

What if voting rights are embraced as climate action?

What if frontline communities help shape policy?

What if we turn to the ocean for climate solutions?

What if our legal system deprioritized corporations?

What if we recognize the rights of nature?

What if . . . ?

10 Problems

- The national climate pledges under the Paris Agreement put the world on track for around 2.5°C (4.5°F) of warming by 2100.

- Many of those (already insufficient) climate pledges are not being met. Without stronger climate policies, the planet is expected to warm 3.2°C (5.76°F) by 2100.

- The U.S. is responsible for emitting more greenhouse gases than any other nation, causing approximately 17% of global warming to date, compared to China at 12%, the EU at 10%, and the 47 least-developed nations collectively at 6%.

- The U.S., China, Russia, Brazil, and India (the top 5 emitters) have collectively caused $6 trillion in global income losses due to warming since 1990.

- Financing for climate adaptation in developing countries is 10 to 18 times below the estimated $194–366 billion needed annually by 2030, and that gap is growing.

- Women represent only 26.7% of national legislators globally. There is evidence that gender equity leads to more effective climate policies.

- In the U.S., fossil fuel power plants are 31% more likely to be sited in formerly redlined neighborhoods, resulting in higher exposure to dangerous air pollution for residents.

- The fossil fuel industry benefited from $7 trillion in government subsidies globally in 2022 (or $13 million per minute), roughly 7 times that year's total investment in clean energy.

- In the U.S., 22% of people are doubtful or dismissive about climate change, higher than any other country, and 57% support expanding offshore drilling for oil and gas.

- More than 8 million registered voters in the U.S. who have environment/climate as their number one voting issue did not vote in the 2020 presidential election, and more than 13 million skipped the 2022 midterms.

10 Possibilities

+ 195 countries, representing 95% of global emissions, have ratified the Paris Agreement, and 97 countries responsible for about 81% of global emissions have set targets to reach net zero by 2050.

+ The Montreal Protocol, signed in 1987 and eventually ratified by all nations, phased out ozone-harming chemicals, and prevented 2.5°C (4.5°F) of warming.

+ Laws recognizing the rights of nature exist in 30 countries, including Canada, Mexico, New Zealand, Colombia, and Ecuador.

+ 33 U.S. states have or are developing a climate action plan, and 23 states plus Washington, D.C., and Puerto Rico have official goals to reach 100% clean energy between 2030 and 2050.

+ 8 U.S. states and dozens of municipalities and tribal nations have filed lawsuits against fossil fuel companies for downplaying their role in causing climate change, and for knowingly deceiving the public.

+ Montana, Pennsylvania, and New York have adopted green amendments to their state constitutions, establishing fundamental rights to clean air, water, and soil, including a stable climate. An additional 15 states are considering joining them.

+ 383 coal plants in the U.S. have been shut down or will be soon. The U.S. is on the path to retire all coal plants by 2030—147 more to go!

+ The Inflation Reduction Act commits $370 billion to climate—the largest-ever U.S. government investment in reducing global warming. Year one of implementation spurred $216 billion in private investments, and 272 clean-energy projects were announced across 44 U.S. states, creating more than 170,000 jobs.

+ Implementing the Inflation Reduction Act could lead to 81% of electricity generation being clean by 2030, and put the country on a path to reduce emissions by up to 48% by 2035.

+ Across 125 countries, 89% of people want their governments to do more to address climate change.

Negotiating and Leapfrogging

Interview with Kelly Sims Gallagher

Climate change is such a thorny challenge in part because it requires global coordination. From air pollution to migration, climate change does not give one hoot about the borders of nation states. This is the United Nations' time to shine. But the COP climate negotiations are, to put it mildly, slooooow. We are now nearly thirty years in, and it wasn't until 2023 that an official COP agreement explicitly acknowledged the need to transition away from fossil fuels.

Climate change necessitates policy change at all levels—international, national, and local. So, what's actually happening (or not) at the UN? What's up with China, the current leading emitter of greenhouse gases? How do we think about fairness when the nations most culpable for the problem (ahem, USA, largest cumulative emitter, and a top per capita emitter) are also the ones with most of the money to fund solutions, and vice versa?

To make sense of all of this, we'll converse with Professor Kelly Sims Gallagher. Kelly and I met in the summer of 2022 at a writing residency hosted by the Rockefeller Foundation.[134] It was there, on the shore of Lake Como, that I re-envisioned this book into the shape and arc you are receiving—after two years of "How can I possibly begin to answer this huge title question I have posed?!" It was a delight getting to know Kelly (ten out of ten, would jump in that or any lake with her again), and there's one particular thing I learned about her that knocked my socks off. She described realizing in the 1990s that China would become a key player on climate, so . . . she moved to China, learned Chinese, started working on climate policy, and ended up in the Obama White House helping to negotiate the first climate deal with China. Talk about seeing the writing on the wall and taking some action.

---

134 The fortune (one of the world's largest) that funds the Rockefeller Foundation was amassed largely via John D. Rockefeller's Standard Oil, which later spawned ExxonMobil and Chevron. The foundation divested from fossil fuels in 2021, and now focuses much of its philanthropy on climate.

Now, Kelly is the acting academic dean and professor of energy and environmental policy at The Fletcher School of Tufts University. There, she directs the Climate Policy Lab and the Center for International Environment and Resource Policy, leading projects on everything from low-carbon growth of developing countries to clean-energy innovation and sustainable agriculture. From Beijing to the White House to academia to the UN, Kelly has many insights to share.

———

Ayana: What would an ideal United Nations climate agreement look like? How would it be different and better than what we've got now?

Kelly: I think the structure that was set up in the Paris Agreement in 2015 is the right structure, after a lot of trial and error. That trial and error started all the way back in 1992, when we adopted the UN Framework Convention on Climate Change—a legally binding treaty that pretty much every country in the world ratified, including the United States. That treaty set out broad aims and principles and goals. From there we need to make much more concrete commitments, which is what Paris was designed to do.[135]

Ayana: What exactly is that framework?

Kelly: It basically said that the aim is to stabilize greenhouse gas concentrations in the atmosphere at a level that would prevent "dangerous anthropogenic interference in the climate system." But how do you define that? Negotiators have spent years arguing over the definition of "dangerous."

*   What's hard about the international negotiations is deciding who's responsible for those greenhouse gas emission reductions. Which countries should do more? Which countries should do less? Not just in terms of emissions reductions, but also in helping less-developed countries leapfrog over stages of development that industrialized countries have already gone through. Should reductions be calculated in terms of per capita income? In terms of cumulative historical emissions? In terms of

———

135 UN climate agreements are colloquially referred to by the name of the place where they were negotiated. So if you want to sound in-the-know, referring to Paris, Kyoto, Copenhagen, and Egypt might help.

total current emissions? The negotiation process has really struggled with defining principles around equity, and also determining how countries divide up this task.

Initially, only already-industrialized countries were to commit to emissions reductions, and those targets were negotiated. So, Japan would say to the U.S., "We think you should do this amount of emissions reductions." And the U.S. would say, "No, we won't do that unless you do this much."

Ayana: Sounds like a mess.

Kelly: It was this complex horse-trading process. And after repeated failures, the approach that was adopted through the Paris Agreement was what's called nationally determined contributions, NDCs. And the principle behind NDCs was that each country determines what it can commit to according to its respective capabilities. That took the pressure off and unlocked a universal agreement in Paris, where every country in the world made a commitment, even if it was a limited one.

The new challenge is that the NDCs are all over the place. They're not even comparable—different baselines, different timetables, different types of commitments. But on the plus side, we finally had a global agreement, and that meant every country was trying to address this global problem.

Ayana: This is what you mean by the Paris Agreement having basically the right structure.

Kelly: Yes, I think it's the only possible way that we can get universal participation.

Ayana: What is the United States' NDC?

Kelly: Our Paris Agreement target was to reduce emissions 26% to 28% below 2005 levels by 2025.

Ayana: Sort of a convoluted goal, but how are we doing?

Kelly: There's a chance we'll achieve that target. It's plausible. The harder one is our 2030 target, which is a 50% to 52% reduction just five years later. The Biden administration is asserting that it thinks it can achieve that, but certainly the policies we currently have in place are not sufficient to get us there. Other action will be needed at the state level and kind of magically through the markets—greater-than-expected uptake of renewables or electric vehicles or whatever.

Ayana: And the Paris Agreement has a mechanism for making increasingly ambitious commitments over time.

Kelly: That's a crucial feature of the Paris Agreement. It's called the global stocktake. And at the 2023 COP in the United Arab Emirates, we will review how much progress has been made. Are countries adhering to their commitments? Are we on track?[136] And that is supposed to then trigger thinking through a new round of commitments—although countries are free at any time to revise their NDCs and make stronger commitments. Every year at the COP, some countries come forward and say, we've decided we'd like to revise our NDC and make this particular commitment stronger because we've discovered it's easier than we thought to reduce emissions in this sector.

The problem is: Many countries are not in fact fulfilling their commitments, for various reasons, some good and understandable, and some not so good and less understandable.

Ayana: Some are bad excuses and some are actual technical or political or financial challenges?

Kelly: Right. Or the commitments are insufficient.

Ayana: Yes, clearly. And global emissions are continuing to increase and break records. So, what do we need to change?

★ **Kelly:** We need a better, more robust structure for tracking each country's progress, and to assist countries with implementation of their commitments. We need more transparency. We need up-to-date reporting from all countries about what they think their emissions are and what progress they made each year. Because right now, in some cases, the information we have about countries is ten, fifteen years out of date.

Ayana: Oh, gosh. Not ideal.

Kelly: It's understandable. It isn't easy to monitor and track emissions, especially for developing countries. But this is one of those cases where if you can't measure the problem, you can't manage it. If you don't know where your emissions are, then you don't know where to start.

---

136 Update: The 2023 global stocktake estimated that the current NDCs set us up to exceed the planetary carbon budget for keeping warming under 1.5°C by about 20 gigatons of CO_2. Aka, we are not on track.

And if you've put policies in place and you're not tracking to see if they're working, that's also problematic because you won't know if these policies need to be revised.

Ayana: How would we better monitor emissions? Is part of the problem that countries are doing it themselves? Do we need an international body to be in charge of that?

Kelly: If we were really serious, we would probably have an international ✱ body, similar to the International Atomic Energy Agency, which monitors a lot of information about nuclear power plants and nuclear weapons, and does accounting of nuclear materials around the world.

And we have the technology to do this now. One of the most exciting things that's happened in the last decade or so is that we're capable of monitoring a lot of the emissions for most of our greenhouse gases from space, via satellites. We can do that in almost real time. But those data are held by certain governments and certain private corporations and it's not all being shared and made useful.

Ayana: This seems like a moment to maybe share info. The stakes are pretty high.

Kelly: Well, this was one of the nice announcements at COP27 in Egypt. Al Gore[137] launched a new initiative called Climate TRACE to gather a lot of these data and make it available to countries so they can pinpoint major sources of emissions.[138] But we need to do much more to support developing countries in translating that data into usable information they can then employ in their policymaking processes.

Ayana: If we had better data on where all these emissions were coming from—which countries, and then within each country, which sources and sectors of the economy—then is the idea that countries would want to do better? Or that we could use that data to shame or peer-pressure them into doing better? Because the NDCs are still voluntary.

Kelly: There's no enforcement mechanism, and it's hard to imagine us getting to one. Maybe when we get serious, we might sanction countries

---

137 Heaps of respect to Al Gore for his decades of important work on climate. Thank you, sir.

138 At COP28, Climate TRACE unveiled data showing that about 5% of total global emissions have gone unreported, much of it from oil and gas operations.

for non-compliance. But right now, the U.S. would never agree to a re-gime like that, I don't think. Nor would many other countries. We should think of the NDCs as an expression of a floor, a minimum, especially for developing countries.

Ayana: Like an under-promise, over-deliver approach?

Kelly: Not necessarily under-promising, but being conservative about target-setting because you want to make sure you're going to be able to deliver. Especially if you don't have good baseline information, it's nerve-racking to make these commitments. And if you don't have a lot of con-fidence that policy X is going to lead to outcome Y, that the policies are going to work, you can see why that would make you conservative in set-ting your target. If we take countries at their word that this is their com-mitment, and we were actually helping them with their policymaking processes, providing advice about which policies might work in their na-tional context, giving them actionable data, then many countries would probably find that it's not as hard as they expected to reduce emissions.

Ayana: Knowing is half the battle.

Kelly: I think so. And to your broader question of "What if we get it right?," a lot of countries would love to find economic development pathways that are cleaner and greener, because there are so many co-benefits. Green industrialization strategies can actually make a lot of money plus create a lot of jobs and reduce pollution. There's a lot of win-win. The nitty-gritty of how you get there is the challenge. Indus-trialized countries (to some extent; it's uneven) are ahead in terms of development of green technologies and industrializing them. And now, of course, China has come from behind and begun to dominate some of these industries. China is a great example of a country with a "develop first, clean up later" strategy.

Ayana: With major public-health implications.

Kelly: Yeah, the acuteness of the air pollution in China got so extreme around 2013 that there was a popular uprising. I remember being there during what was called the "airpocalypse." The pollution was so bad, similar to what we're seeing nowadays in New Delhi, in India. People were leaving the country because their children had asthma and were getting sick.

Ayana: The sky was brown.

Kelly: One terrible photograph that's etched into my memory is a row of mothers lined up to have their babies breathe from oxygen tanks, so they could get fresh air into their babies' lungs.

Ayana: Oh my goodness.

Kelly: It got very severe. And that helped to convince the Chinese government that it was time to clean up. But meanwhile, and much earlier actually, the Chinese government had decided that green industrialization was a smart economic development strategy. And so they started investing in industries like solar photovoltaics (PV), wind, ultra-high-voltage transmission, and electric vehicles. Over the course of the last fifteen, twenty years, they have, pragmatically, developed world-class industries that have challenged existing industries in the U.S. and Europe. All things being equal, if you could choose a cleaner, healthier economic development strategy, why wouldn't you do that?

Ayana: You use the terms "low-carbon development" and "leapfrogging" to describe how developing countries can skip the brown-air apocalypse phase and go straight to renewables.

Kelly: Developing countries still need a lot of support to figure out their competitive advantage in green technology—how can they make this good for their economies, and how are they going to finance this transition? And that brings us to the sticky, tough issue of climate finance.

Ayana: In reflecting on COP27, you tweeted, "Climate finance (in its many forms) is the pivotal issue upon which all future progress rests." What did you mean by that?

Kelly: Well, leapfrogging or pursuing low-carbon development is certainly not automatic. So each government needs to be intentional about the policies it's putting in place to make sure that happens. I think a lot of developing countries now are feeling trapped, between needing to make choices that will help them with their plain old development process, and ones that would be cleaner or more climate-resilient. This came to the fore in Egypt with the discussions about loss and damage.

To give you an example, Pakistan had record flooding in 2022. A third of the country was under water. Imagine, one-third of a country under water. The damages have been estimated at more than $40 billion by the World Bank. Nearly 2,000 people died, many more were injured, a lot of children died; it's totally tragic. And when a country like Pakistan, a less-developed country, experiences an incident like that, it has to finance its recovery process. Somehow it has to get the money it needs through a combination of aid and loans to rebuild.

There was also severe flooding in Nigeria: 440,000 hectares of land were destroyed, affecting food security. Somehow you need to get all of those farmers back on their feet, and restore and rehabilitate their farmland. Typically in that situation countries go to the World Bank or the IMF (International Monetary Fund) and take out loans.

A lot of developing countries, particularly as climate change gets worse and worse, are experiencing more and more severe weather-related disasters and getting into this climate-debt trap, where you have disaster after disaster and you're getting more and more indebted just to recover, and you can't get to the real developmental goals that will help bring the country out of poverty.

Ayana: It's constant triage.

Kelly: And so you're making hasty decisions. In the midst of a disaster you ask yourself, "What's the fastest way we can get power back online?" And you'll conclude, "Oh, let's restart that old coal plant that has been idled." It's a natural thing to do, and we saw a lot of that in Europe actually, as a result of the Russian invasion of Ukraine.

The worry I have is that if we don't help developing countries get out of this debt cycle and debt trap, they're not going to be able to get ahead in making investments in a resilient, low-carbon future for themselves. This is the argument advanced by the prime minister of Barbados, Mia Mottley, known now as the Bridgetown Agenda.

★ **Ayana:** Let's zoom out for a minute and describe <u>climate finance</u>. As I understand it, there are a few major categories: one, financing to help countries transition to clean energy; two, financing for adaptation and resilience, to prepare for the coming climate changes; and three, what's termed "loss and damage" or "climate reparations," i.e., how we deal with the fact that countries that didn't cause this crisis are being disproportionately pummeled by very expensive impacts of it.

Kelly: I entirely agree with your three categories. The fourth category I'd add is just regular development finance, because a lot of developing countries are worried that if we start financing climate stuff, we're going to forget about the development part.

Ayana: So, broader financing for poverty alleviation, infrastructure, and food security.

Kelly: Right, exactly. Those are undeniably important. And there's a possibility that we could have a huge new growth in emissions from developing countries if they're successful in their poverty-alleviation efforts but not mindful of climate as they're going through that process. China is super instructive. In the year 2001, when it entered the World Trade Organization (WTO), its economy took off. Over the next twenty years, China cumulatively emitted more than 200 billion tons of greenhouse gases. That's a lot. I mean, the United States has emitted even more, but in a slower process, over many more years.[139]

It's pretty easy to put together different scenarios where current developing countries could emit that much in the future. The obligation is to help countries achieve their development aspirations, but in a climate-smart way. It's a huge task, and nobody's done it before.

Ayana: There's no country that's done this?

Kelly: I can't think of a country. Can you?

Ayana: No, but I was hoping you would.

Kelly: There's no magic solution, but some countries, like Iceland, were blessed with big geothermal endowments that have allowed them to achieve substantial emissions reductions. So it's exciting forging this new low-carbon development model, but it's also a huge challenge.

Right now, most of the climate finance that goes from industrialized countries to developing countries is in the form of debt, and it's mostly sent to support mitigation. We do very little to support adaptation and resilience in developing countries, and nothing at all has been distrib-

---

139 China has been the biggest annual greenhouse gas emitter since 2006, and accounted for 35% of global emissions in 2023. However, the U.S. still emits more per capita, as do Canada, Australia, Russia, and many other nations.

uted yet for losses and damages associated with climate change, though at COP27 a loss-and-damage fund was established.

Ayana: Given that you have been deeply engaged with the justice elements of climate policy for decades, what did you think when that fund was created?

Kelly: I do feel like the establishment of the UN Loss and Damage Fund was symbolically very important, and satisfying to those of us who have sought climate justice. But I'm also not convinced that big money will ever be mobilized for supporting losses and damages. The reparations analogy is a pretty good one. I think a lot of people in industrialized countries wouldn't feel like they were the ones who caused the climate damages that are being experienced in developing countries. And then how are you going to decide who's the most worthy of funds? There's so much need. Also, most people would rather spend money on prevention of the problem than on cleaning up the problem after it's happened. These will be perennial challenges for this loss-and-damage fund, but several European countries made relatively modest commitments.

Ayana: Shoutout to Denmark for being the first to pitch in a few million toward a multi-trillion-dollar problem.

Kelly: The money raised so far is literally in the millions.[140] The European commitment was small but an important acknowledgment that a lot of countries are already experiencing real losses and significant damages. And some of these losses are existential. Small island states, for example, have lost parts of their territories already, and they're facing complete loss. Or take Pakistan: Parts of that country may never be developed again because it would be too expensive and too difficult.

Ayana: It's a moral question. What is the responsibility of rich countries, whose development caused the climate problem, to poor countries that are dealing with so many of the effects of it?

Kelly: There's no question there's a moral obligation, not only to pay for these damages, but also to help make countries more resilient to climate

---

140 The economic cost of loss and damage in developing countries is estimated to be up to $1.8 trillion by 2050. Initial contributions to the UN Loss and Damage Fund were $430 *million.*

change, and to help them fulfill their low-carbon development dreams. I've spent my whole career thinking about this, and watching this process unfold in what has been an agonizingly slow way. So much of what we're seeing today was anticipated and was predicted and we didn't act in time, so therefore I feel all the more obligation.

But I also think it's in our self-interest to support developing countries. Because if we don't avoid this tsunami of emissions that could come from the developing world, we will suffer too within the U.S. And if we don't help cope with the losses and damages, we're going to see significant amounts of climate migration from these countries. ✳

Ayana: What I'm hearing from you is that we basically have to invent low-carbon development.

Kelly: Yes. But isn't that exciting?

Ayana: It's terrifying. It's an extremely gnarly problem given how little time we have.

Kelly: My optimism is that we do already have almost all the technology we need. And that is something that has become true over the last twenty years, thanks to investments that were made, going all the way back to the 1950s when Bell Laboratories developed the first solar PV cell.

And we also know quite a bit about which policies work—which policies lead to green industrialization, which policies lead to emissions reductions, which policies can help with adaptation and resilience. The missing ingredient is <u>political will</u>. That should be the easiest thing to solve, because that's about convincing people that there's a positive future in front of them and then working toward that future. It means holding leaders accountable for pledges they're making at the COPs and their net-zero commitments.

Ayana: Watching all of this unfold in the U.S., political will certainly does not seem easy to muster, even when there are obvious win-wins in cleaner development—the fossil fuel lobby is so powerful. But say we can get the political will, we still need lots of money for implementation. So let's talk about the mechanisms for financing low-carbon development.

Kelly: There are dedicated climate finance funds internationally, like the Green Climate Fund, which launched at the COP in Paris in 2015. There's an older fund called the Global Environment Facility, which is administered by the World Bank. There is a dedicated fund for adaptation called the Adaptation Fund. And now we have the new Loss and Damage Fund.

Ayana: And where does the money in these funds come from?

Kelly: Largely from contributions from countries. This is public finance. This is taxpayer money from industrialized countries going into these funds.

Ayana: Are rich countries falling short on their pledges? Or are their pledges not even big enough to begin with?

Kelly: Both. It's been particularly hard in the U.S. to honor climate finance pledges.

Ayana: So we're making pledges and just not writing the checks.

Kelly: That's right. Joe Biden committed $11 billion in international climate finance that could have gone to any of these funds but has not been able to get Congress to appropriate the funding. And so the U.S. has not honored its climate finance pledges since the Paris Agreement.

Ayana: This is so embarrassing and so dangerous.

Kelly: But, honestly, there's a limit to how much public money is out there, right? There are only so many taxpayers in the Global North who are able to put up this money. Most of the money is in the private sector. What we need to be doing is steering private sector finance and commercial finance in a cleaner way. We also need to use our development finance institutions like the World Bank, the Asian Development Bank, or the Inter-American Development Bank to get climate mainstreamed into those institutions so that every time they're making a development finance investment, they're thinking about the climate consequences. Which means they may have to rethink some projects. Like, reconsider where a bridge is sited so that it's not going to be flooded. Or instead of building a coal-fired power plant, build a clean power plant.

Ayana: How do we get these development banks to change their funding parameters?

Kelly: It's a big question. A lot of people are calling for it now. This is part of that Bridgetown Agenda that we mentioned earlier. The first thing we need is leadership in these institutions that cares about climate change. And we need to rethink the articles of incorporation, the founding documents for these institutions, and see if they need to be reformed or changed. For example, the World Bank has this very worthy goal, which is to reduce poverty around the world. Could we add an "and climate change"?

Ayana: The two are inextricably linked at this point

Kelly: They are. We also need to look at the lending guidelines, lending ✳ priorities. A lot of concrete and specific policy changes are needed in these institutions. And actually the development banks are further along in doing this than the private sector is. So we also need to ask for this same kind of commitment from our commercial banks. I'm thinking of the JPMorgan Chases.

Ayana: Yes, those biggest lenders to fossil fuel companies.

Kelly: How are they going to be held accountable? The commercial banks are also important for developing countries, we forget about that. They account for a lot of the lending.

Ayana: Who gets to write the rules for commercial banks?

Kelly: This will probably need to be done country by country, with Congress or a parliament developing the rules of the road about what kind of lending you can do, putting a red light on certain types of investments and green light on other types of investments.

Ayana: What about carbon markets?

Kelly: The carbon market idea is one that negotiators have been pursuing for many, many years. The starter initiative was called the Clean Development Mechanism, established during the Kyoto Protocol in 1997. The idea was that industrialized countries could invest in a project in a developing country to reduce emissions, and get credit against their own target for emissions reductions. That of course generated financial flows to developing countries to invest in these projects, but there were lots of controversies about how permanent those emission reductions were. Plus, the unjust thing about it was that the credits for the emission reduction were accruing to the industrialized country and not to the developing country. This gets us into the complicated discussion about offsets.

Another idea that was embraced in the Kyoto Protocol was international emissions trading. The idea there was to allocate permits to emit on a global scale. So China gets this many permits, and Indonesia gets this many, and the U.S. this many. If one country was able to reduce its emissions below its quota, it could sell its surplus permits to a country that needed more. Article 6 in the Paris Agreement sets us up for this, but there are no rules of the road for how to do it. So far no systems yet exist at that global level, though we do have robust emissions trading systems at the regional level. Europe has a European Emissions Trading System that's effective. In the U.S., there's a Regional Greenhouse Gas Initiative in the northeast with a robust emissions trading system for its power sector. California and Quebec have programs. But these have been limited either to countries or regions so far.

Ayana: Let's talk about China, because this is one of your most impressive areas of expertise. When people talk about climate and China, one of the things I often hear Americans say is, "It doesn't really matter what the U.S. does because China's emissions are so large, they'll cancel out anything we do, so why bother?"

★ **Kelly:** China presents itself as a paradox. Because on the one hand, China is the biggest aggregate emitter in the world. But on the other hand, China deploys more new renewable energy capacity than any other country on Earth, and has consistently done that for the last five years; it has arguably done as much as or more than most countries to address climate change. It's hard to hold both of those ideas in your head at the same time. Furthermore, it has brought more people out of poverty faster than any other country on Earth. From a development point of view, that really matters—many countries want to emulate China in its ability to bring hundreds of millions of people out of poverty in a generation.

Ayana: Ideally, emulate without the carbon intensity.

Kelly: Correct. As far as natural resource endowments, like the U.S. China is blessed with huge quantities of coal. So its development process has been extremely carbon intensive because it industrialized on the back of coal. When I first started studying China in the late 1990s, coal accounted for 75% of primary energy supply. As China has grown economically, and started deploying renewable energy in hydro and nuclear, it has brought that proportion down to 60% in 2023. Because of China's enormous scale and size, it's still emitting a lot, but give them credit for that progress.

China also has extremely ambitious plans to construct new renewable energy capacity. And, like it or not, it's constructing more nuclear energy power than any other country on Earth—more than the rest of the world combined. And it is selling more electric vehicles than any other country on Earth.

It's doing all this as a result of deliberate policy. This didn't magically happen. In fact, one of the things Americans love to hate about China is how much planning it does, but China has an extremely comprehensive approach to climate change policy. It's the polar opposite of the U.S. It has short-, medium-, and long-term plans. It has five-year plans. For every industry, every sector, every government agency, there are clear and explicit plans, and those plans incorporate climate change. It uses regulatory tools, efficiency standards, performance standards. Maybe surprisingly to many people, it has a national emissions trading system. So it's using market-based instruments along with fiscal tools.

I would say another admirable feature of China's policy is that it's been stable. There's no volatility; it's predictable. As a result, we've seen clean-energy industries be birthed and grow steadily to the point where they have taken over the world. China's solar PV industry now accounts for 80% of the global market, for example.

Ayana: This paradox you've described is intense. And beyond what China is doing within its borders, around the world it has been the largest *funder* and consumer of both renewables *and* coal power plants.

Kelly: Exactly.

Ayana: What does it look like if China gets it right on climate? What does that trajectory look like?

Kelly: I'm quite confident about their trajectory because their plans are clear. They've made slow but steady progress. They've achieved all the targets for their 2015–2020 five-year plan. So I actually believe that China will hit its goals of peaking carbon emissions before 2030, and reaching net-zero goal by 2060.

Ayana: 2060 is too far from now, though.

Kelly: But, let's start with that, right? What I think China hasn't yet confronted, and neither has the U.S. by the way, is the socioeconomic transition that's required to get all the way to net zero. For instance, there are a lot of coal workers in China. There are whole regions of China that depend on coal. And not just coal but heavy industry, iron and steel.

Where are all those people going to go? What are their new jobs going to be? How is the country going to retrain all of these people? How are these regions going to survive economically? Even though China officially has a policy of no new coal plants, when the economic conditions get hard, like due to the pandemic and the global recession, regions that are struggling turn their coal plants back on and reopen their coal mines because they have people who are hungry and need jobs.

Ayana: Yeah, this gets us back to the idea of a just transition.

Kelly: That idea does not exist yet in China. Yet, ironically, they've achieved what we're trying to achieve in part with the Inflation Reduction Act in the U.S.: a strong manufacturing workforce in clean energy as a result of deliberate industrial policy. But they haven't yet figured out how to cope with all of the workers who are stuck in traditional industries.

Ayana: And there are also human rights issues within those industries, like with solar in Xinjiang.[141]

Kelly: Yes, absolutely. Plus, in the U.S., one of the reasons why our deployment of renewables has been slower is we have a lot of permitting rules and processes that give rights to local land owners. In China, that often doesn't exist. One of the reasons China has been able to deploy renewables faster is that it doesn't give people the right to oppose new solar or wind farms. You can deploy and get things permitted much more quickly and easily in China. And there have been substantial human rights violations. There's a lot of work to do on the social and economic side in China.

Ayana: It's my understanding that, historically, nothing major happens on climate policy globally unless both the U.S. and China are on board. What's your vision for how these two countries could get aligned and collaborate to move climate policy forward?

---

141 45% of the world's supply of polysilicon (a key component of solar panels) comes from Xinjiang, produced with forced labor from Uyghurs and other Muslim minorities.

Kelly: We know they can cooperate because we actually saw them do it in the run-up to the Paris Agreement. I was fortunate to be able to help negotiate the U.S.-China deal in 2014, and the subsequent deal in 2015.

Ayana: This was when you were working in the White House during the Obama administration.

Kelly: Yes. Everybody said we would fail, that it was impossible, because historically the two countries had been at odds in international climate negotiations. Some people called it a suicide pact: The U.S. would always blame China for its inaction, and China would always blame the U.S. for its inaction, and so we were going to mutually commit suicide. But I had an insight from the decade-plus of work I had been doing in China that the Chinese government was doing a lot that Americans didn't understand.

That was right after the peak of China's air pollution, and you could just feel that the government was committed to cleaning up the environment. Likewise, there was political momentum in the U.S. because President Obama had achieved a second term, and he wanted to make climate change part of his legacy. He wanted to do something significant internationally; he had been frustrated by the failure in Copenhagen. It seemed like there was a ripeness, a political window in both countries.

The two countries are so symbolic on this issue as the top two emitters, and as historic enemies in the climate domain. So I had this hope that if they could come together and do some kind of a deal, it would be catalytic for the rest of the world. And pragmatically, although their proportions have changed over the years, the two countries collectively account for about 40% of global emissions. If you could get just the U.S. and China to work together, you're solving almost half of the problem. And Europe was already on board. Actually you were getting to around two-thirds of global emissions with these three regions. Here's where the nationally determined approach was important.

Ayana: NDCs to the rescue.

Kelly: This was the beginning of trying out NDCs. China decided what it could do, and the U.S. decided what it could do. And then the deal was just standing up together and jointly announcing their commitments. The nature of each commitment was different, but they were both symbolically important. Many other countries subsequently announced their own NDCs.

I say all of this to remind people that once upon a time we did have a deal between the U.S. and China. President Trump broke trust with China when he withdrew the U.S. from the Paris Agreement—I give China credit for not following suit and also withdrawing when they had every justification to do so. We need to rebuild the trust between the two countries. They're too big and too important to not work together to address global issues like climate change.

Ayana: Shifting back to the U.S., if we are getting it right, what else would the U.S. be doing?

Kelly: We need more regional planning. I'm part of the National Academy of Sciences committee on deep decarbonization. The report we released a couple of years ago recommended the establishment of a National Transition Corporation that would provide planning grants to local communities and regions for how they want to reimagine their economy and their transition, and then grants to get those plans moving. So that's concrete.

Ayana: Oooh. That seems like exactly what we need—place-by-place designing what it would look like if we get it right, and making a plan to get there. Perhaps we need a little China-long-term-planning energy.

Kelly: We also need the discipline of an <u>emissions budget</u>, a commitment that the U.S. will not emit more than X amount over time. The targets the U.S. set in our NDC have all been set out of the executive branch and not embodied in legislation by Congress.

The UK, New Zealand, and others have actually set an emissions budget for themselves. Then, they track to see if they overspent or underspent on the budget, and hold themselves accountable in a quantitative way. It's time for that in the United States. We're well past this period where we can hope that technology is going to save us. We need to be quite disciplined about our emissions reductions.

Ayana: I want a monthly progress report from the U.S. government of how we are doing relative to our Paris Agreement commitments and beyond in each sector, in each state. What's working? What's falling behind that needs more attention or funding? Is the problem that we don't have enough people who are trained tradespeople who can help us actually make these shifts happen? What are the bottlenecks?

Kelly: What are our constraints? Exactly. We need data so that we can track and analyze the effectiveness of the expenditures, and whether they're leading to emissions reductions or more resilient communities.

Ayana: Like a public dashboard. That would be excellent.

What are the top three things you wish everyone knew about climate policy?

Kelly: There is no silver bullet. It's not like if we just had a carbon tax, we would solve climate change. That's my top one, two, and three. People fixate on, "If we just had X," emissions trading or a carbon tax or what have you, but it's too big and complex. Climate change is generated by all aspects of our economy, whether it's agriculture, individual transportation, our homes, or where we work. So we need a big basket of solutions. And we need to have targeted, specific, granular policies for each dimension of the challenge.

We can embrace climate policy as a living document—an evolving, improving set of ideas. If our planet is built on fluid systems and cycles, why shouldn't the policies we put in place to protect it be the same?

—Maggie Thomas

A Green New Deal

Interview with Rhiana Gunn-Wright

In the 2010s, my work was ocean conservation and ocean policy. For a few years, I led an initiative that supported several Caribbean islands, including Barbuda, in establishing ocean zoning and passing a remarkable set of marine regulations. Then, in 2017, Hurricane Irma hit Barbuda square on with its full Category 4 force. The community was decimated. And I was forced to comprehend that even the best local policies can't protect a place from the onslaught of climate disasters. Before that hurricane, I only really thought about U.S. climate policy insofar as that we didn't have one and that we absolutely should. After it, I expanded the scope of my work from ocean to climate.

Then, in February of 2019, the Green New Deal (GND) appeared on the scene. It had been crafted by a new cadre of climate policy researchers and advocates, still in their twenties. When I read it, my first reaction was: *Holy shit, they really went for it.* Manufacturing and agriculture, healthcare and job guarantees, rural and urban, poverty and race, clean air and clear water, all of it. While I knew it was all intertwined, even I barely felt ready for this. Could America be ready for this? Become ready?

There was a massive amount of press coverage, because it was introduced into the House by Representative Alexandria Ocasio-Cortez—then a brand-new and extremely high-profile member of Congress—along with Senator Ed Markey.[142] The resolution, "Recognizing the duty of the Federal Government to create a Green New Deal," quickly gathered 14 senators and 101 representatives as co-sponsors.[143]

Initially, the resolution was quite well received by the public. Right after it was announced, overall support among registered voters was at

---

142 Senator Markey is an OG of climate policy. See: Waxman-Markey Bill of 2009, which came heartbreakingly close to becoming law, and which would have established a national cap-and-trade program for greenhouse gas emissions.

143 A resolution is not a bill that could become law. It's more of a non-binding vision statement to, as the Library of Congress puts it, "express the collective opinion of [one or] both chambers on public policy issues."

48%, with 28% who didn't yet have an opinion, and over 86% support for nearly every key component of the resolution.[144] Turns out people are into green jobs and a livable planet. That is, until conservative media started putting some aggressive, "They are going to steal your hamburgers!" spin on it. Speaker of the House Rep. Nancy Pelosi also voiced her displeasure with this bold, insurgent policy proposal, deriding it as "The Green Dream, or whatever."

The resolution was forced to a vote prematurely and didn't pass. But the concept, the framework it laid out, had permeated the discourse. It sparked the creation of the House Select Committee on the Climate Crisis and pushed climate to the top of the agenda for the 2020 presidential election. When polled in the lead-up to that election day, 68% of registered voters (and 91% of Democrats) said climate change was very or somewhat important to their choice of candidate. The vision of the Green New Deal was here to stay.

The GND resolution itself is quite short. I had expected it to be hundreds of pages, and was relieved to open the PDF and see that it was only fourteen pages—with large margins, wide line spacing, and a large font, no less. It starts with a compelling set of "Whereas" statements laying out the scientific basis, the United States' responsibility as the largest historical emitter of greenhouse gases, our intertwined environmental and inequality crises, and national security risks as the impetus. I kept wishing (and shouting into the social media void) that more people would actually take a few minutes to read it so our public debate could be about what it actually proposed and how it could be improved, instead of baseless arguing. Alas. Regardless, it was an inflection point in U.S. climate policy. And it was the first time we had something on the table that felt holistic enough and ambitious enough that it could actually transform American society and give us a chance at reining in global warming.

However, there was one *huge* thing the GND omitted almost entirely: the ocean. So I began working to get coastal ecosystems, regenerative ocean farming, offshore renewable energy, etc., integrated into the fed-

---

144 Also, a CNN poll conducted in April 2019 showed that 96% of Democrats "favored taking aggressive action to slow the effects of climate change."

eral climate policy conversation.[145] Through that work I met people like Rhiana Gunn-Wright, the young policy phenom who was a core architect of the Green New Deal. That gave me a semi-inside view into how things played out after the resolution was introduced. I was floored by Rhiana's robust policy proposals, brilliant and relentless press engagement, and creative and effective shaping of what climate justice would look like in policy form. And I was thrilled that a whole new discussion had been opened up about how the U.S. federal government could and should address the climate crisis. Her work helped me understand that policy, at its best, enshrines our shared values.

Rhiana is now director of climate policy at the Roosevelt Institute, a policy think tank, and remains focused on how to make the Green New Deal vision a reality. This interview starts by rewinding to the period before the resolution was introduced in Congress, and explores how we can get it right on developing, passing, and implementing climate policy.[146]

———

Ayana: What exactly is the Green New Deal?

Rhiana: The <u>Green New Deal</u> is, or at least started out as, a vision and framework for how you could decarbonize the U.S. in a decade, while redressing systemic injustices, creating millions of jobs, and re-energizing what some people call the "real economy," which is the production of actual goods and services and not financialization.

Ayana: So, not the Wall Street economy, but the actual making-stuff economy.

Rhiana: Exactly. Some people, like my friend Ben Beachy, who was then at Sierra Club, talk about the Green New Deal as policy at the jobs, justice, and decarbonization nexus.[147] That is the most simplistic explanation.

Ayana: It's been incredible to hear President Biden talk about climate policy in economic and infrastructure terms, because it feels new to hear

---

145 Enter the Blue New Deal. More on that in the following chapter.

146 Implementing environmental policy is my love language.

147 Ben is now working in the White House as special assistant to the president for climate policy focused on the industrial sector and community investment.

politicians talk about it not only as an environmental issue. And I remember seeing him tweet, "When I think about climate change, I think about jobs. Good-paying union jobs that put Americans to work, make our air cleaner, and rebuild America's crumbling infrastructure." And I was like, uh, did you crib this from the Green New Deal?

Rhiana: Bro, yeah.

Ayana: I mean, great that the president adopted that narrative framework. The way we think and talk about climate policy has shifted so quickly. Okay, so what is actually proposed in the GND resolution?

Rhiana: It set out a number of project areas. Building resiliency against climate-change disasters. Meeting 100% of electricity demand through clean and renewable power. Building energy-efficient smart grids. Upgrading all existing buildings—that was a big one—to achieve maximum energy efficiency and water efficiency, including through electrification. Spurring growth of clean manufacturing. Investing in sustainable and family farming. Expanding clean and affordable public transit, including high-speed rail, and expanding EV manufacturing and infrastructure. Funding community-defined and -driven projects to mitigate the long-term effects of pollution and climate change. Removing carbon through low-tech solutions. Protecting ecosystems. There's more, but I forget off the top of my head.

Ayana: So it's a whole set of goals and projects that hit different ways greenhouse gas pollution can be sequestered or prevented, and the societal changes that need to happen along the way.

Rhiana: Right. And that last part is what I actually specialize in: a set of social policy changes and reinvestments and shifts in the social safety net that are needed to support the kind of just transition that the GND vision requires, including empowering people who are traditionally disadvantaged and marginalized.

* For instance, if people have to relocate because they're losing their homes to sea level rise or to climate disasters, or they want to move to be part of some of these new industries, having universal healthcare, not tied to an employer, really helps with that, right? And having things like free college and training and access to childcare helps ensure that everyone can participate in these shifts into these new industries. Because

what we know right now is that, without intervention, traditional clean-energy jobs are quite White and quite male. So if you want these jobs to benefit everyone, especially to be accessible to women, you need accessible childcare.

The fact is, when we have done these sorts of big transitions before, whether it was the New Deal or World War II, certain communities who had a history of being discriminated against were left out. Like, Social Security didn't include agricultural and domestic workers. Similarly, the GI Bill after World War II enabled a lot of home ownership, but also enabled redlining, as did the New Deal.

All of these things—childcare, healthcare, job training—aren't just good to do. They are fundamentally *necessary* so that we do not end up repeating history. Because redlining is one of the reasons that fossil fuel infrastructure is located in low-income, Black and Brown communities. Those communities are more likely to be zoned industrial; it's cheaper to locate there. And, fundamentally, there are people there who have a dearth of political and economic power and social capital, and a history of being discriminated against. It means that you can poison them with-   !
out a lot of consequence.

Ayana: They don't have a lot of power. They don't have a team of fancy lawyers fighting for them.

Rhiana: They can't push back and be heard on the first—or even second or third—try. And the side effect of concentrating all of those downsides in marginalized communities is that unaffected people still believe that fossil fuels are nothing but good for our economy, and good for our energy system, and that they don't really have a cost. They think this because they're not the ones dealing with the health consequences of pollution. If those consequences were equally distributed, if everyone   *
had to deal with the poisonous effects of fossil fuels, I doubt that we would be where we are right now.

Ayana: That's an important point, and a great way to put it.

Rhiana: Whenever you have pockets of people without power, it enables systems to become excessive and not check themselves in ways that they would if everyone had to deal with negative effects equally.

If we get this right, one of the big things I envision is giving power back to folks who have been wrongly divested of it, and getting feedback in real time about how well or poorly our systems are functioning, how they are helping or harming us.

Ayana: I love that the resolution ends with this goal of: "providing all people of the United States with high-quality healthcare; affordable, safe, and adequate housing; economic security; and clean air, clean water, healthy and affordable food, and access to nature." How did you expect people would react to the Green New Deal resolution? And how did that compare to how they actually did?

Rhiana: We expected some amount of excitement about an approach to climate policy and decarbonization that felt in line with the level of change we actually need, with what the science was saying was necessary to deal with the scale, speed, and scope of the crisis. Climate scientists and people who had been working on these issues for decades were like "Finally!" And there was some excitement that legislators were treating it like a crisis and discussing it as a crisis.

We also thought there would be some excitement about taking a different approach to climate policy, one that wasn't a carbon tax and wasn't technical, one that was more based in industrial policy and building real things and actually shifting the shape of our physical world, of our energy system in particular, and shifting the economic policies that were underneath it all. And doing that in a way that felt real and concrete—not just about financial engineering or hyped-up market mechanisms.

Until then, all the talk was about a carbon tax, and how basically just having a carbon tax would lead to renewables and then all these other things would happen as knock-on effects. There was a sense that nothing apart from creating the tax really had to change, even though, if you dig into what's necessary to decarbonize, much less how to do that in ten, fifteen, twenty years or even half a century, it's not possible if all the underlying structures of your industries and economy stay the same.

So I thought there would be some excitement. Definitely thought there would be some pushback. And the response was as expected in type, but *much* bigger in magnitude.

Ayana: Climate policy doesn't usually make that much of a splash.

Rhiana: No, not at all. Since 2009, when the last major federal climate proposal was put forward, there hadn't been much. And during the

2016 presidential election, I don't remember any conversation about climate, and presidential elections are when you tend to surface big issues. So I didn't expect the magnitude of the reaction.

There was an excitement from folks about the connections being drawn, about discussing climate policy and the climate crisis in a way that recognizes the intersections and recognizes how fundamental a shift it is to change your energy source. It's radically different.

Ayana: Yes, I've learned from you that climate policy has to be much more holistic than you would initially think if it's going to be successful, because of all these deeply intertwined threads of how society is currently structured. What were some of those dots that you were connecting that people may not have been connecting before?

Rhiana: One is the connection between the climate crisis and racial injustice and economic injustice. A lot of people did not realize the climate crisis didn't affect everyone in the same way, that some people were hit first and worst, and that those people were largely people of low income and people of color. People didn't realize that fossil fuel infrastructure and pollution are concentrated in low-income, Black and Brown and Indigenous communities, and how that is connected to the climate crisis.

The other dots we connected were between the climate crisis and economic policy. Before the Green New Deal, we were talking about the climate crisis as just a scientific and technical problem, a problem of having greenhouse gas emissions that are excessive. There wasn't an understanding that the energy source you use undergirds your whole economy. When you change your energy source, you're going to have to transform large swaths of your economy. Things will fundamentally have to change.

Ayana: When we no longer have an economy based on fossil fuel energy, that has major ripple effects.

Rhiana: A hundred percent. It has ripple effects between what kind of jobs people have, where those jobs are based, what kind of support you need for those jobs. The Green New Deal helped people understand that it's not just about whether you move to renewable and clean energy, or whether you embrace more climate-friendly agricultural practices, or whether you're switching to electric vehicles. There are also a lot of choices we have to make about how we will structure this transition.

Before we introduced the Green New Deal, people weren't asking a ton of questions about: How will you make a transition to renewable energy equitable? How will you make sure that everyone has access to renewable energy? How do you structure this transition so you actually create more jobs than you eliminate? How do you do this in ways that help drive wealth and ownership and empower groups that have traditionally been disempowered by federal policy, especially during big economic shifts?

The Green New Deal elevated the fact that it's not just *what* we do, it's *how* we do those things. There are important choices about whether or not we reduce emissions and limit warming, but also whether we do that in a way that ultimately helps undo some of the inequities of our existing world instead of replicating them.

Ayana: Yes, if the goal is a certain amount of emissions reductions through this suite of industrial, technological, policy, and nature-based solutions, there are a lot of ways to get from A to B. You can screw people over on the way to B, or not.

Rhiana: Right.

Ayana: I think about the Green New Deal as a vision statement for the kind of world we want to create and the path that we want to take to get there. Was crafting this vision and narrative something you consciously set out to do?

Rhiana: Oh, yes. It was purposeful. Any effort to decarbonize is going to take a suite of legislation, tons of legislation. So a vision is necessary to orient that, and to orient the principles and framework that the work is based on. We wanted to change the direction of policy, but also to change the conversation.

The appeal to narrative was intentional, because the path to policy change actually starts way, way before Capitol Hill. It happens way before any congressperson is usually involved. It requires the development of shared principles and worldview. And all of that's driven by narrative. If you want people to make a different choice, you have to change the story they tell themselves. You have to change the conversations they're having on the regular. And you have to change the boundaries of what's seen as feasible and possible and "smart," and what's not.

It's not enough to change the narrative among policymakers because they don't live in a vacuum—you also need to change the narrative in popular culture. You have to change the narrative of, as trite as it might sound, what celebrities are saying, and what the policymakers' grandchildren are saying.

Ayana: What did you want them to say?

Rhiana: We wanted them to say they liked it. We wanted them to say, "Wow, this makes a lot of sense, this is something worth exploring." And people did come behind the Green New Deal and support it in various ways. We didn't have a lot of money, but—

Ayana: Who's the "we" here?

Rhiana: There was a crew of us that were at the heart of Green New Deal organizing. It was Sunrise Movement, Justice Democrats, and the think tank I was working at, New Consensus. Now that it's grown into a whole movement, leadership is way more decentralized—you have the Green New Deal Network, you have local orgs for a Green New Deal, you have state orgs for a Green New Deal. It's a whole different ecosystem now. But initially it was those three organizations.

Ayana: Three organizations and probably just a few people at each who were really working on this? We're talking about a single-digit number of people who were crafting this?

Rhiana: Maybe double digits, but low, like twelve.

Ayana: An adolescent number.

Rhiana: It was definitely an adolescent number. And, each of the orgs was three years old, tops. We didn't have a lot of money or a ton of formal institutional power or a lot of inside relationships. We obviously had some key connections like with Representative Ocasio-Cortez and with Senator Markey—and their staffs were of course deeply involved with turning what had been a proposal by the team at New Consensus into the congressional resolution. And Sunrise worked to increase the number of political relationships we had, and they grew. But in the beginning, what we had was we were young, we knew the science, we had spoken to economists and experts, and we knew what was cool. And the power of making something cool is really underestimated when it comes ✱ to politics.

Ayana: This is what you were saying about how policy gets shaped well before it arrives on Capitol Hill. Do you mean that culture is shaping what's possible for policy and what politicians will vote for?

Rhiana: It's not just culture, it's more realms of conversation. You have a cultural conversation, which is important in helping folks in power see where the pulse is, see where pressure might be building, see what is seen as cool or not, what's in the zeitgeist. And everybody wants to be cool on some level to somebody, right? Then you have the sphere of movement conversations, which are about how you get a critical mass of the right advocates organizing around an idea they think is important and viable.

Then you have a set of conversations among experts, the people who are considered smart in a given area, the people who policymakers listen to when they want to know "How do I deal with a problem? What's viable? What should I be thinking about?" And of course there's a political conversation that is happening on the Hill, and is happening among policymakers about what's popular among my constituents, who's going to run against me in the primary, et cetera. Our goal was to influence all of those conversations simultaneously, so that we had the best opportunity to push this thing forward.

Ayana: Smart. So, four years into this, has your vision for federal climate policy changed in any significant way? Or are you like, "Nope, we got it right, we gotta make it happen; that's the vision, no edits."

Rhiana: I mean, there's always edits, right? But I do think we got a lot of it right. Our batting average is pretty high, which I'm happy about. Although, and I don't know how we could have done this, I wish that we had been more specific about *how* to do some of these things—like change the electric grid, or reinvest in a manufacturing base—with environmental justice and racial justice at the center.

What I see now is the parts of the Green New Deal concept that have been most successful are about tackling decarbonization through public investment and industrial policy. That's the basis of the Inflation Reduction Act. That was not a conversation that was happening on a large scale before. And the things that have called on traditional visions of men-in-boots—the grid or EVs or clean manufacturing—have been more successful.

The parts that have been the stickiest are the parts about weaving in

racial justice. Too often I see this tension where people want environ- !
mental justice, they want to include it in these projects, but they don't
necessarily think of it as feasible, or they consider it too difficult. They
say the rhetoric without fundamentally including it in how we are actually building out and structuring these various decarbonization tactics. I
wish we had foreseen the ways that some of the racial justice stuff could
be peeled off and left to the side. And I wish we had thought more about
how to knit those together, or maybe offered a bit more of a roadmap for
how you do that.

Ayana: Yeah, I can see that. Okay, can you indulge me in one silly thing?
Do you have the resolution in front of you?

Rhiana: Yeah, I do.

Ayana: Can we read page five together? We can alternate subsections. A
dramatic reading of Section One of the Green New Deal. "Resolved."

Rhiana: "Resolved. That it is the sense of the House of Representatives,
that it is the duty of the federal government to create a Green New
Deal—to achieve net-zero greenhouse gas emissions through a fair and
just transition for all communities and workers;"

Ayana: "To create millions of good high-wage jobs and ensure prosperity
and economic security for all people of the United States;"

Rhiana: "To invest in the infrastructure and industry of the United States
to sustainably meet the challenges of the 21st century;"

Ayana: "To secure for all people of the United States for generations to ♥
come—clean air and water, climate and community resiliency, healthy
food, access to nature, and a sustainable environment, and"

Rhiana: "To promote justice and equity by stopping current, preventing
future, and repairing historic oppression of indigenous peoples, communities of color, migrant communities, de-industrialized communities,
depopulated rural communities, the poor, low-income workers, women,
the elderly, the unhoused, people with disabilities, and youth (referred to
in this resolution as frontline and vulnerable communities)."

Ayana: When I read this, I think: "This is the world I wanna live in. This
is getting it right."

Rhiana: Yeah, me too.

Ayana: Can you imagine a world where we actually do all this stuff?

Rhiana: Totally. As I was reading that aloud, it reminded me that I have to direct my vision back to this, because what I work on most right now is racial justice in urban communities. That's my baseline, that's what I feel most passionate about and where I have the most knowledge. But I'm reminded now that all of this is tied together, all of these places and people are deserving and worthy of exactly what we laid out here. And can I see that? Yes, a hundred percent.

Ayana: Help us see it. What does that world look like? What does it feel like?

Rhiana: So, this is going to be a little bit sappy.

Ayana: Oh, bring it on. I'm here for it.

Rhiana: I recently had a kid, he's one year old. And something I was struck by in having a baby, beyond all of the toil and sleeplessness or whatever, is this incredible feeling of acceptance and welcome—which I know will change as he becomes a teenager. But there's this sense of . . . he is so happy that I am alive.

Ayana: Mmm.

Rhiana: Right? It doesn't matter what I'm producing. I mean, I guess other than milk. But even if I didn't do that, he is just so happy that me and my partner are here and able to be with him. And there's a sense I have for the first time that this is what it feels like to have someone who is grateful just that you exist. And obviously I feel that way for him. So when I think of what if we get it right, you have access to nature, you have access to clean water, and you have clean air, and we have a country, and ideally a world, where the ways you're treated, your surroundings, feels like it's all telling you that we are grateful that you exist.

Ayana: That's such a nice way to think about it. I like the simplicity of that. Just a fundamental respect for each other as human beings.

Rhiana: Exactly.

Ayana: So, you're from Chicago. What would be different in the daily life of someone living in Chicago if we had a Green New Deal implemented?

Rhiana: I'm thinking of this through the perspective of my child when he's older. I imagine a Chicago where all of the housing stock is healthy.

You don't have to worry about whether this house has a lead service line or this house is energy-efficient, because all houses will have that sorted. And if you have a little bit more change in your pocket, you won't have to feel bad about having an electric vehicle or an induction stove or a heat pump because everyone has that, even family members who aren't as well-off. So those things don't feel so much like an announcement of "I had an extra $30,000 to drop," and more of a "Yeah, it's just normal."

Ayana: I'm eager to see whether this new federal funding can actually help create that new normal. And at this point that's a big question mark.

Rhiana: It's definitely a big question mark, and if we're being honest, the mechanisms those investments are happening through, particularly for things like heat pumps or EVs, are individual <u>tax credits</u>, which have a terrible history when it comes to being equitable in terms of income and race. And these tax credits aren't refundable, which means they exclude something like 40% of American households, because that's how many American households don't have tax liability. And that's not even talking about the people who do have tax liability but do not have thousands in capital to front the cost.

That's what I mean when I say we have to try to implement these things as equitably as possible with justice front and center. Even with those legislative limitations, this can push toward a Green New Deal. There's lots of funding in the IRA that perhaps could be used to provide up-front capital, maybe through a grant program. It seems like minutiae, but all of that is important to make sure that we don't have a clean-energy system that replicates the exclusionary patterns of the one that we have now.

Ayana: That's not minutiae; that's where the rubber meets the road. What else do you imagine for a future Chicago?

Rhiana: I imagine if my child has a child there's a choice among fully available, publicly funded childcare options. You won't have to choose based on whether this neighborhood is polluted or more dangerous. And also he can choose whether or not he has a vehicle, whether he has an EV or a bike or whatnot, because the city is planned so that there's more walkability, more modes of transportation are available and public transportation is clean and accessible and affordable, hopefully free.

Ayana: We're heading in that direction in some places.

Rhiana: Yeah. And in Chicago, we have a beautiful lakefront. A lot of it is accessible now, but living near that is expensive. And, because of climate change, maybe people won't want to live there or it won't be as safe. So I do hope that the waterfront, and beautiful places in general, will become more accessible. One of the biggest things I hate most about capitalism is that access to so many of the things that are just nature, beautiful places, are conditioned on whether you can afford to be there or not. And they're so expensive.

I hope that when he goes back to visit the place where I grew up, hopefully that house is still in the family, and being there is a much more joyful feeling, because the neighbors are thriving. People have jobs that pay well or they own their own businesses. He's driving past community solar that's there. And people are walking around with a lot of life in their eyes because they're in surroundings and a community where it feels like people care about you, that the city itself thinks you're valuable, and the people in power do, too. I hope it will be someplace that feels joyful and vibrant.

Ayana: As opposed to what you had to deal with as a kid, which was having asthma because of the air pollution.

Rhiana: Or the way I feel now when I go home. I have these wonderful memories of living there, of my neighbors and the community that we had. But now it's desolate in a way that it wasn't. You can feel that there are no jobs, that the city is disinvested, that this is not a place where the people are valued. You can go into a neighborhood and know that. It takes seconds.

Ayana: You feel it.

Rhiana: And it breaks my heart because I know the kinds of people who live there and how worthy and wonderful they are.

Ayana: What are some of the barriers you see in getting from where we are now to this whole Green New Deal vision actually happening?

Rhiana: Well, the Green New Deal is very much about what we need to *invest in* to build change. But a huge thing that it's more silent about, but is incredibly vital, is divestment from fossil fuels. And it is naive to think that the changes in our energy system, and the market changes that come from shifting to clean and renewable energy, will naturally get rid of fossil fuels. Or that the powerful fossil fuel industry isn't going to create a lot of opposition.

Ayana: They're not going to be like, "You were right, we'll just stop drilling."

Rhiana: Yeah, absolutely not. And it's something we have to think about. How do we divest from fossil fuels responsibly, in a way that minimizes harm? How do we make sure that, when fossil fuel companies wind down, we don't end up in a situation where it's like a fire sale, a mess, pensions going missing, which we've had a taste of, and workers end up holding the really short end of the stick?

Another thing we have to be serious about is the stickiness of White ! supremacy. We like to think about White supremacy as something either external or something that's just about hearts and minds, but in very real ways, it is a tool that we use to solve tough problems, down to where we put fossil fuel facilities. Because we depend on fossil fuels, but they poison us. So we have to be serious about developing alternative tools to help us solve those problems.

Ayana: And where are we siting our renewable energy infrastructure? We are not prepared to answer some of these questions yet. What are we doing with all the old fossil fuel infrastructure? How are we remediating or transforming those locations into something that benefits those communities?

Rhiana: Right. Or even something as fundamental as solar panels and batteries. Renewable technologies, they're wonderful, but they're not without costs, even down to minerals. How are we going to get those ! minerals in ways that don't rely on exploitation and shoving aside treaties and Indigenous rights? How do we do this in a way that's not dependent upon oppression?

The other thing I think about is we can't act like the rise of fascism ! and right-wing violence is not a huge driver in whether or not these policy and societal changes can happen. I mean, there is evidence that environmental stewardship and having a clean environment is seen as important among conservative voters. But the ways that they will go about it, they're not centered around justice. So we have to be serious about how we build and sustain enough power to fight that back.

Ayana: It's interesting because renewable energy is happening on a large scale in conservative parts of the country—rural Texas and Iowa lead the nation in wind energy. There's actually quite a lot of support for green jobs across party affiliations. Same across the political spectrum for clean air and water.[148] But, like you said, when it comes to how we get there, there's definitely not agreement on that.

How do we need to change or shore up our democracy in order to enable this Green New Deal vision to have a chance?

★ **Rhiana:** I think about that all the time. Because a lot of the changes people talk about needing for a multiracial democracy are exactly the things you need for a Green New Deal. Things like abolishing the filibuster, expanding and protecting voting rights—honestly, an abolition of the Electoral College would go so far, so could expanding or eliminating the Senate.

Ayana: Probably the Supreme Court needs to be updated in some way.

Rhiana: Oh yeah. Definitely, the Supreme Court needs to be fixed.[149] All of those are actually things that, I won't say that they're *necessary*, but it'll be very, very difficult to have a full articulation of the Green New Deal without them. Anything that levels the playing field for political participation is crucial because what we do know is that folks who are more likely to be climate deniers are rich, older White men who are centered in our political system and through all of the institutions that I talked about.[150] And the folks most interested in climate action and who vibe most with a Green New Deal are the people who are more likely to be shut out.

Then there are also small-"d" democratic changes, things we have to develop because they don't really exist now—new ways to have procedural justice, like stakeholder input in policy creation and oversight. New ways to create and protect public ownership, particularly of energy resources.

---

148 More than 92% of Americans consider rights to clean air and water to be "essential rights."

149 We'll dive into the role of the courts in the following interview, with Abigail Dillen.

150 Compared to other Americans, conservative White males are nearly twice as likely to deny climate science.

Ayana: What does that look like?

Rhiana: That can look like community solar, more publicly owned utilities, a more decentralized grid, more distributed power resources. Net metering is always nice; it's always helpful for people to get paid for extra electricity they produce and sell back to the grid.

There's also the idea of having publicly minded institutions be anchors of energy production, like hospitals or public universities. I don't know a ton about that, and I have my questions, but it's an interesting idea. Even state public utilities, publicly owned energy resources, all of those things are helpful for both decarbonization and climate justice.

Overall, we need more small-"d" democracy that allows for more public engagement and community control over both how decarbonization and the energy transition are happening, and more generally around public policy. We've seen such an erosion of trust in our institutions and in some ways in civic engagement. Part of the reason for that is it's difficult for communities to have a say in policy because we've swung so much toward experts. I'm all for experts; this is not anti-intellectual. It's to say that having more ways for people to be involved, and have their input be meaningful and not just symbolic, could go a long way to helping rebuild trust in institutions.

Ayana: Are there any cultural barriers that you see, any cultural shifts that need to happen?

Rhiana: We need a renewed emphasis on community and community care, and a scaling back of individualism, of the metric of success being based on individuals or individual households. We are dependent upon one another, upon our physical environments. You can't oppress some groups over in one place and not have that affect the whole eventually. A real cultural shift has to be around recognizing our responsibility to one another, the fact that quality of life depends not just on how much money you as an individual make, but what's happening in your community.

Ayana: Oh, absolutely. And what about corporations? What would you like to see them do?

Rhiana: A shift away from shareholder primacy would be huge. If you believe that your main and highest objective as a company is to make money for your shareholders exclusively, it enables a whole bunch of antisocial behavior because all that matters is your bottom line.

There are similar issues with pensions. The duty of the person or entity that makes investment decisions for pension funds is actually to the health of the fund, not to the pensioners. That can lead to perverse situations where funds are maximizing profits for pensioners through investments in fossil fuels that will ultimately undermine their health and safety, and sometimes even the security of their jobs. We need to move away from that, and have far more thoughtful and rigorous governance of corporations and their boards.

Ayana: Are there places where elements of this Green New Deal vision are already unfolding and solidly underway?

Rhiana: There's that landmark New York State climate legislation, which had a provision that at least 35%, with a goal of 40%, of the benefits of all climate investments have to go to frontline and disadvantaged communities. So that's happening. And there is an offshore wind energy project happening in New York that is in part being driven by UPROSE, a community group. I wanna say they're out of Brooklyn.

Ayana: They are in Brooklyn, I'm proud to say. UPROSE is a great environmental justice group that's been around for decades.

Rhiana: That's a good example of environmental justice and industry and the public sector all working together. UPROSE is helping shape what the project looks like, so that it maximally benefits their community.

Ayana: That's a very big deal.

Rhiana: It's a huge deal. I often hear this refrain that environmental justice advocates don't know how to build things or don't want to build things. This is a clear counterpoint to that.

Ayana: Who do we need on the policy team in order to make all these changes happen? Who do you want to recruit?

Rhiana: I would say some folks in clean-energy industries like solar and wind, deep in them. Because a lot of the data and knowledge about how you would decarbonize different industries is held by the industry. It's not publicly available.

It's also crucial to have more people who are not just knowledgeable about environmental justice, but who have done work with local and state governments and who have the facility to do federal policy. That bench is not deep right now. We need a lot more of those folks. We need policy nerds focused on process and stakeholder engagement.

More broadly, the Green New Deal resolution lays out a bunch of focal areas—agriculture, energy systems, etc.—and it's necessary for experts in all of these areas to be on the team. Because people who know about the grid don't necessarily know about agriculture and people who know about oceans don't necessarily know about housing stock and energy efficiency.

What has to draw together this team is a Green New Deal vision, but also a real commitment to justice in ways that allow us to check whether the ideas are doing harm or not. In all ways, the team should also be porous—they have to be taking in information from outside, they need to have those relationships, and they have to actually be able to go out and talk to people on the ground in communities.

Ayana: And we have to be deliberate about all that—it isn't going to happen serendipitously. What are the top three things you wish everyone knew about climate policy?

Rhiana: Climate policy is much bigger than you think; it's not just "environmental." The Green New Deal is not a single piece of legislation; it's a vision and a framework for how we orient and structure a transition away from fossil fuels. And there is no way to disentangle the decisions we make about climate from the decisions we make about our economy.

To refuse to participate in the shaping of our future is to give up. Do not be misled into passivity either by false security (they don't mean me) or by despair (there's nothing we can do). Each of us must find our work and do it.

—Audre Lorde

A Blue New Deal

With contributions from Jean Flemma

The ocean is singular—one interconnected system of currents and wildlife covering most of our planet. It is tunas and turtles, seaweeds and seahorses, currents and carbonate, mangroves and manganese, beaches and blowfish. It is not separate from us humans and our daily concerns. It is culture, joy, and freedom. Its heat and motion drive our weather and climate.

The ocean is an extremely big deal. And we've pummeled it. Polluted it. Dragged it. Emptied it. Disrespected it six ways to Sunday. And yet. Here it still is. Still nurturing, feeding, and delighting us. Still sustaining coastal communities and buffering the impacts of climate change.

Ten sweet, sweet ocean facts,[151] all possibilities:

1. Ocean-climate actions—like scaling offshore renewable energy, decarbonizing shipping,[152] eating low-carbon seafood, and conserving and restoring coastal ecosystems—could comprise 35% of our climate solution.

   *

2. Forests get all the love, but mangroves and wetlands can absorb 3 to 5 times the carbon per area as a tropical forest.

3. Mangroves can reduce wave height by up to 66%. During the 2004 tsunami in Indonesia, the presence of mangroves is estimated to have saved more than 11,000 lives.[153]

4. Oyster reefs can be as effective (and as cost-effective) as constructed breakwaters, dissipating waves and reducing their height by up to 65%, and can even grow and expand to outpace sea level rise.

5. Coral reefs (gravely threatened by climate change) support nutrition and livelihoods for approximately 500 million

---

151 For sour ocean facts, refer back to the "Reality Check" chapter.

152 If shipping was a country, it would have the sixth-largest emissions of any nation.

153 From 1980 to 2003, over 1.1 million hectares of Indonesia's mangrove forest were destroyed, primarily for conversion to shrimp aquaculture ponds.

people. Plus they provide coastal protection—reefs reduce wave energy by an average of 97%.

6. Coastal wetlands can provide better and cheaper shoreline protection than seawalls. During Superstorm Sandy, although 85% of the wetlands in New York and New Jersey had already been destroyed by development, what little remained prevented $625 million of damage.

7. Every $1 spent on reef and wetland restoration achieves more than $7 in direct flood-reduction benefits.

8. The offshore wind industry could create more than 1 million jobs globally by 2025. In the U.S., the industry could support tens of thousands of jobs by 2030 (more than coal mining currently does), including many union jobs and jobs with skills transferable from the oil and gas sector.

9. Regenerative ocean farms growing seaweed and shellfish can support food security, improve water quality, protect shorelines from storms, and create tens of millions of direct jobs globally. Seaweed can absorb hundreds of millions of tons of carbon every year, and help to reduce ocean acidification locally.

10. Within fully protected areas, fish biomass is on average 670% higher than in unprotected areas. At present, less than 3% of the global ocean is fully protected, while scientists recommend protecting at least 30%, even 50%, of land and sea.

Credit and respect and gratitude where they are due.[154] Thank you, ocean. You deserve better.

_____

154 Bonus ocean facts: (1) Tropical, white sandy beaches can be up to 85% parrotfish poop, because parrotfish nibble on and grind up dead coral as they graze on algae. Protect parrotfish to protect sandy beaches. (2) Octopuses have three hearts and nine brains and can make independent decisions with each of their arms. I love them.

Given all these ways that the ocean can be a significant part of our climate solutions, clearly it should play a major role in climate policy. Coastal ecosystems, offshore renewable energy, ocean farming, and zero-emission shipping all deserve some love. Enter the <u>Blue New Deal</u>.

When we imagine the climate transformation we need, usually the first images that spring to mind are of *physical* transformation—installing solar panels, planting trees, driving EVs—but as the last two interviews have made abundantly clear, underneath that we also need *policy* transformation. So, how do we turn ideas into policies? Policymaking is often a black box, so let's pull back the curtain.

I've been fully in the mix of this effort to put the ocean at the heart of climate policy for five years and counting, so this chapter is not an interview. It's a story, one that includes some of my favorite characters, like Jean Flemma, who contributed to telling it. I'm just going to tell you how things have unfolded. It's been gratifying and even, at times, joyful—partially because of the progress, but mostly because of the camaraderie. Collaboration is key to this whole story. And while this chapter is ♥ nominally about the ocean, it's here as a case study for making change—the steps and tools, the who and what, ingredients for any policy-change recipe.

Welcome to my world. We are going on a wonky journey.

YEAR 1—PRESIDENTIAL PRIMARY

Step 1: Commiserate about the status quo.
February 2019

The ideation of a Blue New Deal started the way many exciting projects start: over a delicious meal. I was feeling bummed about the ocean getting short shrift in the Green New Deal Resolution—just a single, vague reference to the ocean on page 13 (out of 14), in section 4, subjection L—when I met up for lunch with Chad Nelsen, CEO of Surfrider, an ocean conservation nonprofit. We commiserated about the ocean-sized omission. And by the time we were contemplating dessert, we had decided to complain about this in public, aka co-author an op-ed.

Step 2: Write an op-ed.
February to April 2019

In policy land, op-eds (i.e., opinion essays) are a tool for trying to get the attention of people in power and shift public opinion. They can be a key strategic tool, but at this point there was no strategy, just two disgruntled ocean lovers with a Google doc pointing out a void. Then I learned that Bren Smith, a pioneer of regenerative ocean farming and co-founder of the nonprofit GreenWave, was drafting something similar.[155] So we three teamed up and distilled our ideas to four ocean-climate solutions that need urgent federal government attention: (1) restore and protect coastal ecosystems; (2) invest in renewable offshore energy; (3) bolster the "blue economy"; and (4) vastly expand regenerative ocean farming.

Step 3: Get the op-ed published.
July 2019

Easier said than done. We got rejection after rejection from the major newspapers. And then we landed at *Grist*, a wonderful environmental publication.[156] Their editor gave our draft *the business*, at which point it became clear why no other paper had accepted it—it required determined editing to reshape our . . . let's call it un-poignant writing. And *Grist* gave it a great title, "The Big Blue Gap in the Green New Deal." The op-ed was finally published in July 2019—six months after we started working on it! This was during the U.S. presidential primaries, and we hoped it would get candidates' attention. That didn't happen (yet), but the ocean conservation and policy community was into it and embraced the term "Blue New Deal." Now we had a catchy tagline and a concrete concept to back it up.

Step 4: Deepen the analysis with a policy memo.
June to September 2019

As we were trying to find a home for the op-ed, Data for Progress, a progressive think tank that had been promoting the idea of a Green New

---

155 Interview with Bren coming up in the "Transformation" section.
156 *Grist* publishes extensively on climate solutions and possible futures. See: grist.org

Deal through their policy memos and polling data, reached out to Bren and me about creating a memo on fisheries.[157] We teamed up with their senior fellow Johnny Bowman to make the case for three specific areas of policy reforms: ocean habitat restoration, community-based fisheries, and regenerative ocean farming.

Our recommendations included: create a blue carbon fund; include coastal restoration in a future Civilian Conservation Corps; require full supply chain traceability for seafood; pass the Keep America's Waterfronts Working Act; and include regenerative ocean farming in USDA crop-assistance and insurance programs. When it comes to advocating for policy change, specificity is important—that enables your wishes to be granted.

Politicians are often compelled by "How will this affect my constituents? How many jobs is this going to create (in my district)? Is this popular?" So, in our policy memo, "Seafood, Blue Jobs, and the Green New Deal," we highlighted two key numbers:

- The Blue Economy (tourism and recreation, fishing, shipping, and more) supports 3.5 million jobs in the U.S.—considerably more than all the jobs in crop production, telecommunication, and building construction combined.

- 65% of fishermen believe that climate change could leave them "unable to profit" and ultimately "forced out" of their fishery.

Step 5: Make the news.
September 2019

During the presidential primary, climate groups pushed the Democratic National Committee (DNC) to host a climate-specific debate. Young people from the Sunrise Movement even camped out on the steps of DNC headquarters with this request, but to no avail. Since Democratic candidates were prohibited from appearing together on stage except during DNC-sanctioned debates, when CNN stepped in to host, they used a

---

157 Compared to op-eds, memos go into more detail about the what and the how (but are still usually under ten pages), and get published by the organization itself (not in newspapers) if at all (sometimes they are internal documents).

town-hall format. This resulted in an absurd seven-hour marathon of back-to-back forty-minute sessions, with each candidate fielding questions one at a time. Who but climate dorks would sit through this ridiculous extravaganza?! Sigh. But it was better than nothing.

Anyone could submit a short video question, and CNN picked a few to present to the candidates live on TV. I really dislike recording selfie videos, but thankfully Bren, our favorite ocean farmer, submitted a question that made the cut and happened to get directed to Senator Elizabeth Warren:

> **Bren:** My oyster farm was destroyed by two hurricanes. Now warming waters and acidification are killing seed coast to coast and reducing yields. Those of us that work on the water, we need climate solutions and we need them now. The trouble is the GND only mentions our oceans one time. This is despite the fact that our seas soak up more than 25% of the world's carbon. So what's your plan for a Blue New Deal for those of us working on the ocean?
>
> **Senator Warren:** I like that!
>
> **Bren:** How do we make sure that all of us can make a living on a living planet?
>
> **Senator Warren:** So, thank you, I think that is a great question and I think he's got it exactly right. We need a Blue New Deal as well.

I was at a little watch party of climate friends and when I saw Bren pop up on the screen, I yelped. And when I heard Elizabeth Warren's strongly affirmative response, I cheered. I was blown away by this three minutes of fame for ocean-climate policy. But I didn't think anything would come of it, because climate policy, let alone ocean policy, had never been a central issue in a presidential primary.

Step 6: Develop a campaign policy plan.
September to December 2019

Elizabeth "I've got a plan for that!" Warren actually went back to her campaign climate advisor, Maggie Thomas, and asked her to draft a plan for a Blue New Deal. Maggie turned around and called me. When my phone rang that September, I was standing in my mom's kitchen, looking out the window at the vegetable garden in its late-summer abundance. I tried to keep my cool as it dawned on me that I was being asked to help Warren write one of her signature plans.

At that point I didn't know Maggie, but she had previously helped craft climate plans for Governor Jay Inslee's campaign, which were considered the gold standard in the primary. I was thrilled with the chance to team up with her.

Next thing you know, there's another Google doc. I'm in there with Maggie, fleshing out ideas from the initial op-ed and policy memo, and she's reaching out to experts for input and ideas, and we were weaving together this concept for how the federal government could actually make all of this happen—from ending offshore drilling to expanding marine protected areas, from climate-ready fisheries to climate-smart ports, from protecting and restoring habitats to regenerative ocean farming, from disaster mitigation to ocean trash reduction. We covered *a lot* of ground. Three months from that first phone call, the Warren campaign released their Blue New Deal plan.[158]

Step 7: Write another op-ed.
December 2019

As poetic and concise as I thought this plan was, I knew few people were going to dig in and read the eight or so pages of policy recommendations. We needed to promote it. Time for another op-ed, to spread farther and wider a vision for robust ocean-climate policy. With Warren now championing the concept, plus stellar editorial support, this op-ed landed in *The Washington Post*: "Our oceans brim with climate solutions. We need a Blue New Deal."

During a contentious presidential primary, it was exciting to see how

---

158 In this same time frame, the High Level Panel for a Sustainable Ocean Economy, a global group of ocean-climate experts, released a report showing that the ocean could provide 21% of the emissions reductions needed to reach our global climate goals. (That estimate has since been updated to 35%.)

eagerly people embraced Warren's policy plan. Quite likely hers is the only presidential campaign to ever have a standalone ocean-policy plan. When I later met the senator in person, she told me it was the plan that got the most excited mentions in her selfie lines.

At this point, over the course of a year, we'd gone from a few ocean lovers commiserating to two op-eds, a policy memo, some polling data, and a presidential campaign policy plan. Everything was coming up ocean. And then . . . Warren dropped out of the race. I expected that would be the end of the road.

YEAR 2—GENERAL ELECTION

Step 8: Carry the ideas forward regardless of who wins.
June 2020

Reader, it was not the end of the road. After Joe Biden became the Democratic nominee, to consolidate a fractured electorate, he pulled ideas from other candidates' platforms and incorporated them into his own. Enter Evergreen Action, a new think tank founded by alumni of the Jay Inslee campaign. They set to work trying to get the ambitious climate proposals they had put together for the Inslee campaign adopted by candidate Biden and by members of Congress. Maggie Thomas became Evergreen's founding political director. (And I later joined their advisory board.)

One day in the summer of 2020, Maggie asked me to have a phone call with Ali Zaidi, then New York State's deputy secretary for energy and environment, who in his free time was volunteering as a climate advisor for Biden's presidential campaign. Ali asked what elements from the Warren plan I thought should be included in Biden's updated climate plan.[159] I gave him my top three: (1) protect coastal ecosystems to absorb carbon and protect coastal communities, (2) ramp up offshore

---

159 This is one of those moments where I found myself thinking, "Me? You want *my* opinion? Are you sure??" I was sitting on my stoop in Brooklyn, wearing cutoff jean shorts, sweating, and feeling like I really should be wearing something more formal for this conversation.

wind energy, and (3) create a Civilian Climate Corps (an Inslee campaign proposal, formerly known as the Civilian Conservation Corps, that had been carried forward by Warren) to put tens of thousands of Americans to work implementing climate solutions. My overall message to Ali was: Please don't forget about the ocean. The Biden team was gathering ideas from lots of experts, and I had no delusions that what I was advocating for would be particularly notable to them. Then the Biden campaign did something I did not see coming . . .

Step 9: Revel in a new campaign policy platform.
July 2020

The Biden campaign released an ambitious new climate plan, like actually legit, during the general election. And I was thrilled to see that all three of the things I had advocated for had made it in. *Mangroves*, my darling, underappreciated mangroves, got a shoutout.

Even though so many advocates had been chipping away at this stuff for decades, breaking through can often feel impossible, like you need more power and influence or gazillions of dollars or to be on the campaign staff or in the administration to make a difference. Instead, I found myself (my jaded self) crossing paths with all these smart folks who were trying to help from their various positions. The whole thing was pleasantly remarkable.

Breaking through was also happening in Congress. The House Select Committee on the Climate Crisis released a Congressional Action Plan. This was thanks to the work of many ocean advocates and members of Congress who are true ocean-climate champions—including the chair, Kathy Castor, and committee members Suzanne Bonamici and Jared Huffman. And it included significant ocean-climate action recommendations.

Step 10: Publish another memo, this time on offshore wind.
August 2020

Now it was time to prepare for how to make all these ideas and plans into actual detailed federal policy should Biden win. In other words, it was memo time again. This time Urban Ocean Lab,[160] the policy think tank that I co-founded and co-lead with Jean Flemma and Marquise Stillwell, teamed up with Evergreen to publish "A Plan for Offshore Wind Energy in the U.S." The memo laid out some major barriers to offshore wind development: from complex regulatory requirements, to inadequate staffing and resources at federal agencies, to lack of well-designed transmission systems for grid integration. We then offered recommendations for how Congress and the executive branch could remove these barriers and rapidly scale offshore wind: from setting national targets, to jumpstarting American manufacturing, to investing in workforce training, to creating tax incentives, to increasing efficiency of permitting, to bolstering knowledge and research.

Step 11: Conduct polling to show how popular it is.
August 2020

* Politicians often follow instead of lead, only proposing or advancing things when they know there is ample public (or lobbyist or donor) support. With Data for Progress, we designed a short survey to gauge whether voters supported expanding offshore wind energy. And did they ever. Voters support:

  • the construction of new offshore wind farms, by a 48-point margin,

  • speeding up the permitting process for building offshore wind farms, by a 43-point margin, and

  • federal investments in research to improve offshore wind technologies, by a 34-point margin.

And here's the kicker: On all the questions, there was support from a majority (or near majority) of voters *across the political spectrum.* Gotta love a bipartisan climate solution—too few and too far between.

---

160 Urban Ocean Lab's mission: Cultivate rigorous, creative, equitable, and practical climate and ocean policy, for the future of coastal cities. For more, see: urbanoceanlab.org

One interesting additional polling result worth pausing on is the response to this question: "Would you support or oppose reforms that would make it more difficult for wealthy coastal homeowners to stop the construction of offshore wind farms?" Again, across the political spectrum, the answer was "support." This is key because, to date, NIMBYism[161] has held back development of offshore wind energy in the U.S., and here was a sign that people think that needs to stop.[162]

Step 12: Write another op-ed.
December 2020

In policymaking, you have to pursue all avenues for advancement and ✱ hope that at least one will stick. So while some of us were working the communications and campaign avenues, others were diligently working the halls of Congress, where things were in motion. Congressman Raúl Grijalva introduced the Ocean-Based Climate Solutions Act, broad legislation designed to harness the ocean for its climate solutions.

Then Biden won the election. As the presidential inauguration approached, in the spirit of relentlessly pursuing our policy goals, Jean, Dr. Miriam Goldstein[163] (then the director of the ocean program at Center for American Progress, another think tank), and I co-authored an op-ed titled "The Ocean's Heroic Potential Could Be Realized Under Biden." It was published in the *Boston Globe*, the primary newspaper for New England, where ocean issues have long been politically prominent. This piece specifically called for Congress to pass the Ocean-Based Climate Solutions Act, and for Biden to use executive authorities to kick off shovel-ready coastal restoration projects to provide jobs, restore coastal ecosystems, and support coastal communities.

---

161 NIMBY stands for "not in my back yard" and refers to opposition by residents to nearby development—especially when they would support it if it were built farther away.

162 This was also mentioned in the Warren plan: "I won't allow visual and aesthetic impacts to provide a basis for denying federal offshore wind energy permits. The climate crisis is too urgent to let the ultra-wealthy complain about wind turbines getting in the way of their ocean views."

163 Fun fact: Miriam and I both did our marine biology PhDs at Scripps Institution of Oceanography, and she was my first dive buddy for my dissertation fieldwork in Curaçao.

YEAR 3—BIDEN ADMINISTRATION

Step 13: Play the inside-outside game. Personnel is policy.
January 2020

As Biden was building his team, he established an Office of Domestic Climate Policy in the White House and selected Gina McCarthy, former EPA administrator and my collaborator via *All We Can Save*, to become the president's inaugural national climate advisor, directing the office. Ali became the deputy (and when Gina stepped down in 2022, took over as director), and Maggie became chief of staff. This was the first time I had professional relationships with White House policymakers. As Warren says: "Personnel is policy," meaning that who is placed in policy positions determines what policies get made. I considered applying to join that amazing inside team but thought I'd be more useful on the outside, as an advocate and communicator, developing and pushing for various policies and rallying public support. Policymaking is often referred to as an inside-outside game.

Step 14: Celebrate an executive order.
January 2021

One week after being sworn in, Biden signed the Executive Order on Tackling the Climate Crisis at Home and Abroad. This was one of his first executive orders and laid out the administration's climate priorities. It included a lot of the same things as appeared in his campaign plan. And there was one part in particular that I was thrilled to see carried forward:

> Coastal communities have an essential role to play in mitigating climate change and strengthening resilience by protecting and restoring coastal ecosystems, such as wetlands, seagrasses, coral and oyster reefs, and mangrove and kelp forests, to protect vulnerable coastlines, sequester carbon, and support biodiversity and fisheries.

Perhaps this sounds ridiculous, and I know it's just words, but I got a bit emotional seeing coastal communities given specific attention and marine ecosystems individually name-dropped and valued in a presidential order.

And the executive order was just the beginning. Within the first 100 days (a meaningless milestone that is often given lots of press attention) the administration announced some major ocean-climate action commitments, including the goals of deploying 30 gigawatts of offshore wind by 2030; conserving 30% of the ocean by 2030; and reaching zero emissions from international shipping by 2050.

Step 15. Write yet another op-ed.
May 2021

While the president can execute some parts of his policy vision through executive-branch authority, via federal agencies, for many things you need to get Congress on board. At this point I direct you to yet another op-ed, this one published in *Bloomberg* and titled "Congress Must Make Biden's Vision for the Oceans Come True."[164] (This is op-ed #4, for those losing track.) This was again co-authored with Jean and Miriam. Our point was that it was time for Congress to step in and pass laws and appropriate funds so that the work can happen. Because while executive orders are useful, they can be pretty easily undone by the next administration—and this work takes serious money and resources.

Step 16: Don't settle for small victories—keep pushing.
March to June 2021

The Biden administration quickly kicked into gear on offshore wind energy, in March releasing a policy plan detailing how they would ramp up the industry in U.S. waters. Our wind-policy memo was then quickly out of date—the administration was already completing the items we recommended. Maggie, a co-author of that memo, was now working in the White House, which maaayybe had something to do with this. In June, Urban Ocean Lab released an updated memo with an updated set of policy recommendations.

---

164 Note: Writers have little control over op-ed and article titles.

Step 17: Testify before Congress.
June 2021

I received an invitation from the House Committee on Natural Resources to testify on the Ocean-Based Climate Solutions Act. While the thought of doing this was absolutely nerve-racking, it was a valuable opportunity to speak directly to members of Congress and their legislative aides. Jean and Miriam encouraged me to accept, and I was so grateful to have them, with their collective decades of experience as congressional staff, coaching me through it.

I logged on to the video platform (participating remotely was a slight reprieve for my nerves) to offer my remarks. I began my testimony by highlighting some of the ocean facts listed at the beginning of this chapter. Then I focused on three main ocean-climate solutions that at this point will not come as a surprise to you: (1) protecting and restoring coastal ecosystems that absorb tons of carbon and protect us from storms; (2) producing clean, renewable energy from offshore wind; and (3) farming the ocean regeneratively to support a sustainable food system. I also spoke about the need for ocean justice. Then (the scariest part), I fielded questions from members of Congress. I was in excellent company with Marce Gutiérrez-Graudiņš, founding executive director of the Latine ocean conservation nonprofit Azul,[165] and State Senator Chris Lee from Hawaii also testifying in that hearing.

The bill has yet to pass, but hearings can be valuable for advancing issues and educating members of Congress. Some bills are introduced many years in a row before they get enough support to pass. This one was first introduced in 2020, and this second time around had seven more co-sponsors (for a total of forty-six). More important, many provisions of the bill made their way into the Inflation Reduction Act (IRA) and the Infrastructure Investment and Jobs Act (IIJA). Shoutout to Congressman Raúl Grijalva, sponsor of the bill, who was then chairman of the committee and who is a stalwart champion for ocean and climate issues, and for environmental justice.[166]

---

165 More on collaborating with Azul in step 21.

166 Congressman Grijalva also sponsored the Environmental Justice for All Act. More on that in the next chapter.

YEAR 4—KEEP GOING

Step 18. Push Congress to legislate.
January 2021 to August 2022

As much as I appreciate ocean-specific legislation, the megabills working their way through Congress in 2021 and 2022 were the IIJA and the Build Back Better Act (BBB). It was key to ensure that the ocean got some love (and appropriations) in those. There was a full-court press from the whole ocean conservation community—all of us were pushing Congress to include ocean protection, restoration, and wise use in these broad acts. The BBB proposal included $6 billion specifically for coastal restoration, which was whittled down to $2.6 billion by the time it made its way through the legislative gauntlet and emerged in August 2022 as the Inflation Reduction Act.[167] Overall, much of the climate funding remained intact: The initial BBB proposal included around $555 billion for climate and clean energy, and the final was $370 billion. Given how many programs were cut entirely (e.g., universal preschool, paid family leave, expansion of Medicare), this was a significant victory. One particularly sad last-minute cut from the IRA was funding to create a Civilian Climate Corps. And one very bitter pill in the Act was that expansion of offshore wind development was tied to continued expansion offshore oil and gas leasing, an especially brutal blow for climate justice in the Gulf South.

Step 19: Push the president to make a plan.
October 2021

With so much funding moving through Congress, and important ocean-climate policy goals being set by the administration, there was a recognition in the advocacy community that implementation would be more successful with inter-agency collaboration—something that is not sexy but is terribly important for maximizing the benefits of new policies and dollars. So, 118 organizations and businesses co-signed a letter to Presi-

---

167 The IIJA also included $1 billion in funding for coastal restoration and resiliency projects, for a total of $3.6 billion. Unprecedented, and a testament to years of work by the ocean advocacy community.

dent Biden asking his administration to "design and implement an ambitious U.S. ocean-climate action plan."[168] A few weeks later, the president committed to doing just that.

Step 20: Tell the president what you'd like in that plan.
June 2022

Riding high on the commitment by the administration to develop and implement an ocean climate action plan, in June 2022 more than ninety organizations sent the White House a detailed list of policy recommendations: the Blueprint for Ocean Climate Action.[169] It laid out the critical elements of an ocean-climate solutions agenda and advocated for their inclusion in the forthcoming plan.

Step 21: Advance ocean justice.
June 2021—September 2022

Beyond the *what* of policy, there is the *how* of the approach and implementation: It must be equitable; it must benefit coastal communities. While "climate justice" is understood to be key to climate policy, justice is invoked and addressed far less often when it comes to ocean policy. So in June 2021, Urban Ocean Lab helped to launch the Ocean Justice Forum (OJF), a collective of eighteen nonprofits committed to putting justice at the heart of U.S. ocean-climate policy. These organizations represent a broad cross-section of coastal stakeholders—from Indigenous kelp farmers in Alaska, to fishermen in New England, to Latine surfers in California, to Native Hawaiians, to Black folks in the Gulf, to national conservation groups in Washington, D.C.[170]

In September 2022, after four convenings and many, many drafts,

---

168 Urban Ocean Lab was among the signatories.

169 We signed on to that too. Keep showing up.

170 OJF founding members: four steering committee organizations (Azul, Center for American Progress, Taproot Earth, Urban Ocean Lab); ten grassroots, justice, and community organizations ('Āina Momona, EarthEcho International, Green 2.0, Healthy Gulf, Interfaith Power and Light, Native Conservancy, North American Marine Alliance, Salted Roots, United Houma Nation, and UPROSE); and four national environmental groups (Earthjustice, Greenpeace, Natural Resources Defense Council, and Oceana).

this collective released our Ocean Justice Platform.[171] Eighteen organizations co-developing and coming to complete consensus on anything, let alone a policy platform, feels like a small miracle. In the end, we summarized our collective vision quite simply:

An equitable and just approach to ocean policy must:

1. Protect the ocean and the benefits it provides for all.

2. Alleviate the disproportionate burdens placed on ocean-justice communities due to ocean pollution.

3. Promote an economy that sustains the ocean and communities that rely on it.

4. Uplift justly sourced renewable energy from the ocean.

5. Prioritize community social cohesion in disaster response and adaptation investments.

It may shock you to learn that, to go along with the release of our platform, the steering committee members wrote . . . an op-ed, published in *TheGrio*, titled "Ocean Justice Can Help Empower Communities of Color on the Frontlines of the Climate Crisis."

Step 22: Celebrate a White House policy plan.
March 2023

The White House put out the first-ever federal Ocean Climate Action Plan, including more than 200 actions items spanning three goals: create a carbon-neutral future, accelerate nature-based solutions, and enhance community resilience to ocean change. Highlights of the plan include expanding offshore wind energy responsibly, decarbonizing the shipping industry, ensuring climate-ready fisheries, investing in blue carbon, integrating environmental justice, and more. Several key elements in the Blue New Deal memos and many of the recommendations in the Blueprint are now official federal policy.

---

171 Read the full platform here: oceanjusticeforum.info

Step 23: Celebrate the White House Ocean Justice Strategy.

December 2023

So far, more than fifty additional organizations have signed on to the OJF platform. We have briefed the leaders of the Council of Environmental Quality and the Office of Science and Technology Policy in the White House, the federal Ocean Policy Committee (which includes leaders across more than twenty-five federal agencies), and the National Oceanic and Atmospheric Administration. These names may seem like gobbledygook, but in these offices are people making big policy decisions. We needed to get their ears to ensure that as ocean policy is being developed, justice is kept in mind. OJF's influence came from being a collective of united voices—that's how we were able to capture their attention.

Miriam (who was also on the steering committee of OJF) is now the director of ocean policy for the White House Council on Environmental Policy. In fact, upon arrival in that role, she helped crank out the Ocean Climate Action plan and then turned her attention to the development of a federal ocean justice strategy. And when the government sought public comments on that latter effort, nearly 95% of the more than
* 16,000 comments mentioned the OJF platform. Policy documents may sound boring, but putting things in writing can be powerful. In December 2023, at COP28, the U.S. government announced an official Ocean Justice Strategy, and it includes key elements of our OJF platform.

Step 24. Create a Blue New Deal for coastal cities.

January 2024

This chapter has been focused on federal policy, but a lot of ocean and coastal management decisions get made at the local level. So Urban Ocean Lab created a climate-readiness framework for coastal cities (aka guidance for creating local Blue New Deals), along with an open-access resource hub to help ensure that coastal city policymakers and community leaders have the information they need to develop their own customized, local plans for ocean-climate policy.[172] And we released a

---

172 See: urbanoceanlab.org/resource-hub

funding guidebook to help cities identify and access the federal IIJA and IRA funds they qualify for.

Twenty percent of Americans live in coastal cities, where nearly 60% of those residents are people of color, 51% are renters, 26% are immigrants, and 16% live in poverty—all higher than the national averages. So getting ocean-climate policy right in cities isn't a "coastal elite" issue, it's about 1 in 5 Americans, a diverse cross-section of our society. Local governments are often the ones making decisions about planning for sea level rise, coastal development, port operations, and more. Having strong federal policy is key, but the local level is where it ＊ gets real.

———

That's how far we'd gotten when this book went to press. We are five years, five op-eds, four policy memos, two polls, two executive orders, one funding guidebook, one ocean justice platform, a White House action plan, a White House ocean justice strategy, one city-level policy framework, and innumerable meetings, calls, emails, and texts into the work of bringing the elements of a Blue New Deal to life. And for decades before that, so many people had been working on these ocean-climate justice policy ideas, waiting for and working to create the political moment when politicians would finally pay attention. And just as with the Green New Deal, the reality is that the concept must be implemented across many pieces of legislation, executive actions, and agency regulations. There will never be just one thing we can point to and say, "This is it, we did it."

A few lessons emerge from all of this: ＊

1. **Write it down.** As the saying goes, talk is cheap. The memos and the op-eds and the reports—that's what starts to make it real. The pen is indeed mighty.

2. **Team up.** Join forces with other organizations, with people you actually enjoy, with people across the country with different areas of expertise. Change can be accelerated through deep and broad collaborations. Also: Make friends with staffers in Congress and in the administration.

3. **Tell a compelling story.** Make your case—publicly and privately, in the press and on the internet. The message and the messengers both matter. Make it catchy; make it stick.

4. **Don't quit on big ideas.** No matter who is in office, keep pushing. Relentlessly, assiduously pursue change from every angle, at every opportunity. Pry open windows of opportunity.

Changing policy, getting from ideas to implementation, is big and nuanced work. It's endurance work. It's collective work. It's strategic work. And, clearly, it's work that is far from over. So I'll be keeping up the drumbeat, passing the baton, and all other applicable metaphors.

Bonus: Remember the Civilian Climate Corps? Well, even though that didn't get funded by Congress through the IRA, the Biden administration (ahem, instigator/champion Maggie Thomas) cobbled together existing funds and in September 2023 announced the creation of the American Climate Corps, and started recruiting the first cohort of 20,000 young people for training and placement in green jobs. And one of the featured areas of work is: "Rebuild coastal wetlands to protect coastal communities from storm surges and flooding." Yay!

You will be in positions that matter. Positions in which you can decide the nature and quality of other people's lives. Your errors may be irrevocable. So when you enter those places of trust, or power, dream a little before you think, so your thoughts, your solutions, your directions, your choices about who lives and who doesn't, about who flourishes and who doesn't, will be worth the very sacred life you have chosen to live. You are not helpless. You are not heartless. And you have time.

—Toni Morrison

See You in Court

Interview with Abigail Dillen

When I was in college, I took a history class on the U.S. Supreme Court of the Civil Rights Movement era. The Warren Court ended legal segregation, established Miranda rights and the right to free counsel, and declared laws against interracial marriage unconstitutional. That was a transformational period in U.S. law, and taking that course was transformational for my understanding of the great power of our judicial system. Many of us only think about this third branch of government when it's doing things like ruling on the outcomes of our elections or our right to abortion, but in state and federal courts across the nation, judges are also deciding a wide range of cases that determine how much pollution, habitat destruction, and harm to biodiversity is allowable.

Our legal system can enable us citizens to hold individuals, corporations, and our own government accountable when they fail to adhere to environmental laws and regulations. We can sue. Such lawsuits are often brought by nonprofits, and often specifically by Earthjustice, whose ace tagline is "Because the Earth needs a good lawyer." They've got around 200 lawyers serving 1,000 clients, pro bono, from environmental justice groups like Rise St. James to the American Lung Association to the Sierra Club.

Attorney Abigail Dillen has been on the Earthjustice legal team since 2000, when she started working on cases in the northern Rockies to protect grizzlies, wolves, and salmon. She went on to lead their coal program, winning cases against the EPA to prevent pollution from coal plants, cut funding for them, and block permits to build new ones. We met in 2020 through Mary Anne Hitt, who was then director of the Sierra Club's Beyond Coal campaign and had teamed up with Abbie to shut down coal plants across the U.S.—383 coal plants down, 147 to go![173]

---

173 Mary Anne's essay in the *All We Can Save* anthology tells the remarkable story of this wildly successful campaign.

My conversation with Abbie, now Earthjustice's president, is about getting our legal system right for addressing climate change: our foundational laws, the regulations that implement them, the gaps that need filling, and how to sue the pants off governments and corporations when they run roughshod over waters, air, and wildlife. We talked about the big cases (and big wins) on her mind, spanning pipelines and permitting, environmental justice and the rights of nature, and the power of the laws we already have. I'm so glad she and the whole crew at Earthjustice are on the case.

———

Ayana: Let's start with the big picture. What is the role of the legal system and the courts when we think about how to get it right on climate?

Abbie: Laws are meant to embody the values of the people who live in the society. They're how we set some rules, make agreements together, and hold each other accountable to them. Our system is broken in some important ways, and our courts have certainly never been perfect and are in particular peril now—especially when it comes to protecting and restoring our planet in the next ten to twenty years as we have to do. But strong laws and strong adherence to and even love for them is part of how we will get it right.

Ayana: A social contract for how we want to operate, and the rules that we will abide by.

Abbie: Yeah. The last time in our lifetimes that people in this country realized that we were overtaxing our planet, that we were setting the rivers on fire, that the air was unhealthy to breathe, we were able to come together in a bipartisan way as a country and invent the strongest environmental laws in the world. That was the 1.0 version of what we need now and it was strong.

Ayana: You're talking about the early 1970s.

Abbie: Yep.

Ayana: It was then that major pieces of legislation—the Clean Air Act, the Clean Water Act, the National Environmental Policy Act, the Endangered Species Act—came out of a bipartisan Congress. On the first Earth Day, in 1970, something like 10% of Americans were in the streets protesting. That led to the creation of the Environmental Protection Agency, the EPA. It's nice to hear you describe that set of foundational laws as having been the strongest in the world. Is that still the case?

Abbie: No, it's not. I'm happy to say we have incredible examples emerging in other parts of the world and in some of our own states. For instance, in countries including Indonesia and South Africa, constitutional entitlements recognize the right to a clean and healthy environment as a human right, and the UN recognized that right in 2022. New York has adopted it recently, and several other states in the U.S. include it in their constitutions as well. What is especially exciting is we are seeing more and more test cases successfully vindicating that right.

We are also seeing the emergence of legal frameworks that recognize the rights of nature. In New Zealand and Australia, among other places, you have rivers protected as living entities with rights that the government and water users must respect.

Ayana: The Ganges River in India, too. And similar frameworks are emerging across South America. In Ecuador and Bolivia, protection of nature has been enshrined in the constitution. And in Colombia, a court case brought by young climate activists resulted in rights for the Colombian Amazon.

Abbie: Yes, exactly.

Ayana: This concept of rights of nature or rights for nature is something quite new within the Western legal system. Giving official legal rights to a river or a forest can be hard to get your head around.

Abbie: It's not a new idea, but it has been regarded as fringe in the U.S. because our whole legal system is premised on the rights of the individual, on the right to own property—which can include water, air, the right to pollute up to certain limits. In our social contract in this country, the limits we impose upon each other are around trespass.

Ayana: Mm-hmm. "Don't come over here. This is mine."

Abbie: If anything is sacrosanct, it is the right to property. So much of the law is built around protecting that right. That is completely incompatible with a natural system in which there are limited resources that everyone needs. And so there have always been laws premised on the idea that there are some essential resources that should and must belong to everyone.

The idea of a public trust—stewardship for future generations—is an old idea in the law. Of course, it's fundamental to an Indigenous worldview, but it also finds its way into American law in various ways, including through the precautionary principle that influences some statutes. But the overwhelming worldview in this country has been, "I've got what's mine, don't mess with me."

Rights of nature challenges that. It no longer assumes humans as the rightfully dominant force. Instead, it puts us into kinship relationships with all of the species on Earth. Ideally, it puts us into right relationship with all of the systems that are sustaining us and all other creatures.

Ayana: So the role of the legal system and the courts and lawyers, if we get it right, shifts from thinking about the rights of individuals to justice for the collective, from protecting property to prioritizing healthy ecosystems and life on Earth.

Abbie: Justice is a practice. It isn't this enlightened state that we will inhabit in any ecstatic way, but if we get it right we'll have different norms and ideas about what constitutes justice. If someone is overstepping and throwing a bunch of toxins in a river, the punishment, the social opprobrium will be different in a more advanced society. It won't be a modest fine in the course of business as usual. But even now, if we reliably enforced the laws that we had on the books, we would make it economically implausible to pollute the way that companies do now.

Ayana: Just by enforcing existing laws.

Abbie: Just by enforcing existing laws. The Clean Air Act as it stands, the Clean Water Act as it stands. We have the statutory framework we need to keep our air and water clean, to protect the most fragile and imperiled species, to get a handle on our chemical exposures, to actually make agreements as a society as we're about to build so much new infrastructure, and to analyze the best way to do that. We have an incredible legal framework in place to do those things.

If we funded our government agencies and accepted as a society that we wanted our laws to be enforced as they were written, really in keeping with their spirit, it would be a dramatically different world. And most of all, if we decided to resource that enforcement equitably so that you didn't have a different standard for a White, wealthy community as opposed to a low-income community of color, it would be night and day.

We wouldn't have a Cancer Alley in Louisiana.[174] We wouldn't have a chemical corridor through the Ohio River Valley.

Ayana: I was coming to this conversation assuming we didn't have the legal structure that we needed. But you're saying these bedrock environmental laws already do give us a lot of what we need to hold polluters, ecosystem destroyers accountable.

Abbie: We have a solid foundation and we can't take it for granted. As environmental laws become more and more important, they're becoming political targets because they're having real economic consequences. They can be used as tools that dramatically curtail what the fossil fuel industry and other polluting industries want to do. And so we have to protect that foundation and build upon it.

Where we need help is recognizing more explicitly what the climate crisis, the extinction crisis, and stark environmental injustice demand of us. For example, we have to reform our water laws, which are the zenith of conferring private property values on something that is not only a collectively critical resource, but one that's dramatically diminishing.

We do not have the water laws that we need, and I don't think we have the full suite of ecosystem protections that we need. The existing laws in these areas were built on preventing extinctions, preventing the worst kind of loss. We're at a point where we need to be focusing on restoring and protecting abundance to avoid wholesale losses that threaten our whole web of life. There are legal inventions that all of us need to get our heads around and bring into being. But we don't need to throw up our hands until we get there. We have an extraordinary set of laws to enforce now.

Ayana: It's interesting to hear you specifically call out water laws, because too rarely do we think about the water cycle as a key part of climate change and how we need to address that crisis. If you could upgrade our water laws, what would you change?

Abbie: Well, I would strengthen the Clean Water Act, which protects wetlands, rivers, waters, and streams. It's a law with correctly ambitious

---

174 The eighty-five-mile stretch along the Mississippi River between New Orleans and Baton Rouge is nicknamed "Cancer Alley" because of its concentration of more than 200 petrochemical facilities and associated high rates of cancer.

goals and many strong provisions to reach them, but it needs strengthening to overcome relentless efforts to chip away at it.

I would want to change the water entitlement system, the water law in the West. It evolved out of a Wild West regime where you could just say, "I'm using this water and therefore it's mine." And then going forward, you would have what's called a "prior appropriation." You would be able to appropriate running water to yourself.

Ayana: That is ridiculous.

Abbie: That's what water rights are. They're an appropriation of the most public resource we have. And of course, water is not just for people, it's supporting all life.

Ayana: Okay, our water law definitely needs to change. What about the atmosphere? Do we have, in the U.S., the legal structure we need to address climate change as far as controlling greenhouse gas emissions?

Abbie: Let me answer that question two ways. The first way is to reframe ∗ the question to say, do we have the courts that we need to regulate carbon emissions? Because the Clean Air Act, even though it was written back in the '70s, contemplates atmospheric protection. It's expansive enough to let responsible, smart presidential administrations create the regulations we need. But the act is not written with precise tactical clarity about implementation. These laws don't specifically say, "This is the exact way that the EPA is authorized to design a program." And obviously, that would be impossible for a big group of people in Congress to legislate because they don't have the time or capacity to go deep on those kinds of details.

Delegation is what makes government work, but conservative judges are now advancing their power to decide what the limits of executive authority are. Most students of law would read our capacious statutes to allow a sitting president to respond to the environmental crises that we face now. But we have a conservative supermajority on the Supreme Court that's willing to override Congress and the president and aggregate power to the courts to decide what the president can do with broad statutory authorities.

Ayana: When it comes to air pollution, the Clean Air Act provides the broad mandate and the Environmental Protection Agency then writes regulations and figures out how to enforce them under the direction of the presidential administration.

Abbie: Correct. And the EPA has a direct obligation from Congress to keep evolving those regulations, to reflect the most current science and best available technology to control pollution. So if we have an airborne crisis that's driving us into disaster, yes, that's in the EPA's wheelhouse. There should be a clear-cut pathway for the Biden administration, for instance, to regulate polluting industries. But in a deregulatory landscape and with the Supreme Court that we have, a brand-new doctrine was introduced in 2022 in the *West Virginia v. EPA* case.

Ayana: Ah, yes, the major questions doctrine.

Abbie: That was the majority of the court embracing a radical and new idea. And the premise is basically, if a judge thinks a policy is significant and has big consequences, then the assumption is that Congress would have specifically spelled out the executive branch authority to execute it. It's contrary to three co-equal branches of government. If everyone in government were to apply the major questions doctrine, government as we know it would grind to a halt.

Ayana: Because Congress doesn't have any granular expertise to enable them to spell out every detail. That's not what they're for.

! **Abbie:** And that's true even of a very functional Congress. This is a cynical power grab by a court that wants to shrink the role of the president, and, as is becoming increasingly clear, the role of Congress. This is a court that's aggregating power to itself to advance the ideological agenda of the conservative majority of justices.

Ayana: So, to put it bluntly and broadly, what do we do given the anti-environmental supermajority on the Supreme Court?

Abbie: We navigate the courts with care, and we recognize that the judiciary is still the weakest branch of government. The Supreme Court does not weigh in on the vast majority of decisions that any government, whether it's the federal government or a state government, makes. When they do weigh in in ways that are harmful, we can <u>lose loudly</u> and we can build momentum for a congressional fix.

Ayana: "Lose loudly" is a nice turn of phrase. I like it.

Abbie: We say it a lot at Earthjustice and we're learning to do just that. We live to win in court, but when we lose, losing loudly is the way to make the case to the public for new laws or more enlightened action by our elected leaders. And losing loudly is essential to build pressure for court reform, which we may need to see in our lifetimes. I increasingly think we will.

Ayana: What do you mean by court reform?

Abbie: It's not written in stone that the Supreme Court has justices who ✳ are appointed for life. The way we do it now is not constitutionally required. You could have staggered terms, for example. What you want is a stable institution that is not locked into the agenda of any one political party at any given time. So if each president could appoint two justices, that would be a very different world. People also talk about rebalancing the courts, appointing more justices. The key is shaping an institution that is more responsive to and protective of our democracy.

Ayana: Each president getting to appoint two justices makes me imagine a scenario in which the court doesn't whiplash between extremes, or have a dominant partisan view, and we could get back to a court that is a bit more apolitical.

Abbie: Exactly. For instance, we as a society made a decision that energy reliability is important. So our Federal Energy Regulatory Commission (FERC) is always bipartisan. There are always two Republican commissioners. There are always two Democratic commissioners. And then the fifth tiebreaker is generally from the sitting president's party. We recognize that's a way to create stability, buy-in, reliability over time. And we could do that with our Supreme Court, too.

Ayana: Right now in the U.S., when it comes to courts and climate, is there good news?

Abbie: There's so much good that's happening. For many obvious reasons, we focus on Congress and on the Supreme Court, but the vast majority of what happens is in district courts and in state courts. And lots of states are accountable to voters who overwhelmingly want to see ambitious climate action. One of the things people aren't paying enough attention to is the bipartisan gains we're making on biodiver-

♥ sity protection, even in red states. People can really get together around protecting <u>wildlife corridors</u> and habitat and places that everyone feels deep love for.

Ayana: For example, the Great American Outdoors Act of 2020, which had eighteen Democrat co-sponsors and ten Republicans.

Abbie: Yes, exactly. And there are other major points of social agreement across partisan divides. People may not agree on climate change, but they widely agree on clean air and water and curbing pollution that poisons people and the environment. At Earthjustice, we are harnessing the strong health standards in our laws to go after the biggest polluters, and those polluters are usually driving climate change as well as disease and premature deaths. In this way, we're forcing the retirement of dirty power plants, blocking construction of dirty new petrochemical plants, and stopping other big fossil projects. That's a model people know about, using the law to stop stuff.

Then there's a whole other wonderful tide of work to compel investment in clean energy and climate solutions. Our monopoly utilities are required to provide energy to people at "just and reasonable" rates, and so we can go to all our state energy commissions and literally have a trial to determine what's the most economically prudent way to meet our energy needs. Even in the reddest of red states, we can succeed in making the winning case for clean energy based on economic grounds.

Ayana: Because renewable energy is usually less expensive than fossil fuels now, you can make the financial case.

Abbie: Yes. And we might find big industrial energy consumers like Walmart taking the same view because they want to save on electricity costs. On top of that cost advantage, many states are starting to factor into their decision-making the so-called non-energy benefits, meaning the climate benefits, the health benefits, the societal benefits of transitioning away from fossil fuels.

As people are catching up to the reality of climate change, we're finding a willingness to adapt to it across all kinds of different laws. It's going to change how insurance regulations work. It's going to change how the Americans with Disabilities Act works. What are healthy accommodations for people in a world that's more challenging? How do buildings have to be constructed to protect you?

Ayana: Are there a few specific legal cases Earthjustice is working on that illuminate some of these points? What are you excited to win on, or, if need be, lose loudly on?

Abbie: One case that's captured a lot of imaginations has been our suit against the U.S. Post Office. When the Biden administration decided they were going to build out their next-generation fleet with fossil fuel mail trucks, we were able to bring a lawsuit that stopped that bad decision in its tracks.

Ayana: I remember when the administration made that decision, I was like, "Really?! We're not gonna electrify mail trucks?" It seemed like a no-brainer.

Abbie: I know, if we can't electrify the mail trucks, what can we do? But now we're going to have the biggest electrified fleet in the country.

Ayana: How did that happen?

Abbie: We sued the post office for failing to consider electrification as a much better alternative, for the full suite of reasons you can imagine. We were able to slow them down, and ultimately the Biden administration responded to the lawsuit by making a commitment to procure almost all electric vehicles by 2026.

Ayana: Amazing. Thank you. This makes me a little wistful that I did not become an environmental lawyer because I would love to write those briefs with you.

Abbie: You would love it the most, Ayana.

Ayana: This is a great example of not just stopping the bad stuff, but pushing the good stuff. But we do still really need to fight the bad stuff.

Abbie: Let me give you an example of that. ConocoPhillips has a gigantic project in the western Arctic. It's a massive oil and gas drilling proposal called the Willow project that would open up a whole new swath of the Arctic to drilling.

Ayana: A truly terrible idea.

Abbie: It's horrible. Just to give you an example of how bad it is, they would have to put chillers in the permafrost to keep the oil operations stable enough because it's melting so quickly there.

Ayana: If ever there was a sign that we should not be doing something, it's having to air-condition the permafrost to keep it cold enough so that their heavy oil and gas extraction infrastructure doesn't sink into the mud.

Abbie: Correct. Imagine the energy that air conditioning is using. And the permafrost as it melts is releasing methane, a super pollutant when it comes to warming the atmosphere. It's a poster child of the worst possible idea.

Ayana: And this project is happening in the Arctic, the fastest-warming region in the world.

Abbie: The Trump administration in its last days railroaded the project through. Our lawyers pulled a month of all-nighters litigating this thing, and were able to ultimately win a lawsuit to stop it in its tracks. That sent it to the Biden administration, and we thought, "Thank goodness, this is a climate-conscious administration, there's no way that they're going to let this project go forward." Well, they approved it, and we're back in court. It's a heartbreaker.[175]

Ayana: We simply cannot build major new fossil fuel projects. But at the same time, this is a big deal economically in Alaska. People are thinking about the jobs and their tax base and revenue for the state. I get that it's tricky, but we've gotta find another way.

Abbie: The economic concerns are so real. But we're going to spend trillions of dollars over the next decade to transition our economy. Alaska should be a first priority. It has disserved most people in the state to have survival tied to the booms and busts of the fossil fuel industry, and it's past time to break that vicious cycle.

Ayana: Okay, so suing the post office, trying to block the Willow project, and is there a third case example you want to share?

Abbie: There's a great example of what communities can do when they bring the law to bear. We have a gigantic petrochemical industry in this

---

175 Post-interview note from Abbie and me: Lest we over-focus on the negative, there have also been *many* heartening energy decisions under the Biden administration. For example, hundreds of billions of dollars have been allocated to climate projects under the Inflation Reduction Act (the largest climate investment in history), dramatically restricting pollution from power plants, strengthening emissions standards for cars and trucks, overhauling energy efficiency standards, jump-starting the offshore wind industry, and investing billions for solar power in low-income communities. A tremendous combination of investments and standards.

country. And the oil and gas industry, in its heart of hearts, knows that we're going to have to stop burning oil. For them the question is how else to use this big asset under the ground, and one big answer is to use oil and gas as the feedstock for petrochemicals, which are then in turn the feedstock for plastics.

We now have an increase in plastics demand being driven by the fossil fuel industry.[176] About midway through the Trump administration, we caught on to a gigantic play to build out a whole new generation of petrochemical infrastructure in South Texas and in Louisiana, where the chemical industry already holds enormous sway, as of course the oil and gas industry does.

In the face of a new build-out, the communities that were already ground zero from many of these facilities got organized, and they have been fighting these permits through every venue. They've been working on it in their city councils. They've been challenging state agencies. My favorite example is the Formosa Plastics case, where we have the privilege of representing Rise St. James, among others, in a powerful coalition of groups in Louisiana.

Ayana: Describe this scenario, because it's a great example.

Abbie: Formosa Plastics is a multinational corporation that wanted to build a sprawling new billion-dollar plant in St. James Parish. Rise St. James was started by an unstoppable woman named Sharon Lavigne, who was determined to make a different future for St. James. It's a parish that was settled by formerly enslaved people. There are residents who are deeply connected to these lands where their ancestors are buried. They have fought tooth and nail to be on these lands, to farm them, to continue their generational through-line in this place.

Sharon Lavigne and other leaders brought their community together and they made a deeply compelling case to the court, going up against the Louisiana Department of Environmental Quality. And the decision from the court is one of the most moving that I've ever read. In its first few paragraphs it recognizes the history and the current stakes for the community of St. James Parish. It engages with the reality that is so

---

176 Annually, the U.S. plastics industry produces carbon emissions equivalent to 116 averaged-sized coal-fired power plants.

often overlooked: The community is so deeply tied to this place, they have a right to live there and breathe healthy air and to live free of cancer. The court got that and ruled against the State of Louisiana for failing to address environmental injustice specifically, and overturned the permit for the plant.

Ayana: I love this. So no Formosa Plastics plant will be built.

Abbie: No Formosa Plastics project.[177] And likewise, this amazing coalition has stopped the Louisiana methanol plant that was also proposed in St. James Parish. What I wish people knew is that the law provides a path to go up against the most powerful, well-resourced forces *and win*, when you have the facts and justice on your side. Not always, but it is possible. I want people to know how few individuals were involved in stopping these gigantic projects and how it changed the common sense of the industries who are proposing them to the point that we're not seeing new proposals of this kind.

Ayana: That's so great to hear.

Abbie: There are laws to protect people's health and the public interest. If we enforce them and get organized, we can stop the bad ideas and make room for the good ones.

The postscript is that the fossil fuel industries are recognizing that petrochemicals may not be their perfect play in the United States. They're saying, "Hmm, there's a lot of money for hydrogen in the Inflation Reduction Act. Let's use this glut of cheap gas to create fossil hydrogen hubs and to market hydrogen as the next green solution." They always come up with something new.

Ayana: You're out there playing Whac-A-Mole with these fossil fuel companies.

Abbie: We are. But Whac-A-Mole assumes that it will go on forever.

Ayana: Someone's gonna run outta quarters. Is this what you're telling me?

---

177 Actually . . . bummer update: Since this interview, an appeals court overturned the ruling that denied Formosa the permit. So Earthjustice and the Louisiana environmental justice groups have taken this case to the state supreme court. Outcome not determined by date of this book's publication.

Abbie: Someone's gonna run outta quarters. It's not gonna be us.

Ayana: Maybe this is a terrible analogy.

Abbie: The long view is that people won't be resigned to so much pollution that is killing us when there is obviously a better way forward. How many friends do you have that have asthma or have kids who do? Why do we settle for this? When I think about "What if we get it right?" I see a world where we'll be so much healthier. Imagine how much less cancer there will be.

The more that we create and enforce laws that require us to look at the best science and the best way forward, the better chance we'll have to stop settling for this totally outrageous present that takes a totally untenable toll on us.

Ayana: It starts with that initial kernel that's often just a few people who ✳ don't accept the status quo who say, "Wait, why are we putting up with this?"

Abbie: Yes. Yes.

Ayana: When you think about what the world looks like when we get it right, what else comes to mind?

Abbie: I don't want to minimize how hard it is to imagine a different future in the communities that are so tied to fossil fuels or other polluting industries. That's where the jobs are and it ripples out to everyone. So, I don't want to minimize the project of imagining a new economy and a new way forward.

Ayana: You have to imagine the just transition along with it.

Abbie: You do. And I understand why people who are at the frontlines of this transition have a lot of distrust around the words "just transition." It hasn't been laid out. The project of the next few years is regionally envisioning what a different future is, and being clear-eyed about the economic imperative to achieving that vision.

When I think about getting it right, I imagine your zip code having ❤ no special correlation to your health, much less your lifespan.[178] I think

---

178 Average life expectancy is 20 to 30 years shorter in some disadvantaged U.S. zip codes.

about making room for trees, birds, and pollinators, about living in places with ground that's permeable so the water can be absorbed instead of flooding, about living more comfortably in buildings that are using natural airflow. About what it would be like to be liberated from oil and gas so that our prices and our politics aren't set by dictators. So that we don't have wars driven by oil. So that we could have a humane immigration policy in the U.S. and around the world. Climate change is driving migration, and that reality has to shift how people are allowed to move for their survival.

Ayana: What would be your top three priorities if you were able to wave your magic wand and change elements of our legal system to make sure we had all the tools we needed to address the climate crisis?

Abbie: In no special order: First, I would make fair courts a voting issue in this country.

Ayana: What does that entail?

Abbie: We understand as a country that our civil rights depend upon
* fair courts. We do not understand as a country that our climate trajectory, that the future of life on Earth, is connected to the courts. We have to be ferocious about keeping courts open and transparent and making sure that there are fair judges sitting in the seats.

Ayana: What else is on your list?

Abbie: I would love to pass the Environmental Justice for All Act.

Ayana: Yes, please.

Abbie: This was a bill put forward by Representative Donald McEachin, who we lost in 2022, tragically, and Representative Raúl Grijalva. They worked with communities around the country to understand why communities of color, lower-income communities, are strapped with the burden of pollution in this country. The bill addresses all of the holes in the laws that we've been living with for all of my career. It would be transformative.

Ayana: And what's the third thing you would wave your magic wand at?

Abbie: Third would be to get everyone in a position of power to come together to reform our water laws. And to come into it with a mindset that is truly radical. What we have in place is never going to work for the future. That's assured. The discussion about water is mature in some places, like on the Colorado River, but it's nowhere close to where it needs to be nationally or globally given the crisis that we're walking into.

And I would add one more, which is not only restoring our Endangered Species Act, but adding an additional legal framework to protect and restore ecosystems and species. We have pieces of this legal authority in the Clean Water Act and the Clean Air Act, but we need a more holistic set of laws.

The common core of all of these ideas is law that recognizes the rights ✱ of the collective. We need to move from a system that reifies individual property rights and shifts to not only account for, but truly protect, the common good of public interest.

Ayana: For so long the environmental movement has been focused on stopping bad things, making sure companies are denied permits for pipelines, et cetera. But now we need to build a bunch of good things—the whole infrastructure for renewable energy—which raises the question of whether and how permitting processes might need to be updated. What needs to happen next on this front?

Abbie: We can't afford any more new long-lived fossil infrastructure. The scientists of the world are very clear on this point. We can't develop vast new gas fields. We can't be locking ourselves into piped gas to power this country. And so what we need is policy that facilitates building of the new clean infrastructure, the whole new clean economy, *without* facilitating a new wave of fossil fuel infrastructure that is underway now. The U.S. is the biggest oil and gas producer in the world.[179]   !

Ayana: That is not a good thing.

Abbie: And by the way, all of this clean energy is going to need critical minerals. So we're talking about a resurgence of mining in this country. There's an important argument that we should do that mining here because we can do it with greater environmental safeguards. Coming back to my list of things that I would want to change if I had a magic wand, I would also use it on our mining laws. As we embark on a new wave of mining, I think we can do better than the General Mining Law of 1872.

---

179 And the biggest in history, now producing nearly 14 million barrels of crude oil *every day*.

Ayana: That's a great point. And while the impact of mining for materials to make batteries should absolutely be minimized, it's important to remember that the alternative is to keep drilling for fossil fuels, which are top-to-bottom harmful. We need to be honest and clear-eyed as we consider the trade-offs.

Abbie: One of the most powerful tools we have to evaluate decisions like this is the look-before-you-leap requirement in the National Environmental Policy Act (NEPA). That requires the government to do an environmental review of projects and gives the public time to say, "What are the climate impacts of this? Here is important research or local knowledge you need to consider."

At the same time, we do want to get important new infrastructure built. And one of the most important bottlenecks is how we regulate electricity transmission in this country—figuring out who pays for this expensive infrastructure, figuring out the siting (it doesn't belong everywhere), modernizing the grid, and battling incumbent utilities that are trying to run the show to benefit their existing fossil assets as long as they can.

* As we're building a trillion dollars' worth of new infrastructure, we need to do thoughtful planning, and we need to ensure that communities are consulted in a real way. Not just a public meeting far away from where working people are, and at six P.M. when they're trying to pick up their kids and make dinner. Not just sending a form letter to a Tribe.

We've starved our agencies of resources and we haven't created an expectation for serious outreach and planning at the outset to get buy-in to a master plan. If we had a good master plan in place, we could get projects that are in line with it permitted faster.

Ayana: And if we want to move fast, we need to hire people and build capacity within the federal government offices that are making the permitting decisions.

! **Abbie:** Exactly. It's so cynical for people who've wanted to shrink the government and starve the agencies to then say, "Oh, they're not moving fast enough. This means we should get rid of the review requirements."

Ayana: So frustrating. But overall, I'm relieved to hear that our legal system might be up to the task of climate.

Abbie: We have strong foundations, though we do need some new laws. But in the short term, ambitious regulation and enforcement pursuant to existing laws could take us so far. Don't mistake me; I have big critiques of the law. But if I had to choose between new laws and different ✶ politics, I would choose different politics. We need political will behind us.

Ayana: What is the nerdiest, most esoteric thing we need to do to get it right on climate in terms of our legal system? Is there like a subsection 309B of something that we need to rewrite?

Abbie: We need an army of law nerds. I love working in an organization of the most amazing law nerds in the world. I'm thinking now about all the answers that would pour forth if I emailed that question to our team and crowdsourced answers. I would get like two hundred different answers because it all matters. We have to sweat the details. I can't pick just ✶ one; I can't choose.

Ayana: Okay, okay.

Abbie: But the question reminds me of something we have to stop doing. We have to stop thinking that there's one answer. We have to recognize that climate change touches everything. We have to unleash people to work on climate wherever they are, on everything from a provision of flood insurance in federal law, to a provision determining what would make an energy efficiency program economic in the eyes of a public utility commission. It's everywhere. We have to bring climate reality to every corner of business-as-usual in the United States and around the world.

Ayana: Okay, but I'm not going to let you punt on this last question: If you could sue one person or entity, who would it be and why?

Abbie: I'm a big fan of the government, but I also think the government is the most important entity to sue. Because they're so cross-pressured into short-term decisions that are politically motivated or born of resource constraints. Our ability to sue the government, to force them to do the right thing, is everything.

COMMUNITY FOREMOST

Where do we go from here?
Chaos or community?

—Dr. Martin Luther King Jr.

What if we redefine home?

What if we step back from the rising waters?

What if climate mitigation and adaptation are just?

What if extreme weather doesn't mean catastrophe?

What if we prize interdependence over independence?

What if we develop new traditions for this new era?

What if we take care of each other?

What if . . . ?

10 Problems

- Indigenous peoples have been forced out of their historical territories and onto lands that are more exposed to climate risks like drought and extreme heat.

- Air pollution from burning fossil fuels is responsible for 8.8 million premature deaths annually—1 in 5 deaths worldwide. People of color in the U.S. are exposed to 14% more air pollution than White people.

- By 2070, up to 3.5 billion people may have to move as the areas where they live become too hot for normal human life.

- There is now a billion-dollar weather/climate disaster once every three weeks in the U.S., causing financial hardships for 1 in 10 Americans. In the 1980s, there was one such disaster every four months.

- Sea level rise could displace up to 410 million people globally by 2100, including 13.1 million people in the U.S.

- The number of affordable housing units in the U.S. exposed to coastal flooding is expected to triple by 2050 as sea level rises.

- White communities hit by hurricanes and floods have received approximately three times more FEMA buyouts per capita than communities of color.

- About 36,000 U.S. public schools lack adequate cooling and ventilation systems to deal with extreme heat—and heat waves are increasingly leading to school closures.

- There is a 1% increase in heat-related workplace injuries for every 1°C above 21°C (70° F).

- More than 50% of young people globally are worried about climate change and express feeling sad, anxious, angry, powerless, helpless, and guilty. Over 45% say it negatively affects their daily life.

10 Possibilities

+ Indigenous peoples (less than 5% of the world's population) manage or have rights to around 25% of the world's land area, which is home to about 80% of the Earth's remaining biodiversity.

+ Immediately reducing greenhouse gas pollution and supporting green and resilient development could reduce the scale of internal climate migration by 80% in developing countries.

+ Issuing a warning 24 hours before an incoming storm or heat wave can reduce damages by 30%.

+ 11% of Americans have considered moving to avoid the impacts of global warming, and roughly 75% are hesitant to buy homes in areas with high climate risk.

+ Removing 1 million U.S. homes from flood-prone areas and relocating residents would cost $180 billion but save over $1 trillion in property damages over the next 100 years.

+ Over $29 billion in U.S. federal funding has been allocated as part of the Justice40 Initiative, which directs 40% of benefits of federal investments to disadvantaged communities.

+ Hispanic, Black, and Asian Americans are highly concerned about climate (70%, 65%, and 65%, respectively, compared to 47% of Whites), yet are perceived as being less concerned.

+ Cooperatives account for over 33% of the electric utility industry in the U.S., serving more than 18 million homes, businesses, and schools, while supporting nearly 623,000 jobs.

+ Union representation is higher for workers in the clean-energy sector—transmission, distribution, and storage (18%), solar (11%), and wind (12%)—than the private sector overall (7%).

+ When advocating for climate action, love for protecting future generations is the biggest motivator, 12 times more popular than job creation or economic growth.

Disasterology

Interview with Samantha Montano

Every so often, at a fancy party, when the discussion turns to climate, a rich person with a vacation home near-ish to the coast laughs that climate change is great for them because, *Soon I'll have beachfront property!* I hate this "joke." What a cruel dismissal of the hardship ensuing globally, and what naïveté about the destruction of infrastructure and livability associated with changing coastlines, theirs included.

What used to be a once-in-a-hundred-years storm is now an annual event. We live in a different climate scenario than even a decade ago. Even if we stop emitting greenhouse gases tomorrow, we'll still have to contend with a fired-up planetary system. Extreme weather is spreading—places where people never needed air conditioning are now experiencing dangerous heat waves.

To sustain ourselves, we must sustain nature. This is true every day, but natural disasters lay it bare. Fundamentally, we cannot innovate our way out of our dependence on nature. But we can respectfully adapt. While extreme weather events are a certainty, we can make them less disastrous and less deadly. We can recover from them more quickly and more justly. We can be coordinated. We—as individuals, businesses, communities, and all levels of government—can be prepared and resourced before disasters strike.

This conversation with emergency management expert Samantha Montano makes very clear both that we are nowhere near ready, and that there are so many things (often simple and not so expensive) we can do to get ready. Samantha, who calls herself a disasterologist, brings clarity and wit to many things. I was introduced to her work a few years ago by Kendra Pierre-Louis,[180] and I started following her on social media. Until then, I had not fully appreciated that emergency management is an entire academic field with a rigorous social science literature.

---

180 Yes, the journalist Kendra in the "Planet Is the Headline" chapter. She has great sources.

Experts are diligently analyzing what went wrong and right in various emergencies, and how we can do better. When she announced her book *Disasterology: Dispatches from the Frontlines of the Climate Crisis*, I immediately pre-ordered it, eager to fill in some of my blind spots. It's a page-turner.

How we deal with disasters is not merely tactical—although we are going to get in the weeds here on what's up with the Federal Emergency Management Agency (FEMA) and flood insurance and go-bags. It's also a matter of justice: Who is most impacted, and who has what they need to get back on their feet? Over and over again, we see poor communities and communities of color get pummeled and stranded. Sometimes recovery remains incomplete years or even decades later. It's heartbreaking and gut-wrenching to watch this history continually repeat itself.

Samantha's first experience with a disaster was in the aftermath of Hurricane Katrina in 2005. For me as well, that was the storm that made unbearably clear the injustice, insufficiency, and even absurdity of our emergency management systems. Though we don't know exactly when or where, another hurricane will hit Louisiana. Same goes for places contending with floods and fires and storms. Same goes for the slower-moving disasters of rising sea levels and temperatures—which, you'll recall from our conversation with Dr. Kate Marvel, compound to worsen storms. But since we know, more or less, what's coming, we can craft policies and protocols and budgets for getting the response and recovery right. We are far from helpless.

———

Ayana: Your work in emergency management started with an eye-opening trip to New Orleans after Hurricane Katrina. What did you see there? What stood out for you?

Samantha: I had grown up in Maine, and when Katrina happened I was still in high school. So everything about the catastrophe was very far removed from my life and my experiences. I had never been to New Orleans before. I had never dealt with a hurricane before, but I had this opportunity with my high school to go down to the city to volunteer for a week to help with some of the initial gutting-out and rebuilding efforts. When we got there, it quickly became clear that the amount of help that was going to be needed to even begin to rebuild the city was at a scale that is really hard for anybody to comprehend who wasn't physically there themselves.[181] It was going to require everyone pitching in. It was a DIY, do-it-yourself, recovery. And so I was there initially just for that week, but ended up deciding to move to New Orleans and lived there for a little over four years.

Ayana: That's a big decision. You went from "Oh, this scenario is even worse and more intense than anyone could imagine" to "I'm going to move to New Orleans."

Samantha: Some would call it a little dramatic, but I grew up in a family where we were all oriented toward helping others, and this was the biggest example I had ever experienced of a lot of people really needing help and me being in a position to be able to provide that help.

Ayana: So you went to college down in New Orleans, and spent years helping with recovery work in various forms. You were physically helping with the rebuilding, like framing and sheetrock and roofs. I'd expected that stuff would have been done by contractors or state and federal support. I hadn't realized just how much of the rebuilding was done by volunteers.

Was there something revelatory in your experience in New Orleans that clarified for you, broadly, what's broken and what you wanted to help change?

Samantha: During my first two years in New Orleans, I wasn't paying attention to other disasters. I had a one-track mind on Katrina recovery. But then in 2010, the BP oil disaster happened along the coast. And it was in the comparison of Katrina to BP, and later to the Joplin tornado and the Tuscaloosa tornado, that I started to see, "Oh, what's happening in New Orleans isn't a unique situation. The problems we're facing in New Orleans are representative of what is happening in communities all across the country every time there is a disaster, every time there is a catastrophe." Seeing the problems in disaster recovery, and emergency

---

181 Hurricane Katrina damaged more than one million homes across five states and displaced roughly one million people. Over $120 billion in federal funds was spent on the recovery effort.

management more broadly, pushed me from doing physical hands-on work, literally hammering nails and putting houses back together, to thinking about bigger questions of policy change.

Ayana: You've described Hurricane Katrina as being beyond a disaster. You call it a catastrophe. How do you differentiate between those two?

Samantha: Disaster researchers struggle with this question a lot and where we've landed right now is that when we think about bad events that happen, they fall along this spectrum where you have emergencies on the low end that are handled mostly locally. In the middle you have *disasters*, like a major tornado, those classic disasters that we think of. But then you have *catastrophes*—the events that are so big that they completely overwhelm not just one city, but an entire region. You need resources coming from outside, oftentimes internationally. You need leadership coming from the outside. By all those markers we have in disaster research, Katrina stands out as a classic example of a catastrophe in U.S. history.

Ayana: What are some of the systemic problems that create the scenario for a disaster or a catastrophe?

Samantha: Disasters are a manifestation of all of the inequalities in our day-to-day systems. They create a moment where the repercussions of racist housing policies and classist transportation systems, for instance, all come together and erupt. The impacts from the event itself are disproportionately spread across the population. And then as you move into the recovery, the aid that is coming into that community often falls along the lines of race, class, and gender, and those inequalities deepen.

A second systemic problem is the way our laws and policies lead to bad development decisions that worsen the effects of these disasters. How and where are we building new neighborhoods? Are we, as in Houston, allowing developers to pour concrete over bayous and other natural flood-protection systems, while not fixing our drainage systems to accommodate major rainfall events?

Then you have a third set of systemic problems, which is the emergency management system itself. In 2005 especially, we had huge problems within FEMA. A large part of that stemmed from the post-9/11 policy changes that completely turned FEMA specifically, but really the entirety of the emergency management system, upside down and on its head with the creation of the Department of Homeland Security and putting FEMA within it.

Ayana: One Katrina anecdote in your book was a particular gut punch for me. You wrote that "the storm was at the end of the month, which meant, for many, bank accounts were empty." To think that not having enough money for gas could be what keeps you from evacuating, just, come on, there's gotta be a better way.

Samantha: It's so frustrating. Evacuation of a major city is an incredibly complex endeavor. But we've done a million evacuation studies. We know why people aren't evacuating when they're told to. And we know that not having the resources to do so is one of those major factors. This issue of somebody having a car but not having money for gas, that is a completely solvable problem. What you do is you go to the gas station, you buy up all their gas, and now everyone's getting free gas to evacuate. You start handing out gas cards. Problem solved.

Ayana: You are the co-founder of Disaster Researchers for Justice. What would "disaster justice" look like?

Samantha: We have 300+ folks who have signed up to be part of the group, and one of our main topics of conversation has been examining what exactly we mean by "disaster justice." Each discipline seems to have a slightly different perspective on this. To me, from an emergency management perspective, disaster justice looks like preventing disasters ✱ and eliminating risk as much as possible. And, if we fail to do that, then being effective and equitable in how we do preparedness, response, and recovery.

Ayana: What is the biggest shift we need in order to make headway toward better emergency management and toward disaster justice?

Samantha: The overarching conceptual change that's needed is moving ✱ from an individualistic approach to one that is community-based. Going all the way back to the 1950s, the federal government has emphasized that it is the responsibility of individuals to prepare for disaster, and that they're largely on their own. The government will put out these lists of supplies that you should have in your emergency kit, tell you to make sure your family has a plan, and tell you to do a fire drill with your family. And those are not bad things to do. Everybody should do those things.

But that is an incredibly limited perspective of what it takes to actually be prepared for a disaster. It also ignores one of the key findings of disaster research from the past hundred years, which is that when we respond to disasters, we do so collectively.

Ayana: So, don't worry so much about my go-bag. Worry more about improving community resilience and social cohesion.

Samantha: Right. The problem that causes you to get stuck in your house for a week after a terrible hurricane is not that you did not put enough water in your coat closet. The problem is that developers have built neighborhoods in highly vulnerable places. The climate is changing. Your local politicians did not call for an evacuation in time. There are all of these much bigger reasons that you are in this situation. Again, yes, do what you can as an individual (just as with climate change), but that is not by any means going to be enough. Real preparation is addressing risks before the disaster happens and then making sure we have a collective response to the disaster carefully planned out.

Ayana: There are so many things we could do to manage risk and decrease vulnerability. And it seems like a lot of them are pretty simple and often relatively inexpensive. What are some of these key solutions?

Samantha: It's a long list. In terms of minimizing our risk overall, there are the big-ticket items like mitigating climate change or preventing new development in high-risk areas. Then there are some other things that are lower-hanging fruit but still complicated, like updating building codes across the country and investing in our local emergency management agencies. One thing most people don't know is that most emergency management agencies across the country only have a part-time emergency manager—who often doubles as the fire chief. In some communities the only emergency manager they have is a volunteer, which is even more alarming.

Ayana: Oh wow. That is not ideal.

Samantha: You cannot do the amount of work that you need to do to be a successful emergency manager working part-time. It is not possible. So, one of the most obvious things that needs to happen is that we need more support across the country for local emergency management agencies.

Ayana: And where does that funding come from? Or where should it come from?

Samantha: Excellent question. Currently, a significant portion of funding for local emergency management comes from FEMA through the Emergency Management Performance Grant (EMPG) program, which is a relatively small annual allotment by Congress. The funding is given to states and territories, and then they can choose to pass the funding along to county and municipal emergency management agencies. That small fund needs to pay for everything from building emergency operations centers to the salaries of emergency managers.

Ayana: When you say "small fund," what are we talking?

Samantha: In 2023, the EMPG funding was $355 million.[182] This relatively small amount of money is propping up the entirety of the emergency management system across the country. There is some additional funding through FEMA and local budgets. But we're talking about many agencies across the country that are trying to operate with, in small places, a couple thousand dollars, maybe in the tens of thousands of dollars. They simply do not have the money, the resources, the staff, to do what we know they need to be doing.   !

Ayana: We don't even spend a billion dollars total on local emergency management nationwide?!

Samantha: It's worse than that, actually. On multiple occasions, there have been threats to take this money away. So the lobbying in Congress has focused on making sure this funding doesn't get taken away rather than saying, "We actually need a lot more money than this." And there's certainly a question here about the responsibility of the federal government versus states versus local. I don't particularly care who funds these local agencies, but better funding for local emergency management is a precursor for us to be able to do really anything else.

Ayana: A study I've come across shows that for every dollar the federal   ＊
government spends on mitigation of disasters (through, for example, restoring coastal ecosystems, my favorite option), we save $6 on response and recovery. It's a knock-your-socks-off kind of stat.

---

182 For scale, the 2023 Department of Defense budget was $816 *billion*.

Samantha: And that's the average. If you break that down by different hazards, it actually can be much higher than that. For earthquakes it's something like for every $1 you spend on retrofits to meet building codes, you save $12 or $13 in response and recovery. So you're talking about pretty astronomical savings based on cost alone, let alone lives saved. The financial argument is certainly there for doing more mitigation.

I will say the Biden administration has put significantly more money toward our hazard-mitigation programs through FEMA. There's also some mitigation funding tucked into the Bipartisan Infrastructure Law that is going out through some other federal agencies like the Department of the Interior. So there is more money at the federal level in recent years going toward mitigation. It's a work in progress.

Ayana: Paint us a picture of how it would work if we did emergency management right—from prevention and preparation through response and recovery. Let's say a hurricane is coming to Florida, which will happen at some point again soon. What does this process ideally look like, before, during, and after?

Samantha: First of all, long before that hurricane ever even formed we will have invested in local emergency management agencies that work on community preparedness. So people have evacuation plans. There are public-assisted evacuation plans for people who need it. The legalities of calling for evacuations have been figured out ahead of time. Shelters are open, and people know which shelters they should be going to. Ideally, there also would've been changes to Florida's building codes. Development along the coast would have stopped.

And then in terms of a federal response, FEMA would be taken out of the Department of Homeland Security and returned to its status as an independent cabinet-level agency, as it was pre-9/11. And in doing so, the funding for the agency will have increased so that they have a well-trained staff and enough people to meet the needs across the country.

Ayana: Especially as those needs are going to grow because of the climate crisis.

Samantha: Yes, and you would see that the staff at the FEMA regional level has good relationships with the state of Florida and local emergency managers. They would all be working in a well-coordinated way and communicating well with one another.

You would see that contracts for doing various response-and-recovery ✱ tasks have been worked out ahead of time and funding is going to companies that are paying their workers livable wages and are not profiteering off of the disaster. You would see that our national disaster nonprofits are well funded through donations and grants and able to deploy to Florida to assist local nonprofits in those efforts that the government is overwhelmed by. You would see that you have strong community-led groups that are familiar with how response-and-recovery works, and that they have the resources to begin doing the grassroots organizing that needs to happen.[183]

You would see that every community had a pre-existing recovery plan, so that once the storm did pass, they were able to immediately jump-start the recovery and close the window for opportunists to come in and make negative changes for the community. And you would see that in the recovery process the community was doing even more to mitigate future risk, asking questions about which areas should or should not be rebuilt, about how to rebuild in an even safer way than we had before. That would be a good start.

Ayana: FEMA is obviously at the core of a lot of this nationally. It's an acronym that is very loaded at this point. A lot of people have deep disappointment or frustration or anger when they think about what FEMA is or isn't doing. If you were the head of FEMA, what is the first thing you would change?

Samantha: The first thing is getting out of the Department of Homeland Security. The other thing is shifting the organizational culture. FEMA notoriously is not a place where people want to work. It's a place that burns through staff quickly. And to be able to do this work, you need people who are passionate about this field, who are extremely well trained, and who have a depth of experience and a depth of formal education for how to do this. You need people who are from communities that have been through these disasters. There are certainly some people at FEMA who are fantastic and check all those boxes. But the overall organizational culture just does not seem, and this has been true for de-

---

183 As I heard a climate justice leader say recently, "Community organizers are first responders."

cades, to be supportive of people who are looking to make changes and to better the system.

It's a place where you're encouraged to follow the checklist and just get to the end of the day. And that isn't what we need right now. Sometimes during an actual response, that's what you need, but the rest of the time you need people who are thinking big, who are thinking about how we can be transformative in our work. You need people who are willing to go to work every single day and potentially get fired because they've spoken up and said what needs to be said. And I understand that's a big ask, but we need to find a way to create space in the agency for that in a way that has never really existed there.

Ayana: Yes, this is a time for courage.

We've talked a lot about the role of government, but I know you've also spent a lot of time thinking about the role of communities. Can you talk about what it looks like when communities get it right on disaster response and emergency management?

Samantha: There are tons of examples of community groups across the country that are working on these efforts. What always comes to my mind first are the mutual aid groups on the coast of Louisiana and in southeast Texas. They are in communities that have been hit by one disaster after another in the past several years, and in response they've built up organizations and a network of people with a depth of disaster-specific knowledge.

Ayana: When we think about community organizing, community-level response, neighbors helping neighbors, one of the things that people often need is help navigating actually getting the resources that are available, that are supposedly allocated to them in terms of recovery. And you describe this frustrating phenomenon of people compiling binders full of paperwork.

Samantha: I came to call them "recovery binders." When I lived in New Orleans, I would go out every weekend to do rebuilding work throughout the city with various rebuilding organizations. The homeowners almost always were right there alongside you helping rebuild the house and directing how they wanted things. So we spent a lot of time with the folks who owned the houses that we were rebuilding. And over lunch or taking a break in the afternoon, they'd pull out this ginormous three-ring binder, and it was packed to the brim with all of this paperwork. And they'd start going through it and showing you, "Oh, here's my correspondence with the insurance agency, and here's the back and forth with FEMA, and here's this grant program that I found."

This is *years* after Katrina has happened. They had been accumulating all of these documents you have to keep track of as you try to navigate this incredibly complex recovery system that we have across the country. I had thought it was just a New Orleans thing. Then as I started to go to other disasters all around the country, I'd go to a community meeting or be rebuilding a house, and out would come the binder.

Really what that binder represents is the complexity and the amount ! of time that it takes to try to simply get the resources that you need to even begin the physical rebuilding process, let alone any other aspect of recovery, like mental health or economic recovery, which also needs to happen.

Ayana: This is heart-wrenching because after living through a disaster, a catastrophe, the last thing you want to do is be bogged down in paperwork and not be able to move forward. On top of the fact that the money is usually not enough to rebuild your house anyway, or to relocate. I learned from you that the system we have here in the U.S. for recovery is called a "limited intervention model." What is that, and what might be better?

Samantha: Basically it's saying that the government is going to be limited in their involvement in individual and household recovery. It's saying: "You're primarily on your own; we might give you a little something."

Let's say your entire house is destroyed in a disaster. You have three primary sources of aid that you can go to. First, you can rely on yourself. If you have insurance, if you have an ample savings account, if you are independently wealthy, relying on those resources is the best option. Second, you're also going to turn to your social network. So, help from family, friends, co-workers, neighbors—that could be financial help, that could be a couch to sleep on while you're displaced. And then the third source is institutional aid. This could be from nonprofits, which is great if you can get it, but is often quite limited.

And then there is government. For the vast majority of disasters, the only help you would be getting from FEMA is through the Individual Assistance program. And getting that help is a little bit complex. The first barrier is that you have to live in a county that actually gets a presidential disaster declaration and is specifically approved for individual assistance. If you break through that barrier, then you can apply for that assistance, which is incredibly limited. The number changes a bit year to year. Last I looked, the most you could get was around $42,500 from FEMA. Total. And only about 1% of the applicants receive that. The average amount people were getting between 2010 and 2019 was actually $3,522.

Ayana: Hard to rebuild a house for that.

Samantha: That maybe is enough to rent an apartment while you rebuild, but it's not enough for somebody to recover. And so what happens is you have individuals who are having to pull from all of these different sources and piece it together. This is why it's a do-it-yourself recovery. This is why you have a recovery binder. Because the system is intentionally set up in a way that limits how much help you're going to be receiving. And the disparities in who is going to be able to navigate this system are pretty clear. Anybody who is low-income, anybody living paycheck to paycheck, even people who are middle class, are really going to struggle with this approach.

This is where the strain that is being put on the nonprofit sector is incredibly concerning. Our system was designed with the expectation that nonprofits were going to fill that gap, and many of them are trying, but they aren't capable of doing it at this point. Our need is so great. And this is where the overall economic well-being of individuals across the country has this huge effect on how recovery plays out. Do people have enough money this year to pay for flood insurance? Do they have enough money to be building a savings account? And for most people at this point, the answer to that is no, they don't. The entire way this system has been structured sets people up for getting stuck in recovery for a really long time. It takes a long time to navigate this process.

Ayana: Could be a decade.

Samantha: Years and years, which also increases that community's and that individual's vulnerability for the next event. This is what's happening in Louisiana and southeast Texas, where people cannot get through one recovery process before the next disaster happens. They're then more vulnerable to that next disaster.

And then, and this is where I think a lot of the frustrations with FEMA　✱ actually come from, this limited intervention model is not what people expect. People expect that FEMA is going to come in and that they're going to help you fully rebuild, that they're going to be handing out checks. And that is not what FEMA is even trying to do. There's this fundamental misunderstanding about what FEMA is, what FEMA does, what Congress allows FEMA to do in recovery. That is creating this tension, this animosity between communities and FEMA, which just overlays everything.

Ayana: What a mess. Can you imagine a world where we no longer need recovery binders? Are there places where this better kind of emergency management is already happening?

Samantha: Other countries have other models. Every single one of them comes with its own set of problems, and I don't think any of them would work in the U.S. given the way our government functions. So the answer to that is no. But I do think there are some tangible solutions that would have a huge effect on how recovery unfolds. The first would be making some changes to the National Flood Insurance Program (NFIP).

Ayana: I was going to ask you about this. Wonky but critical. What do we need to change about the National Flood Insurance Program?

Samantha: Well, you need a lot more people to sign up for it.[184] It's pretty　✱ limited now in terms of who is actually required to carry an NFIP policy. You need to have a federally backed mortgage and be in a certain flood zone. So one of the major changes we can make is to change that requirement, require more people to buy into NFIP. That would have two effects. One, that's going to bring more money in, in terms of premiums year to year. Two, that's going to provide a much bigger pool of money for people when they get to recovery.

Ayana: And I understand there's also an issue with the flood maps themselves?

---

184 Following this interview, I signed up for it. Samantha Montano, flood insurance influencer.

Samantha: Generally, the thing that I say about the flood maps is that they are wrong. FEMA is in the process of trying to fix the flood maps, but the biggest flaw is that the majority of them do not account for climate.

Ayana: Like sea level rise and more intense rainfall.

Samantha: Yeah. They also don't account for new development, which changes your flood risk. Like anytime a bunch of woods get knocked down and paved over, that changes your flood risk. Basically, we're not updating the maps as quickly as they need to be updated. When you look up your flood map for your community, it's a pretty common occurrence to find that the most recent map is from like the late '90s or the early 2000s. And of course, your flood risk is going to have changed in the past twenty years.

Ayana: If people had accurate information with which to assess their own risk and vulnerability, then they might make different decisions about where they would buy and build. And the fact that they would have to pay a lot for insurance, or maybe can't even get insurance, might be a deterrent from living in some of these most dangerous places.

Samantha: Potentially, yeah. Often you hear people say, "We bought this house five years ago; nobody ever told us it was in a flood zone." That's another change we can make. We can pass laws that require flood disclosures. Often people are buying a house and if it hasn't flooded there recently, they would have no idea that it's in a flood zone. Nobody would bring that up to them at any point, and if they didn't know to independently, on their own, go check that, they would have no idea.

Ayana: And if they do check and the maps are out of date, that's also not that useful.

Samantha: Right, exactly. One of the main theories with the creation of NFIP was that it's a tool to make people move out of high-risk areas. And they've tried several times to say, "We're not going to provide flood insurance for this community or this property because it's too high-risk." And unfortunately what we see happen sometimes is that people will say, "Okay, I'll live here without flood insurance." And sometimes that is because they are wealthy and they like their beach house and they're fine to just pay for damages on their own.

Where we run into equity issues is when this happens in communities that are low-income and own their houses because they've been passed down from generation to generation and they don't have the resources to move somewhere else. This is where we start getting into some overlap with buyout programs and managed retreat and how to approach that in an equitable way. All of these things build on one another and are connected to one another. Again, that's why this needs to be thought of as a system. If you make one little change to the flood maps, there can be this domino effect in all of these other programs that we use to mitigate our risk.

Ayana: What is the nerdiest, least sexy, most esoteric thing that we need to do to turn your future vision of excellent emergency management into reality?

Samantha: Fund local emergency management. There needs to be ✳ money for emergency managers to join the International Association of Emergency Managers, and there needs to be money for them to go to conferences and talk to other emergency managers across the country. They need to have the money to hire a specialist to come in and do public-awareness and public-education campaigns. Frankly, you need to bring in people who can write response plans, write recovery plans, who are good at facilitating community meetings. All of these skill sets are needed for effective emergency management. And all of that starts with funding positions for people to be able to have these as their full-time jobs. That's not as cool-sounding as, "Here's two billion dollars to build this crazy seawall," but it's more important, arguably.

Ayana: What can people do to help make all these changes happen? It sounds like advocating for that funding is critical, and showing up and being a volunteer in the recovery phase is really useful, and supporting the nonprofits that are doing good work, as that is the stopgap measure we have. Perhaps showing up at community meetings and commenting on different proposals? There's also getting involved in mutual aid. What else?

Samantha: One thing I want everyone to do is google what their local ✳ emergency management budget is. That is the thing I think can radicalize people. When you see how little money is going to emergency management, especially compared to the law-enforcement and fire budgets, that's gonna make you want to go to your community meetings. That's going to make you want to go ask your city council, "Why are you not putting money toward this?" They're using climate change and disasters in their campaign speeches, but they're not following through, they're not putting money toward that. And emergency management budgets are the biggest, most obvious example of that.

Ayana: Excellent civics homework assignment. Okay, what are the top three things you wish everyone knew about climate solutions and emergency management?

Samantha: One, mitigating climate change itself is not going to stop all disasters from happening. Two, climate change is a factor in many disasters, but it is not the only factor. And three, absolutely we need to be mitigating climate change, but we also have to be looking at these other issues like development and social inequalities to be able to make tangible differences in how communities experience these events.

Ayana: Last question. I know you're a Taylor Swift fan. If there's one thing you could ask Taylor to do to improve emergency management, what would it be?

Samantha: I've truly never been asked this question before.

Ayana: This is the most important question I've asked today.

Samantha: You know what, she has started doing this, but what would
✱ be very useful is for her to use her power to get people to vote in *local* elections.[185] Because so, so much of our emergency management policy is happening at that local level. Who your mayor is really matters when there's a disaster. That is the person who's going to ultimately decide evacuation orders and make decisions about how your community recovers. So voting in local elections, high priority.

---

185 On National Voter Registration Day in 2023, 35,252 new voters registered via Vote.org—and there was a 1,226% jump in registrations within the hour after Taylor Swift shared the link on Instagram.

The sea rises, the light fails, lovers cling to each other and children cling to us. The moment we cease to hold each other, the moment we break faith with one another, the sea engulfs us and the light goes out.

—James Baldwin

Diasporas and Home

Interview with Colette Pichon Battle

Climate change is already forcing tens of millions of people to relocate every year. Whereas disasters (especially poorly managed ones) cause punctuated increases in migration, increasing sea levels, heat, drought, and crop failures will cause a *persistent* upward trend. We are not prepared—in terms of policy, infrastructure, or culture—for these shifts.

It is critical to note that people of color disproportionately bear climate impacts, from storms to heat waves to pollution. (Although climate change is coming for us all.) Fossil-fueled power plants and refineries are disproportionately located in Black neighborhoods, where storms and exposure to air pollution and toxins cause great harm to public health. If we get from climate disaster to climate solutions while, as usual, screwing over poor communities and communities of color, that's not success. Inconvenient though it may be, our climate crisis is intertwined with our racial-inequality crisis.

While you may have heard about these disproportionate impacts, you probably don't hear much about how people of color actually care more about the climate crisis than White people do, and are more inclined to want to be part of the solutions. In the U.S., 65% of Black people and of Asians, and 70% of Hispanics are concerned about climate, compared to 47% of White people. That's tens of millions of people who already care deeply about the environment and are down to pitch in on the massive amount of climate work that needs doing. Imagine the additional and huge contributions all these folks could make if unburdened from the dangerous distraction that is racism. And goodness do we need all of that energy and ingenuity. I simply don't see how we win at addressing the climate crisis without elevating Black, Indigenous, Latine, and Asian leaders.

The person who has most helped me see the connections between all these dots is Colette Pichon Battle, a Black lawyer and organizer from rural Louisiana. She co-leads Taproot Earth, an organization that "builds power and cultivates solutions among frontline communities advancing climate justice and democracy." As it did for Samantha, Hurricane Katrina sparked Colette's current work. While living in D.C., Colette watched that storm rip through her home community, then moved back to help with recovery. She has grown into one of the most powerful, compelling, and grounded voices in climate justice.

I met Colette in 2019 at a Green New Deal event organized by Rhiana Gunn-Wright.[186] She was excited to meet a Black marine biologist and I was excited to meet a wise organizer from Louisiana who was thinking deeply about how to shape the future. Her energy is very much, "What's next? What can we create? Who's the crew we need?" As she put it to me,

> I'm looking for others who see that the win requires us to change a little bit. We gotta grow a little bit. Our planet needs us to get out of our ego and into a collective decision-making process that leads to some collective solutions.

Naturally, I have since lured Colette into ocean policy work—together we sit on the steering committee of Ocean Justice Forum. She has also recently expanded her geographic scope from the Gulf South to the global Black diaspora. Here she shares with us the possibilities she is working to unlock, and helps answer one of my big questions: What does "home" mean in the context of the climate crisis?

———

Ayana: How does your geography, your rootedness in the Louisiana bayou, shape the way you think about the climate crisis?

Colette: I'm always grateful when people remember that I'm from the bayou. My maternal family, the Pichons, are from the triangle of watersheds—Bayou Liberty, Bayou Vincent, Bayou Bonfouca—in southeast Louisiana. I know it, and everybody down there knows it, knows what family line I'm coming from.

It's such an important landscape for me, and it has a magnetism for tropical storms. Those storms look for warm. The Atlantic is not warm. The Gulf is warm, and that's where storms end up. Louisiana is a smaller state, but it's a top receiver of federal disaster dollars from FEMA. Be-

———

186 The same gathering where I met Brian Donahue.

cause that's how many disasters we have, not just from storms coming up the coast, but also rain events that come because of our particular topography.

This geography has put me in the climate fight. This was not an "Oh, let me go work on climate change" situation. This was a "Katrina has destroyed us. What do we do?" situation. I'm in a place that has an authentic standing in a global conversation. And I have a duty to this place that I love. Every day I go to work and I get to root my work in love because it's rooted in this place. There's a calling that applies to the storm looking for that warmth that also applies to my life and this movement.

Ayana: One of the hardest things about being rooted deeply in a place is seeing that place change in ways you can't stop. So this is a question I wish I didn't have to ask, but when we think about sea level rise and hurricanes, there is a hard truth that some places will no longer be safe to live in, and our lives will be transformed, and our connection to these places will be transformed. There are some hard discussions we need to have about climate-driven relocation, and some impossible decisions that somehow have to be made, because not making them is also setting a path.

Colette: Impossible decisions. And we have to keep rooting people not in the devastation that is the climate crisis, but in the opportunity that is this liberation horizon that we're on.

Ayana: What would it mean to take care of communities in the face of our climate reality? What does it mean to transform coastal communities in a way that is just?

Colette: It's the right question and it's the hardest one. Because it's the question that actually acknowledges that people are part of an ecology. And an ecology, a whole bioregion, is being reshaped or destroyed or changed in this climate crisis.

The hardest thing is to hear communities in South Louisiana question whether or not they should leave or move or relocate, because rationally you think, "Of course you should get out of harm's way." Then I met some Indigenous folks who don't share that opinion, who say, "We're not leaving. This is our land until the end." And that sparked questions in me: Is my job to get people ready for relocation, for a very difficult separation and goodbye to something they have never contemplated having to lose, which is their place? Or is my job to help develop the ability within leaders and residents to exercise their own self-determination and autonomy?

At Taproot Earth, one of the things we're bringing into the climate justice movement is the creation of self-determined, self-governed formations that practice democracy.

Ayana: What do you mean by formation?

Colette: Some people like to say coalition, some people like to say organization, but we're not that tight. Think of a flock of birds.

Ayana: Like a murmuration.

Colette: These formations are people coming together, agreeing to move together, like Gulf South for a Green New Deal, or the Just Transition Lawyering network, or Taproot Noire, a global network of Black climate leaders. And we are helping them practice decision-making and governance. Because what you're up against in the near future is twenty-two thousand more problems. You're gonna have to know how to bring more people in and how to make decisions together.

And the need to practice "facilitative leadership," that's not what the funders or national climate leaders want to hear. They want to hear how you're going to reduce CO_2 emissions. But the problem isn't just emissions. The underlying problem is our society has checked out of even the notion of democracy. This is a moment for democracy to either get a second breath or it's where democracy is going to die. We've got leaders who lack the courage to represent their constituencies. If you ask folks, "Excuse me, do you really think we should have a combustible plant in your parish?" the people will say, "No." But then their leaders say, "Put it in Cancer Alley."

There are communities fighting for their lives right now. We need folks to stand in solidarity with us, to get their legislators to get behind some of what our communities are asking for. Because our legislators down here are paid for by oil and gas. But they cannot stop us if we move together.

Ayana: It's been wonderful to watch you tap into the collective wisdom of diaspora as part of this liberation work.

Colette: With our creation of Taproot Noire, we're hoping to connect the Black diaspora to create solutions that can be replicated throughout the Black frontlines globally, because otherwise I'm not sure we're gonna survive this. This is what we learned in Katrina.

Ayana: Help is not coming.

Colette: It's not coming. I'm doing this work because I've seen the failures of the system. And somebody has got to care what happens to Black people. My hope is to bring together some of the best and brightest people who love their place the way I love my place.

And inside of that, we will see beautiful solutions that may or may not be invested in by the mainstream financial markets, but that are replicable by other folks who are networked. So if there is a solution for desalination in Cape Verde, that knowledge ought to be transmitted. There should be an information network so people can start getting ready and saving themselves. I hope to save Black people with this network.

Ayana: What does this collective decision-making and formation-building look like practically?

Colette: Especially in the U.S., people have been able to start their own nonprofit and do their own thing and basically just move without anybody. We have re-created a society of individualism inside of our nonprofit structures that have good intentions, but they're not moving together and we're not winning.

In a formation like Gulf South for a Green New Deal, we say: "You got an organization, I got an organization, and we're gonna sit down and figure out how to move together inside of our state." And then we connect to other states and as a region to see what bigger things we can do together.

At our Climate Justice and Joy event there were over a thousand people from all five Gulf states, and from across the nation, actually. The South wanted to bring people in, not into this conversation of what we're going to lose, but into a conversation, rooted in culture and tradition, around what we could save together. It was a narrative shift.

If you looked at it from a distance, it might look like a big fun party— which we are not opposed to here in South Louisiana, just so you know. But we had sessions where people talked about offshore wind, about language access in climate policy, about the role of art and culture inside of narrative- and message-building. It was beautiful. There were people from Standing Rock, from the labor movement in California, and all of these folks are saying, "We recognize the Gulf is an epicenter and we know we've gotta be in alignment."

Ayana: Having joy as an organizing principle is really helpful.

Colette: Exactly. We didn't even start with words. We started with drums. You feel the power words will never give you, which is rhythm, which is everybody moving together, freedom and being free in this moment. If we're going to go have a conversation about climate solutions, let's root it in joy and abundance as opposed to fear and scarcity.

Ayana: And community.

Colette: And community. That is what we have to acknowledge, Ayana. We live in a society—not just the U.S. but all of the Westernized nations—where individualism has been awarded, rewarded, and advanced. But the only way we're going to make it through this next phase of our planet ✱ is together. We're gonna have to remember what it is to actually be in a community. It's odd for people whose daily interaction is on a technological device, physically alone while pretending you're together. We've forgotten how much energy it takes to just be in a room with a lot of people. It's work.

Ayana: Taproot Earth's vision is "a world where we can all live, rest, and thrive in the places we love." How do you imagine the transformation to get there?

Colette: We have to get people back into the game. Residents across the Gulf and Appalachia and the Black diaspora have been told that they have no place, that they're not wanted, not valued. And we have to say, "Actually, that's incorrect, whatever you've been taught for generations. Get in here. We need you and we love you. And whatever you have to ♥ contribute is good enough." To transform the South, to get beyond the notion that we have no voice, no right, and no value, is a real piece of work. That transformation will be invisible, but it will be the most powerful.

We also have to have a conversation about why it is that our leaders feel it's okay to sacrifice the people for the profits of multinational corporations, many of which aren't even headquartered in the U.S. Why are we fracking gas to send to Europe? Why are we cutting down old forests in Mississippi to send to Europe?

Ayana: For their wood-pellet stoves, that they pretend are net zero. Just the most bananas.

Colette: It's crazy. And we are allowing that kind of greenwashing to happen. We have to see the South not as the poorest place, but as the place most rich in resources. And those resources are being extracted by corporations.

We need to uplift <u>justly sourced renewable energy</u>. That's a term Taproot Earth uses because some folks believe all we need to do is switch over to the solar panels and windmills and that will fix it. And I blame the liberal climate movement for making this all about CO_2 emissions. It is not. It's about control of natural resources, management of collective resources.

We cannot say "Let us make more windmills" without understanding where that material comes from. We cannot say "Let's make EV batteries" and not understand that Indigenous communities are saying "No more cobalt mining, this is not what we do with the Earth." And we have to be clear that a lot of the source materials will still have to be extracted, and are actually fueling war and conflict and violence against women. We have to put "justly sourced" in front of anytime you say "renewable energy." If we don't say that, we simply replace the oil and gas sector with the renewable energy sector and maintain that same philosophy of extraction that is the problem.

We have to take our power and our Earth back. And when we win, we have to come up with processes that allow us to steward that power in the right way, as my Indigenous brothers and sisters would say, for the next seven generations at a minimum. All of our policies should be rooted in that.

Ayana: Let's talk more about traditional knowledge and <u>Indigenous wisdom</u>. I know that's a big part of the way you think about the problems and the solutions. How does this fit into your evolving worldview of where we should be heading?

Colette: Basically, when you're raised with a lot of Black women in the South, the response to any question is: "The older Black women have the answer." They know what's wrong with the baby. They know what season the flower's supposed to grow. They know lots of things. They know what they were told. And they've been told things for generations, so it's this aggregated knowledge.

Ayana: And your heritage is Indigenous as well as Black.

Colette: My grandmother was part of the Choctaw Nation and my grandfather is a Louisiana Creole. And what both of them knew most of all was our bayou. They knew our land. They knew what grows there. They knew what storms should make us afraid versus the storms just coming through because they're coming. And after Katrina, the women in my community knew how to clean the soil from the flood.

And I remember this because at the time the Red Cross was giving everybody bleach to clean with. And I was saying, "We cannot use that, it's gonna run off into the waterways." And at that point, none of the grocery stores were open and everybody was fishing. So anything going into the waterways was going to end up back in people's bodies.

But the one piece of information everybody had access to and everybody knew and did not require you to have money or a network outside of Louisiana was knowledge of greens and sunflowers. All of the women in my community, they all still have gardens, and they knew to plant greens and sunflowers and to wait one rotation before you ate out of that garden again.[187]

I didn't know this was called <u>traditional ecological knowledge</u>. I thought that was just Miss Mary's instructions for what we needed to do. I'm often challenged on this traditional knowledge, especially by scientists, no offense.

Ayana: None taken. I want to circle back to this big question around place. Where do we go? Or do we stay? I'm curious to hear more about how you conceive of "managed retreat" and if you have a word or a term that you like better, because that one is not good.

Colette: It's not good. My visceral reaction is that while organizers are actively rebuilding people's ability to see themselves as self-determined humans, the last word we would ever bring into that space is "retreat." It

---

187 Sunflowers can naturally extract poisonous chemicals and heavy metals from contaminated soils.

is, I believe, indicative of the system rooted in a philosophy of victimization and ignoring people's power. It is not a catalyzing framework. It's not helpful.

I think self-determined relocation and climate migration are stronger terms and allow people to have dignity inside of it. We have to be careful with how we label some of these next phases of what's coming. And we have to understand that inside of this label is deep loss. If you mess that label up, you will turn this deep loss into something tragic.

Ayana: And bureaucratic.

* **Colette:** And bureaucratic. South Louisiana can teach the world how to say goodbye to something. If you've ever seen a second line, there is respect, honor, joy, and a release. It is a ritual rooted in African tradition. Unfortunately, since Katrina we have had this terrible gentrification and commodification of our culture. Now they put on a second line just for tourists to party. It's hard to watch. This is meant to be a ritual for particular celebrations and jazz funerals, a ritual for saying goodbye. Because the release and grieving must occur in order for you to go forward with any kind of productivity.

Now is the time for us to take these cultural traditions, give them the honor they deserve, and figure out how they can be brought into climate solutions. And as we get to the stage of relocation, we're going to have to see some rituals and some ceremony.

Ayana: Absolutely. I've started to call that "climate goodbyes."

Colette: It is a terrible goodbye. People are going to have to move. And it's sad and they get depressed and they lose their community and they lose all of their networks. My mother left Louisiana after Katrina and still lives in Texas to this day. Eighteen years. Because some people cannot return after a storm. Luckily, we got some family around that can get to her relatively quickly. But she used to live in a place where she knew everybody on the street. And they knew exactly who she was.

* It is a terrible loss. And loss can turn into anger, rooted in fear. We have to go back to the cultural traditions that allow us to acknowledge that things shift and move all the time. The traditions that tell us that loss is part of life and there are things we can do to honor that loss and ways to heal from it.

I visited Jade Begay[188] in New Mexico and she said, "If you look at the long trajectory of a lot of these Indigenous communities, they've moved before. Migration does not need to destroy the culture. But you do have to do it right. And you have to do it in a way that is honorable."

Ayana: There are a lot of goodbyes that are gonna need to happen real quick. Do you have a vision for how we could start to approach this question of how we let go of places but still stay connected to them?

Colette: When the land goes, when it's not livable anymore, how do we honor it? The answer to that question allows for people to say goodbye. Especially in South Louisiana, it's not unheard of to honor places, from old trees to battlefields.

I went to Egypt for the climate talks and got to go to the city of Aswan and sit around a fire at a ritual of coffee with the Nuba people—also called Nubian, but the correct term, I learned, is Nuba. And this one dude had his hand drum and sang this song, and I watched the men around the circle get teary-eyed because he was singing what is now called the National Anthem of Nuba. He was singing a song of a land that they'll never see again. It was so powerful, I'm tearing up even right now. I was sitting next to ancient people who can tell me about land they lost when damming the Nile River started. The entire circle of people knew that song, and they sang it with him.

I'm understanding that the power of culture, music, and tradition ♥ will not just heal people in this moment of tragedy and current loss, but it will allow for the healing to continue down the road, because everybody longs for their homeland. We don't have to die when we lose our land. It was such a beautiful moment.

Ayana: The goodbye anthem. And the moving-forward anthem.

Colette: That's right. And it's not something that you do once. It's something that is sung over and over again, whenever the spirit moves, so that people who still need to heal a little bit can heal. It gave me a lot of hope for South Louisiana.

---

188 The following interview is with Jade.

Ayana: What does "home" mean in the context of diaspora and the climate crisis?

Colette: My gut reaction to the question of home is it's a painful acknowledgment. Because I come from a place that will be taken by the sea in the next fifty years. And we are aware. We have the maps. So what does that mean when our land is no more? Everybody who knows me knows I'm from South Louisiana. What happens when we don't have that anymore?

Ayana: To me, it feels like we have to make and remake home, we have to embrace that it is a decision, a process, and a commitment. There's a home that is geographic and there's a home that is cultural and there's a home that's where your people are.

Can you imagine what getting it right for self-determined relocation looks like?

Colette: Well, this one's complicated because my normal answers have now led me to understand violations in current law. So I'm going to get a little lawyerly on you.

Ayana: Please do. Bring it on, Colette Pichon Battle, Esquire.

Colette: Esquire—boo, boo. Esquire.

So, imagine you are trying to move a group of people into a new place. If you try to say that this new place is for a group of folks with a shared racial identity, and that federal dollars should go to support that group of folks, some would say you are now violating the Equal Protection Clause. Because equal protection says you can't ever do one thing for just one group of people.

Equal protection was rooted in challenging White supremacy, and dismantling oppression. But now it's a legal barrier when you're trying to preserve a culture that only exists because of that intact community.

Post-Katrina, you could see both the power and the attention the Vietnamese community could get because they were moving back into their homes together. They got their lights turned back on. Because all of a sudden, hundreds of Vietnamese Americans had moved back to New Orleans East, and you can't have that many people without the power being on. And the different Black folks who had dispersed all over the South did not come back as one group.

Now, there's lots to unpack there, but what is interesting is you can see the power when people move together. But you can also see how the other groups that are not inside of that move would feel slighted or left out, and there are legal parameters in place so that doesn't happen. We're actually going to have to get new laws.

Ayana: And because of sea level rise, we're talking about whole neighborhoods, maybe even whole counties considering relocation. So what would these new laws be?

Colette: You are speaking my love language here, talking about changing laws. Because the truth is, all of the laws we have on the books at the federal government level and even at the state level are not ready for this new climate reality. They don't even acknowledge it. Once people finally get their heads around this climate crisis, they're going to understand we're not just talking about disaster, we're talking about whole societies being transformed.

Which creates an opportunity to build a whole new society, an opportunity to do things a different way. This could be brilliant. If we have the courage to let go of the hierarchies that exist in society, which many of us benefit from, you could actually fix this whole thing.

Think about what it's going to be like to move, to migrate. We need comprehensive immigration law reform. And it can't be to the benefit of one country and the detriment of another. We're going to have to understand migration as a part of a new climate reality. We're going to need new housing laws. We're going to need new disaster laws. The disaster laws right now say FEMA will only help if you build back in the same place.

Ayana: In the same risky place. So illogical.

Colette: That's real. Another piece of law we need to rethink is property ownership. What does property ownership mean when your house is on land that will have no market value? What does it mean when you have enough money to buy the high ground but others don't have enough money to get out of harm's way? How do we make this fair?

Ayana: The folks in the bayou, who are losing their homes into the sea, did not cause the climate crisis.

Colette: Accountability is a word we seem to be afraid of in this country, except for when it's thrusted on Black and Brown people. But when we talk about multinational corporations and leaders, somehow accountability is something we walk right around.

Ayana: If you, with your powerful pen, could rewrite a few laws to get us on the right track toward adapting to this new world, where would you start?

Colette: The first step is for this nation to honor the sovereignty of the Indigenous people of this land. We are not living on land that was uninhabited. We are not living in a place that wasn't already being cared for, stewarded, used, and lived on. We have to recognize the sovereignty of all Indigenous nations.

    The second thing is we have to protect the natural resources of this nation, giving rights to water, to clean air, to land, to trees, rooting our constitution in human rights and in the rights of nature.[189]

    Third is we have a constitution that's rooted in property rights, because the right to vote was only for White men who owned property. We need a new constitution that is for everybody, every human, rooted in humanity, and that gives rights to the basic things that are required for life. Water is required for life. Housing is required for life. Food is required for life. Energy is required for life.

Ayana: Along the legal side of things, what is the nerdiest, least sexy, most esoteric thing we need to do to step forward toward this vision of getting it right?

Colette: We need to repeal all tax abatements for any polluting industry anywhere, at the state and federal levels. All of the refineries and polluting facilities must start paying the property taxes that they don't currently have to pay because some politician waived them for the "jobs" that they would create. All of those tax laws need to be repealed. And all of the money they will then pay in property taxes (I personally believe this should go back to the date of first abatement, but we could go back five years or whatever), that goes into the state coffers. And that goes into education, healthcare, and housing first.

---

189 Abbie also raised this and the issue of private property in the "See You in Court" chapter. Lawyers focused on climate justice are in sync on key changes that we need.

Tax laws are the unsexy but sexy legal thing that could help. Because right now these industries are not just making millions and billions, they're making millions and billions and leaving environmental hazards that then have to be dealt with by public dollars, and that can't even always be cleaned up.

And then when people get sick with cancer and all these other things, the healthcare system has to take care of that. So this company that didn't have to pay taxes, got all of its profits, paid a couple of people from your community—who by the way are now sick and have to have cancer treatments for the next twenty years—and then got to leave without cleaning anything up or taking any responsibility for the economic impact they have on our healthcare system, yes, stop giving them tax breaks.

Ayana: I'm here for it. Let's rewrite the tax code.

If we're able to apply all the lessons you've been sharing, how does that change the way people are living in the Gulf South a decade or two from now? I would love to hear your dream.

Colette: The future of the Gulf South could be brilliant and beautiful. But first we've got to stop the drilling for oil and gas in this region. It's ! gotta stop. This is not something that should be slowly tapered off. It has to stop now. That will affect you. Some things are going to cost more. I'm gonna need you to just pay the extra price. We need justly sourced renewable energy. If we're going to do it, let's do it right. And we can <u>consume less</u>. That should be part of the solution.

And new investments should be in sources of energy that don't harm the planet or its people. Seeing the sun, the water, the wind in this region makes me know that renewable energy is something we should be abundant in. The dream is not just to shift the source of energy, but to actually ✳ reimagine the systems of energy so that there is not a monopoly around energy generation, energy transmission, and energy management, but instead, the community collectively owns its own energy and controls it.

That looks like houses that can run the next day after a hurricane because you will always have a sunny day after a storm. So why are we going five and seven and fifteen days without power when the sun is out? We should be investing not in the profits of utilities, but in the ability for people to have power come back online, have their refrigerators working with their medicine, have their baby stuff working that needs a plug of electricity.

Ayana: Homes and communities that can power themselves.

Colette: That can power themselves, and that can control and manage and govern themselves. The dream is to have a system that doesn't have to extract from the ratepayers or the users, but gives them the control over their own energy system.

We see a lot of investment in offshore wind. That's coming. That's beautiful. That's great. Except we need to know that the fishermen, a large piece of our economy here on the Gulf, are not going to be negatively impacted by the building of these things. We've gotta make sure the mining of the raw materials we use to build these wind turbines are not boosting war in South America and in Africa. We've gotta have justly sourced materials for the renewable energy that will be community controlled. This is part of the vision not just because it sounds good, but because this is how we repair past harms. This is actually how we go forward in solutions in an equitable way.

And I believe sovereignty of Indigenous nations is the paramount step to all of this. I dream of a time when all of the Tribes of Louisiana, Mississippi, Alabama, Florida, Texas, and Puerto Rico are recognized as the sovereign nations that they are. And we work with them.

Ayana: How do we get there? What needs to happen now to set ourselves up for having this possible future?

Colette: My gut reaction is, are we ready? Lord have mercy, are we ready?

We live in a society that keeps us ever fearful of what the other person is going to do to us. And it breaks down our ability to build trust, coalition, and alliance. So we're going to have to address power. We're going to have to confront power.

! The economic system of this nation is the root cause of both the climate crisis and the enslavement of African people, of both the housing crisis and the incarceration of Brown people. It is all profit-driven. All of it. None of this is rooted in humanity. And we need to protect elections from voter suppression. There's a system in place here to silence the voices of as many people as possible.

* The shift that we're going to have to make is around courageous leadership. We absolutely need to see it in the electoral space. And we need to see preachers say, "Enough is enough, it's time for us to take care of our fellow man and woman and community." It is now time for us to do a collective calling-in for our survival. And we're going to need these folks who claim to hold the moral center of our hearts and our societies to actually understand that if you are neutral, says Bishop Desmond Tutu, in the face of injustice, you have chosen the side of the oppressor. We have to understand what moral leadership is and who's worthy of informing our ethos.

The answer to all of this is community. Knowing people and seeing people's humanity and building from there.

Ayana: What are the top three things you wish everyone knew about climate?

Colette: Climate change is real. Change is inevitable. Justice and equity are not.

This Living Earth

Steve Connell

Believe about the beginnings, what you will;
This much we know for certain . . .

At some point, darkness released its grip on light
the night / cracked / wide
the universe unclenched its jaws—
and life began to talk.
The very synapses of existence
began firing at once, moving as one thought;
since the beginning, it has been this way,
when all life on Earth communicated without words.
And still today . . .
    in the twisting swirl of a million butterflies,
    the schools of fish fast twitching in the surf,
    and the delicate reaching up of vines;
    in the shared mind of a stampede pounding,
    or beneath ground, in the sound of mycelial nerves extending

. . . this resounding language of connection is observed.
But at some point the harmony was broken—
a dissonance occurred . . .
and slowly the new song of humankind was heard with words like:
    "manifest destiny" . . . "industrial revolution" . . . "dominion over all"
    . . . "survival of the fittest" . . . "urban sprawl" . . .

And as the song of life became a solo
a new thought began
    "Perhaps the world was not alive after all—
    but we were simply alive on it"

And with that thought—
the world became a fixed thing—

it simply was;
as the sun or space or rocks;
as if it could never be diminished;
as if our actions could never cause it harm;

. . . but we were wrong.
And now we have evolved to become both destroyer and savior—
    our behavior is our greatest threat,
yet hope resides in our resolve;
in our collective compassion and formidable talents;
the way we rise to any challenge
once we accept we have no choice

And that is why, *if,* from this moment forward, in our voice,
you once again hear a harmony with the Earth;
then from out of destruction our rebirth comes,
and isn't that what life does best . . .
    It goes on;

when it seems like it is finished, it survives;
that is why in the most hostile conditions in existence
something thrives—

from toxic soil, a mushroom grows
in a desiccated river bed, water flows
in a still smoldering forest, a flower leans into the wind
and in the midst of this vast expanse . . . spins / a planet / comprised of:
    promise. spark of light, oxygen, and bone.
    photosynthesis, salt water, honeycomb, hydrogen atom,
    skin cell, hair follicle, matted fur, acetone,
    slice of wing, wind current, opposable thumbs, cyclone, dorsal fin,
    deep breath, ozone, atmospheric river, heart valve, limestone

All life is connected
Nothing lives alone
On this one and only living Earth: home.

Building Indigenous Power

Interview with Jade Begay

Jade Begay's work, like Colette's, makes one thing poignantly, unequivocally clear: Community is the answer. No need to put "What if" before those four words, or a question mark after them. There are no solitary solutions, as much as it pains my introverted soul to admit it. Ultimately, we rise or fall together. But what does it really mean to build community as we build a better world? How should we pull from the past to create the future, harkening back to a time before over-extraction and frayed ecosystems? What if we view adaptive change as a necessity and a welcome iteration?

There are, of course, a multitude of answers, and Jade is touching many of them. Jade is Diné and Tesuque Pueblo, rooted in New Mexico, and has a layered strategy for change that is grounded in community and in the soil. At the time of this interview, she was serving as director of policy and advocacy at NDN Collective, an organization building Indigenous power through both nonprofit initiatives and investments in Indigenous entrepreneurship. Jade works on climate justice solutions at many levels and time scales—she's as likely to be at an adobe brick–making workshop as a meeting at the White House.

Jade and I met briefly in 2019, at a workshop on climate storytelling, but it wasn't until early 2022 when NDN Collective published a gorgeous anthology that she edited, *Required Reading: Climate Justice, Adaptation and Investing in Indigenous Power*, that I became familiar with her work. Since then, I've come to appreciate the clarity and creativity of her strategic mind—on everything from voting and federal policy, to communications and financing, to what I call climate apocalypse skills.

There's a key statistic from a World Bank report Jade often cites that seems to underpin her approach: 80% of the world's remaining biodiversity is in areas managed by Indigenous peoples. That is a bulwark against extinctions and ecological collapse. When you put together real solutions that meet real needs with a commitment to stewarding nature, it is a small step to the concept of Land Back, a growing movement that advocates returning control of lands to Tribes. More broadly, environmental advocates (including Leah and Colette in earlier chapters) are increasingly uplifting the importance of <u>Indigenous sovereignty</u>. This refrain exemplifies what I see as an expanding understanding of justice (and its benefits) in environmental work.

This conversation with Jade clarifies that the work is big but the ways to chip away at it are clear and delightful. Getting it right is a process.

———————

Jade: I love this question, "What if we get it right?" This question is in the territory of thinking with abundance. It's a joyful prompt.

Ayana: Ooh, I like thinking about it that way. Okay, let's start with the mission of NDN Collective, "building Indigenous power." What does that mean to you? Why is that the fundamental driver of the work?

Jade: For me it means strengthening our ability to live and thrive in our own self-determination. Our communities are underserved. There are all these gaps with resources, with infrastructure. So we need to invest in ourselves and grow Indigenous-led enterprises and land projects. Build-   *
ing Indigenous power is really just growing leadership.

Ayana: And growing skills, too, it seems. Before we met, I remember seeing, on your gorgeous and informative Instagram account, a workshop NDN put together on making adobe bricks. And I was excited to know that there is an organization thinking about everything from shaping climate policy to teaching traditional skills we need to adapt to the changing climate.

Jade: That is definitely a part of building Indigenous power—reclaiming our knowledge systems and having the right conditions to be able to practice, learn, and share that. The adobe workshop was part of our Climate Preparedness Program. We spent three days learning all about making adobes, which are mud bricks. That material goes back thousands of years. Indigenous peoples in Africa, South America, Latin America, and North America all use it. If you build an adobe home and if you do it double adobe, which is two bricks side by side, it's very insulating in the winter and very cooling in the summer. It's really low-impact. The house I grew up in was adobe, so I had this experience of not needing to struggle in the seasons. I grew up warm and cool as the seasons changed.

Ayana: That's remarkable. And it makes me wonder how adobe could be part of the work that's needed to properly insulate the tens of millions of buildings and homes in America.

Jade: Absolutely, that's exactly the point. During the workshop, I didn't even prompt a climate conversation, it emerged naturally. And the workshop served a dual purpose. One, we're teaching people practical skills on how to build with these sustainable materials. Two, it's a way to grow our community. As we're literally having our hands in the dirt together, we are also growing relationships and a community of knowledgeable and trusted folks that we can call upon.

Ayana: I'm enamored with your multi-layered approach. In addition to climate preparedness, there's also your work supporting entrepreneurs and engaging in federal policy. Let's talk about NDN's financing work.

Jade: We offer loans meant to drive resources and help build capacity in our communities and territories and nations. That's through NDN Fund, a CDFI.

Ayana: What is a CDFI?

Jade: <u>Community development financial institution</u>. There are now more of these in Indian Country, but it is definitely still a growing need to have our own financial institutions to hold our wealth and our resources, and then be the ones who distribute them.

We also have NDN Holdings, a for-profit entrepreneurial arm. Through that we can partner with communities and individuals on growing Indigenous entrepreneurship, which we all know is an important aspect of building regenerative models. They can't all be nonprofits.

Ayana: For sure. And then there's your policy interventions. What are you working on in terms of big-picture policy?

Jade: One thing we're doing this year is political trainings. We want and need our people skilled in navigating environmental and climate policy, in knowing how to anticipate what's to come and how to land our wins. That's something that we often get burnt out by—we fight big fights, and we struggle really hard to stop a pipeline or some other project, but national policy isn't implemented into law, and so we see zombie pipelines emerge again.

Ayana: You mean how we keep fighting these smaller fights about individual projects because we're not winning legislative and regulatory victories on the federal and state levels?

Jade: Right. And so, with partner organizations, we've been doing these trainings to get people savvy in this policy world, and try to make it fun.

We throw out a policy problem and ask, "Who's your next call? What's your next email? Who do you reach out to to get something going on this or to figure out a strategy to take down a bad policy? Who are you building this effort with?" Hopefully these trainings will also drive people to be more engaged in things like the White House Environmental Justice Advisory Council (WHEJAC), which I serve on.

Ayana: What is this White House council doing? Can you peel back the curtain for us?

Jade: So, the WHEJAC is a group of about thirty external experts who have been brought in to provide recommendations to the White House on how to best address current and historic environmental injustices, primarily through figuring out how we can strengthen monitoring and enforcement of environmental laws.

There's one particular area of work I'm proud of. We were tasked with developing, another fun acronym for you, the CEJST, the Climate Economic Justice Screening Tool.[190]

Ayana: Oh, yes, I have explored this on the internet.

Jade: It's a cool tool to help people determine whether their community is eligible for Justice40 funds. Justice40 is the program the Biden-Harris administration created to drive 40% of sustainable development and economic benefits into communities that have been overburdened by environmental injustice.[191]

Let's turn back to this question of what if we get it right, because I know you have a strong vision of ways you can see the world becoming

---

190 The tool is available at: screeningtool.geoplatform.gov

191 Passing the Environmental Justice for All Act would also help address the issue of some communities having to deal with multiple pollutants and hazards, like oil refineries, plastics and chemical manufacturing, etc.

different and better. If all of your work at all of these different levels comes to fruition, what does that world look like and feel like?

Jade: A lot of people I meet who happen to be Indigenous and are in this work share this vision that when all is said and done, and we've won all the necessary fights and all the things have worked out, all the strategies have worked, that we're just fully living in our self-determination, living in our homelands or someplace in the world, and we're in deep relationship with the natural world and engaging with our cultures in a way that is peaceful and beautiful, and nothing else is pulling us from that.

In my work, when I see individuals from Indigenous communities around the world, people from the Amazon, the Arctic, my own homelands, and they say, "I wish I didn't have to come to the COP, or go to D.C. and lobby. All of this organizing work is pulling me away from learning my language and contributing to that revitalization." Or sometimes, in the summer when they have to travel to do advocacy work, people are missing their ceremonies, like their community's yearly hunting ritual for caribou or salmon. And that's sad. It cuts into the soul because we want to be immersed, and build with our people, and spend time in relationship with the natural world, especially when we know that the natural world is changing so fast.

Ayana: Hearing you, it strikes me that a big part of getting it right is that we won't have to waste any more energy and time convincing people about the things that need to change, and instead we'll all be doing the actual work of mitigation and adaptation.

Jade: Though I wish there would be a point where we no longer had to work on policy, there will always be resources to distribute. But, yes, the top answer is that getting it right would look like being able to just be in our homelands, and truly participate and live our lifeways without disturbance.

Ayana: You mentioned living in relationship with the natural world. For people like me, who have spent most of their lives in big cities, tell us what that means for you.

Jade: It depends where exactly you are. But that's also a mysterious question because things are changing so rapidly. It's drier here now, and things aren't going to grow like they used to. So I'm probably going to have to buckle down and learn how to develop healthy soil that can retain more water. And I'm probably going to have to do a lot of research into seeds, and build some type of seed bank with my community or with my family. And we're going to have to learn which seeds withstand hotter temperatures and which ones are more drought-resistant.

We're probably going to have to look into better ways of mitigating wildfire. In 2022, we saw the worst fire season in New Mexico's history. The forests were overgrown and so they burnt quickly. That could have been prevented if we had better management or co-management with Indigenous peoples.

Relationship with the land is also the relationship with my body. I'm going to have to figure out how I survive these drier, hotter summers. How do I make it so that I am comfortable, because heat stress is a big deal.

Ayana: And there are some places where it will be too hot for our bodies to adapt. There is a limit to the heat we can take, and some places around the world are now already pushing that limit.

Jade: I'm in the market for a house and I'm thinking adobe needs to be the way, because that's going to help me regulate myself during these hotter summers. That's an important thing for Black and Brown and Indigenous people to think about. The CDC released a report sharing that Indigenous people have more heat-related deaths than any other population in the U.S.[192]

Ayana: For you in particular, in New Mexico, in the Santa Fe area, what does that part of the country look like, say twenty or thirty years from now, if we're well on our way to getting it right?

Jade: I don't think I will be in Santa Fe proper. Personally, I will be up north. Northern New Mexico is more abundant with water. That will be a huge driver in where I end up. In many ways, I'm already preparing for this five-, ten-, twenty-year plan, because I know what's coming. I have

---

192 "Observed differences in heat-related mortality across racial/ethnic groups can also be associated with social vulnerability, which often tracks with factors leading to heat exposure (e.g., less green space and more heat-absorbing surfaces), health disparities manifested by lower income, and absence of structural adaptations such as air conditioning."—Centers for Disease Control and Prevention, 2020

the goal of being a little bit more self-reliant, having a small homestead where we're making tools, we're growing and making our own food. And of course, it's all in relationship to the people around us. I won't be too far from my community. And we, my immediate family, will be sharing workshops and sharing space where we can help build these skills, with young people especially.

Ayana: You've mentioned in our personal conversations how much you think intergenerational living and community is going to be a part of this future as well.

Jade: Oh, absolutely. I moved in with my father during the pandemic, and I don't think I'm the only one. There was a trend of recognizing how important it is to be close and the consequences if you are far. I remember the amount of gratitude I had in early 2020 that my parents were both about ten miles from me. Just knowing that they were close by. I could check on them.

As part of my vision, in two more years I'm giving up flying, so that's a big change. I'm not giving up cars or trains or bicycles; I'll be able to get around. I'm just good on the jet fuel and airplanes. I never really liked them anyway. So, I'll be pretty much localized. I'm excited about that.

Ayana: This is a very literal grounding. You're going to stay on the surface of the Earth and put down roots in your community, literal roots in the soil with seeds. This is a big change for you because I've always known you as someone who is in the room where it happens.

* **Jade:** Yeah. That is going to be a big commitment, but I'm ready. I hope we can popularize this, normalize these types of commitments, these types of <u>devotions</u>. We're going to need to get devoted to our place where we want to localize. We've gotten so comfortable with these lifestyles based on fossil fuel consumption. I don't think we're going to be able to keep doing the things we got comfortable with and used to doing. Make a bold commitment. It doesn't need to be exactly like mine. When I was thirty, I made a commitment that I would stop flying in five years. Yours could be something different, but I would love to see more of these commitments.

In the climate movement, we avoid blaming the consumer, but I really do think there's a power in being a consumer. We do create demand. We can make bold commitments to reduce energy consumption or reduce the consumption of fossil fuels. We can embrace that and claim it, instead of avoiding it and saying "No, it's just the corporations."

Ayana: One of the challenges to quitting air travel is that the U.S. is so big and our train system is so abysmal. So if we give up flying, we're so much more constrained than we would be, say, in Europe, where the veins of the continent are train lines. Getting high-speed rail across America is going to be a total game-changer.

The other part of what you're saying, or at least as I'm hearing it, is that you won't *need* to travel as much. It's not that you're sacrificing it. It's that you're building a life where that's just not a part of it. Even if we do get it right, there will be so much change, all of this extreme weather we're going to have to deal with. Where do we want to be for that? Who do we want to be with for that? I think about this a lot. What does community look like to you if we are effectively coming through the climate crisis?

Jade: One of the reasons I'm excited about northern New Mexico is that there are already a lot of people out in that direction who are savvy being off-grid. Out that way there are people who will be able to help me out if my solar panels aren't working, who might be knowledgeable about the aquifer, the land itself, navigating where to harvest what in the summertime or in the fall time. Not to mention it's surrounded by Indigenous communities. So I'll always be close to my community, my pueblo, my Tribe. Having that nearby is one-hundred-percent a priority.

Any community in the near or far future is going to have to localize. ✱ People who know that this transition, these adaptations are coming, and who are entering those calculations into their life plans, those are the people that I want to be around, that I'm attracted to.

Ayana: I feel the same way.

Jade: Back to the climate-preparedness stuff, another reason why we're hosting these workshops—getting people skilled up in adobe-making, or holding cultural burns to learn good fire—is to figure out: *Who are my people?* Who are the people who are interested in this stuff? Who are the people who want to spend a week roughing it in the woods to learn about this practice? Who will come from all corners of the country to learn about mud?

Ayana: The number of mud pies I made as a child in my backyard in Brooklyn is a very high number.

Jade: So you're one of 'em.

Ayana: I'm definitely with you in spirit, from across the country.

Jade: In the community that I want to build, we're learning and sharing these skills.

Ayana: There's something so gratifying about creating something that's real, physical in the world.

Jade: Exactly. It's a part of being human. I also think that's my "What if we get it right?" I just want to be human in the world for as long as it's safe and livable.

Ayana: What do activists need to be doing right now to open the doors to this possible future?

Jade: If I could prioritize one objective for us to really hold on to in our work, it's prioritizing and uplifting Indigenous rights. We need something like FPIC—free, prior, and informed consent—as the law of the land in the U.S. and across the world, which has been the goal of Indigenous communities for decades.

Ayana: Meaning that Indigenous communities have a say in what happens in their places?

Jade: The simplest way to break it down is that it's the right to self-determination. If FPIC was in place, that would mean Indigenous peoples have legal authority to give or withhold their consent over lands they manage. Say, for example, there's a mega-extraction project that's going to threaten lands, drive up carbon emissions, and do all this destruction. Indigenous peoples could use FPIC and flex it to stop the project. We would see a lot fewer harmful projects happen in the world. And we would probably see the protection of biodiversity go up in a big way— Indigenous peoples protect so much biodiversity.

All of these things have the impact of driving down emissions or creating the benefits for carbon sequestration. And we would see much stronger protections around traditional ecological knowledge. The adobe-making, the knowing about soil, knowing about seeds, seed banking, that's all traditional ecological knowledge. It is crucial for adaptation.

Ayana: Free, prior, and informed consent would be a game-changer.

Jade: Even if the U.S. would honor Indigenous rights, that would go a long way. There's so many things to advocate for, but I wonder: What if we put our efforts into prioritizing something like Indigenous rights? What if we went to the COP and that was a goal of everybody who went there? It doesn't have to be the only goal. It has to be a part of the environmental movement's goals. We need to get commitment from governments on this.

Ayana: Related, NDN Collective works on Land Back, this idea of returning land to Indigenous peoples.

Jade: This is definitely another way we can get it right on a large scale. Land Back is a movement to reclaim land and culture, or, as people in the Land Back movement have said, to reclaim everything that was stolen from us. It's not an eviction notice. That's important to say, because sometimes you say "Land Back" and it triggers people and people think they're going to get kicked out. Actual reparations of land is important, but it's much more than that.

Ayana: How should we think about the practicalities of this?

Jade: There are actually a couple of ways people think about Land Back. ✱ It's co-management or Indigenous-led conservation. Some of the elders I work with and receive guidance from, they just want to see Indigenous management of lands. This could be public lands, Bureau of Land Management, BLM lands, national parks, national forests. The idea is that if we're able to manage these lands, that would help make sure that the lands are well kept. And that means jobs for our people and bringing our people back into the land.

Indigenous peoples and nations have different agendas and priorities, but for the most part, people want to keep those lands intact and not overdevelop, or if there is development, make sure that it's going to be regenerative, it's going to be development that can serve the local and regional economy. It is climate justice, actually, because it's (1) the reparations of the land, and then (2) creating healthy, regenerative, sustainable conditions for people to live and work.

Ayana: What is the nerdiest or least sexy thing we need to do to get to this future vision? It can be specific and granular.

Jade: Engage in public hearings and public comment periods with federal agencies.

Ayana: Show up. State your case. Participate.

Are there places where you're already seeing the broad transformation you describe underway, this future unfolding?

Jade: Definitely. Through the work at NDN, I see it every day. Our team is driving resources into communities—to solar farms, to kelp projects. They're getting land back. They're working on conservation models, land-trust models. It's so exciting to see people thrive once they're resourced. And that is one of our mottos at NDN: When Indigenous peoples are resourced and networked, they have the power to change the world.

Proto-Farm Communities

Olalekan Jeyifous

What if, back in the 1980s, when climate science started offering warnings of where we were headed, we had begun to put far-reaching policies in place? The following works offer a protopian, sustainable community in '90s Brooklyn—a future that is, as Olalekan puts it, "decolonized, decarbonized, draped up, and dripped out." These photomontages posit a variety of advanced green technologies for the production and dissemination of food, seeds, and freshwater, set within an alternate world in which the government combats climate change through a market-based system of "mobility credits" that curtail movement of the poor and working classes. Olalekan imagines a fugitive network of farming societies that he refers to as the "Proto-Farm Communities of Upstate New York" (PFCs). These works build upon a vision of resilience and resistance that references the tradition of maroon communities, recast through the lens of an Afrofuturist, ecological, and solar-punk-meets-salvage-punk world brimming with Black joy.

PFC Catwalk

PFC Compound

PFC Agrotech

PFC Rainwater Harvest

PFC Poultry Farm

PFC Movie Nite

PFC Oyster Farm

PFC Farm Fintech

PFC Brooklyn Depot

PFC Seed + Water Drone

Massive change is happening and it will be inevitably increasingly hard and painful, and many of us will get lost along the way. The only protection against the worst-case scenario is also in community.

We need each other. Either way and always.

—Mia Birdsong

TRANSFOR-
MATION

The time has come for us to re-imagine everything . . . to become the new kind of people that are needed at such a huge period of transition.

We must transform ourselves to transform the world.

—Grace Lee Boggs

A Note on Hope

People often say that hope is important as motivation to address the climate crisis. But I have something to admit: I have a tenuous relationship with hope. I don't even like the word. It seems so passive, like wishful thinking. Like, "I hope that works out." Or, "I hope someone does something about that." And *that* vibe certainly isn't going to get us anywhere.

My aversion has been solidified by being asked, ad nauseam, What makes you hopeful? How do you stay optimistic? That always snags me because I am neither. I'm a scientist; I'm a realist. I immediately think: *Fuck hope. Where's the strategy? What are we going to do so that we don't need hope?*

It took me far too long to realize that what people were really asking was "Can you please give me hope? I need some." And that version of the ♥ question, that vulnerable and true version, that "Tell me how you, how I, can keep going" version, I totally get. We, those of us trying to turn things around, are not robots. We are humans with heaps of emotions and attachments and fears.

The fact that I am not prone to depression is an enormous gift, especially as someone who absorbs loads of bad climate news daily. (In a sense, my professional success has been tied to my ability to communicate about the collapse of the ecosystems I have studied. How messed up is that?) So I ease my anxiety with productivity, by completing action items. That's not to say I haven't been a puddle of tears on the NYC subway while reading UN climate reports, because I'm picturing the human suffering and biodiversity loss implied by those graphs. I feel all of this, deeply. I can absolutely relate to the sense of grasping for a reason to keep at it. And it would be easy, after reading all these chapters, to focus on the complexities and challenges and just throw up your hands.

After all, what is hope? The dictionary definitions of "hope" and "optimism" both include the *expectation* of a positive outcome. A positive outcome is a wild thing to expect given the scenarios we face. But the definition of "hope" also includes the word "desire," something I have in abundance. I want climate solutions *so* badly. And while it would be foolish to *assume* that our story on this planet has a happy ending, every day I wake up, and I think more and more of us wake up, and consider what we can do to manifest that desire, to nudge ourselves closer to a healthy and safe and restored and resplendent life on Earth.

For who are we to give up on this planet or one another? We simply do not get to quit. Also, how do we keep moving forward despite the intimidating odds? It's normal to grapple with all of this, and the truth is, you don't *need* hope at all. As philosopher Joanna Macy has put it:

> It's okay not to be optimistic. Buddhist teachings say feeling that you have to maintain hope can wear you out. So just be present . . . And when you're worrying about whether you're hopeful or hopeless or pessimistic or optimistic, who cares? The main thing is that you're showing up, that you're here, and that you're finding ever more capacity to love this world because it will not be healed without that.

Here's the thing: Octopuses and rainbows and music and dinner parties and love and snow flurries and the aurora borealis all exist! The world is full of delights even as it may also be spiraling toward conflagration and deluge. So when people perceive me as hopeful, I think what they are actually seeing is that I am joyful. And thank goodness for the human ability to decouple hope from joy.

My concern is that hope is insufficient. So, I encourage you to, in the words of Terry Tempest Williams, "make vows to something deeper than hope." If not hope, then what? Truth, courage, and solutions. Love. Collaboration and community. And all the sweetness along the way. That's what can get us there. *Possibility*. That's a word I can wholeheartedly race toward. I find motivation for action in glimpses of what could be, and in values instilled by my parents that say it is my responsibility to try, without any guarantees of success. The force that propels me is a simple and deep desire to be useful.

But if you are into hope, let me stop raining on your parade, and let's embrace author Rebecca Solnit's definition:

> Hope is not a lottery ticket you can sit on the sofa and clutch, feeling lucky. It is an axe you break down doors with in an emergency. Hope should shove you out the door, because it will take everything you have to steer the future away from endless war, from the annihilation of the Earth's treasures and the grinding down of the poor and marginal . . . To hope is to give yourself to

the future—and that commitment to the future is what makes the present inhabitable.

Yes. I can roll with a <u>catalytic hope</u> like that. If by "hope" you mean simmering passion to implement climate solutions, then despite a muffled cringe at your word choice, I am with you. Because what could be more depressing than just passively watching the world burn and melt and crumble? No, thank you.

As my friend Frank says, "Having hope allows you to be ready. If you're not hopeful, you're not looking for solutions." As Paola puts it, "Hope is a propellant." As my friend Boris says, "What we need is mega-pragmatic utopianism—utopianism plus a lot of detail. We need to go all-in on that." As Greta Thunberg challenges us, "Hope is something you have to earn." And as Katharine Wilkinson and I concluded in *All We Can Save*, describing our notion of "can" in the book's title:

> "Can" speaks to sheer <u>determination</u>. This shit ain't over yet. Possibility still exists, as documented in data-driven analysis of climate solutions and temperature trajectories, and as imprinted in the persistence of life despite all odds. We are a miracle. Our task and our opportunity is to face a seemingly impossible challenge and act in service of what is possible.

It's worth repeating: This shit ain't over yet. And while I don't have any assurances for you—hot damn, the world is a wreck and the future uncertain—I am overflowing with motivation to work toward a better world, even knowing it won't be a perfect world. A world with mended landscapes and renewable energy and clean air and climate justice is possible. And that is worth a shot.

An Ocean of Answers

Interview with Bren Smith

What if we are, right now, in the process of getting it right? In fits and starts, the transformation is underway. Yes, there's an astronomical amount of necessary work and change ahead, and humans will keep innovating because that's what humans do, but we truly already have most of the solutions we need. We have a full toolbox; the fundamental challenge is pace. Now it's about speed and scale. What we have on our hands is a race against the clock. So we can't allow ourselves to get caught up in loops of analyzing and diagnosing the problems, raging at the culprits and the long odds. Instead, we must slide into a Möbius strip that loops between making plans and implementing them, circling in the virtuous orbit of solutions.

In most industries, pursuing climate solutions means there is a lot to undo, rethink, and redesign. But where whole new industries are emerging, we have the chance to start somewhat from scratch and avoid disastrous mistakes of the past—the corporate loopholes, the mistreatment of workers, the absurd inequality, the environmental "externalities." So for this last interview in the book, we focus on one such new industry: ocean farming. Not of fish (which has lots of downsides), but of seaweeds like kelp, and of bivalves like oysters, mussels, clams, and scallops.

As an ocean person, I have a predilection toward ideas emerging from marine biomes, and I'm enamored with this new/old tool in our toolbox. Various permutations of ocean farming have been practiced by Indigenous peoples around the world for centuries if not millennia—from fishponds in Hawai'i to shellfish farming in British Columbia to seaweed farms in Japan. But as an economic sector in the U.S., it is still nascent.

Most of what I know about this industry and its enormous potential, I know via Bren Smith, former fisherman, current ocean farmer, and cofounder/co-executive director of GreenWave. I've been following Bren's work since we met at a food conference in 2016. We have since teamed up on shaping the Blue New Deal,[193] and, through Urban Ocean Lab, on policy recommendations for this burgeoning ocean farming sector. A few years ago I went from superfan to board member of GreenWave, whose mission is "to replicate and scale regenerative ocean farms to create jobs and protect the planet." Filter-feeding bivalves, photosynthesizing seaweeds, thriving coastal communities, sustainable seafood, green (blue!) jobs, bolstered biodiversity, natural fertilizer, carbon absorption . . . what's not to like? I'm in. What if we build this industry fairly and well? What if we get it right from the beginning?

Ayana: What is a <u>regenerative ocean farm</u>?

Bren: It's a pretty simple structure. It's anchors, lines, and buoys, and from there we grow a bunch of different species of seaweed and shellfish. It's replicable because it's pretty cheap to build. And that's important, because it's a way folks like me coming out of the fishing industry can shift from being extractive to growing things that breathe life back into the ocean. That's the underlying principle.

Ayana: You are super into polyculture. What species are in the mix at a regenerative ocean farm?

Bren: On my farm, and ocean farms all around the country, it's a mix of seaweeds (mainly kelp at this point), plus mussels and oysters. Some folks are doing clams and scallops. Mother Nature loves polyculture. If you grow mussels next to kelp, you get a 38% more productive kelp yield.[194] ✳

Ayana: Wow.

Bren: And if you surround an oyster farm with kelp, you get something called the "halo effect," where seawater acidification rates drop inside the farm.

Ayana: So those sweet little oysters can actually grow their shells.

----

193 Flip back to the Blue New Deal chapter if you skipped it because it seemed too wonky!

194 This seems to be because the water-filtering power of mussels makes the water more clear, which means more light gets to the down-current kelp.

Bren: Exactly. Polyculture and regenerative ag are not just for people in Brooklyn to love, they also make sense from a farmer's perspective. We're mimicking and stealing from Mother Nature.

Ayana: I don't think she minds. She's like, "You don't even need to cite your sources, please just do it."

Bren: She's been scolding and giving stern looks here for quite a while.

*   As we rethink the ocean as an agricultural space, the key is asking the ocean what we should grow. And when you ask, it says, "Why don't you grow things that don't swim away and you don't have to feed?" And the beautiful thing is, that's a good way to run a small business, because you don't need fresh water, you don't need fertilizer, and you don't need animal feed. It's low overhead. You're also ensuring you're a steward of the ocean and a piece in the puzzle of climate solutions.

Ayana: Where did this concept come from?

Bren: People around the world have been doing versions of this for hundreds of years; a lot of it builds on Indigenous practices. I didn't invent anything here. Anybody who tells you that they invented anything, don't believe them. Individual inspiration is a total myth. It's about borrowing and stealing, it's about networks, it's about learning together. It's an organic, collective process.

Ayana: Iterating, improving, the trial and error of it all. Why is regenerative ocean farming an effective climate solution?

*   **Bren:** Three reasons. One, it's regenerative. Shellfish and seaweeds breathe life back into the ocean. Kelp is the sequoia of the sea—it soaks up carbon and nitrogen, provides habitat, supports biodiversity, and produces oxygen. That soaking-up of carbon through photosynthesis helps address acidification. And what we're doing as farmers is creating structure underwater, a place for little critters to come hide and thrive. We're planting it and harvesting it, but the kelp and oysters are also sending their seed and spores out into the environment, replenishing the area around it.

Two, it's replicable. That isn't something you usually think about when it comes to climate solutions. You usually think of environmental benefit, but replicability is important. Twenty acres of ocean, a boat, and $40,000 to $50,000 is all you need to start a farm. That means people all around the country can build their own farms, and this is already happening.

Three, it's scalable. A World Bank study says that if you farm not even 5% of U.S. territorial waters, you could create millions of jobs, absorb the carbon equivalent of 20 million cars, and produce protein equivalent to 2 trillion hamburgers. So we're not stuck in this small, beautiful model.

Ayana: It's not a climate solution if it's just a tiny, niche thing.

Bren: Exactly. The reason I'm confident in regenerative ocean farming is the climate winds are at our back. Whoever's doing zero-input food, whoever's not using fresh water, fertilizer, or even land, is gonna have the most affordable food on the planet at some point, because the price of everything is going up. The climate economy is gonna push our zero-input food to the center of the plate.

Ayana: Did you have a specific moment when you realized the work you're doing or the challenge you're facing is actually about climate?

Bren: For many years I made my living chasing wild fish. But we, humans, got too good at it—that model of large fleets chasing fewer and fewer fish, and going after just a couple kinds of fish.[195] I witnessed the collapse of the cod fishery and thought, "Oh my God, my life is governed by the health of the oceans and there aren't going to be any jobs on a dead ocean." That was my 1990s experience.

Then I went to work at the salmon farms—with the feed required, and the disease from growing them in such high densities, and all the fish shit, it's not sustainable. And we don't want to just shift from wild harvesting to farming those same few species.

So I remade myself as an oyster man. And then I got hit by Hurricane Irene and then by Hurricane Sandy. After the first one, you're like, "Ah, you know, hurricanes happen." Two years in a row, you're like, "Whoa, this isn't some slow lobster boil. Climate change is here now. We've gotta deal with it." I thought climate change might drive me off the ocean, into the hellish world of all the terrible things I've had to do on land in my life, and I'd end up in deep, deep depression.

At that point, it became clear that if you're going to farm the ocean, you want to be a willow, not an oak. When I first built this farm, I had

---

195 This overfishing emits over 200 million tons of CO_2 annually, and is largely enabled by $22 billion in government subsidies every year, which the UN says must be eliminated. I agree.

these big buoys, and a storm came in and just lifted the farm and moved it. And I thought, wait a second, this thing's gotta bend. So now, the buoys are smaller, and when the storms come through, the buoys sway underwater and then pop back up.

Ayana: Oh, Bren, I wish everyone could see your dance moves right now. Your buoy-impersonation interpretive dance is A+.

Bren: I look like the inflatable tube man.

Ayana: Haha, you totally do.

Bren: And polyculture is about resilience too. It enables farmers to have a crop for all seasons. Folks can be out there working year round, not just a three- or four-month period. And there will be seasons where things fail and I need to make sure there's another crop on deck that will keep me going. If kelp fails, I have oysters and mussels.

Ayana: So this is not just a climate and ecological solution, it's an economic one, too. In the U.S., the effort to build a significant regenerative ocean farming sector is new, but growing quickly. What would this industry look like in 2050 if we build it in the best, most resilient way?

Bren: The biggest question is, What does scale look like? I think there are three different ways we can scale, and we can do them all at the same time.

One is what I call the reef model, where you have twenty-five to fifty farms in a region. You've got a shared processing hub, a hatchery, and then rings of entrepreneurs and value-added producers. Then you replicate that up and down the coast.

The second is a full reimagining of marine protected areas into blue carbon zones. We need people at work in these zones, breathing life back into the ocean, growing shellfish and seaweeds. We need a just transition for fishermen so they're actually working in the zones. There would be underwater reforestation with kelp, so you could have artisanal fisheries and ecotourism because fish are attracted to those areas. That's the marine protected area of the future, to my mind, rather than a Teddy Roosevelt form of conservation, which isn't gonna work.

Ayana: My take on that is that fully protected parks are important, but static conservation, like drawing a line around a place and keeping everyone out, is no longer enough as *the* conservation solution. In a lot of places, we need to protect *and* restore and adapt.

Bren: And we can't leave people out of the mix. You can't do half a climate solution—draw down carbon but don't lift up communities. The beautiful thing about this moment is we can link both. And, quite honestly, if our strategy is just to draw down carbon, we're never going to get the policy and politics right, unless people benefit and see themselves in these solutions. ✱

The third way to scale is blue economy zones. Let's farm on offshore wind farms. When I see a wind turbine, I see an amazing anchor.

Ayana: It's a lot of infrastructure you could grow seaweed and shellfish on.

Bren: Right. And you can share boats, you can share workers. You can share working waterfronts and all this other infrastructure.

Ayana: You mentioned the vision of having millions of people working in this sector, and GreenWave itself has a goal of training 10,000 regenerative ocean farmers by 2030.

Bren: When we first launched GreenWave's Farmer Training program, more than 8,000 people signed up. Since we couldn't reach them all with our hands-on training, we built an online Ocean Farming Hub, and that's now been able to reach almost 7,000 people with our learning network and farmer trainings. We also had over 2,000 people sign up for our six-week Kelp 101 course. That interest, that's not about kelp. That's about building a better future. That's about agency. It's about having a self-directed life, being a regular person and jumping into this solutions space.

Ayana: It's so cool that there is so much interest. Especially since this farming is not super easy. There's a lot of problem-solving, there's a certain rigor to it. What does it take to become a successful ocean farmer?

Bren: Boat handling, boat safety, and other maritime skills are key. And then you need to know how to stage growth.

Ayana: You mean, the business side of scaling a farming operation?

Bren: Yeah, so many people come in like, "Okay, I'm gonna grow this stuff." But first, you gotta ask the ocean. It sounds crunchy and hippie-ish, but it's true. You need to put a couple lines in the water and see what grows. And then build your business around that, modulating your expectations and your plan. Start small and gather information about what kind of business partner your patch of ocean's going to be. And then quickly scale from there.

It's hard to grow food underwater. You have no control over what the seawater does. You can't see what you grow. But you counteract that through shared knowledge, through doing blue-collar innovation. To deal with volatility, and to develop a blue thumb, you need knowledge networks. GreenWave does a lot of farmer-to-farmer exchange. The community is so key. The ocean changes day to day, year to year. So farmers need folks they can ask questions about what's happening on their farms. Even more than going through a training program, having a national or a local group, where you can be constantly getting feedback so that you can correct, self-correct, is way more important than any formalized curriculum.

Ayana: Where is regenerative ocean farming taking off so far?

Bren: There's a bevy of activity throughout New England and New York. The Maritimes, up in Canada, has a lot of activity, and also British Columbia and Alaska. The Pacific Northwest is in formation. And we're seeing so much activity globally now—in New Zealand, Australia, Scotland, Greenland, Iceland.

In New England, where I'm based, there are about forty seaweed and polyculture farms. There are probably a hundred or more shellfish farms that can be diversified. We have the marine protected area offshore in New England, which is highly controversial with fishermen. So let's invite 'em in and turn them into re-foresters, right? Rewilding can be part of the new fishing. And then there's a wind farm off of Rhode Island, and there have already been discussions with those wind farm companies about embedding regenerative farming on there.

When I started, there were a handful of us doing this. Now there's so much activity. The network impact has been stunning. My job is kind of done, right? I can go back and farm. My harvest actually starts tomorrow.

Ayana: No way, not done yet. And you are not just *hoping* for this network effect; GreenWave is deliberately nurturing that network and that information exchange in a bunch of ways.

Bren: Yeah, we've dialed in on four areas of programming. One is farmer training and support. That includes the online Ocean Farming Hub, in-person meetings around the country, harvest workshops, and the community, the learning network.

The hub is the seed-to-market curriculum and training program—eighty how-to videos, everything from knot-tying, to gear lists, to getting a permit, to how to start a community-supported fishery, a CSF. It's everything I wish I had twenty years ago when I started.

Ayana: Amazing. Full toolkit plus community.

Bren: The second piece is our Kelp Climate Fund, which is the subsidy to pay farmers to plant more kelp and allow us to collect data on impact.

The third piece is market development, and that's where we're working directly with buyers. Right now, we have over sixty companies in our network. Through our Seaweed Source digital platform, farmers and buyers can initiate forward contracts. We believe that farmers' crops should be sold before they grow.

And then the last piece is farmer-owned infrastructure, working with farmers so they're able to supply their own seed and processing capability. This is a huge challenge. Right now, kelp needs to be processed within twenty-four hours of harvest. Stabilization and processing that can extend that and enable you to harvest all your kelp and then sell it over, say, an eight-month period is important for farmers. They can own part of that value chain to get a better price, and have the ability to stabilize and not have that chaos.

Ayana: More like apples.

Bren: Yes, exactly.

Ayana: Big picture, once this industry is built, what's the journey—from kelp seed through harvest to processing to final use—when we get it right?

Bren: The most important thing is to keep community and people and climate together in the solutions. If we get it right, we'll have many farms ✱ dotting our coast. Many farmers will own their own farms, but also own seed banks, nurseries, and processing facilities in a horizontal development, where everybody's benefiting. And all this would be grounded in farmer co-ops that are embedded in local communities. Then further up the chain, farmers are selling to larger companies, companies that are creating plastic alternatives and foods and fertilizers and things like that. There's a real opportunity for farmers to have a lot of ownership in this sector.

Ayana: It's exciting to think about revitalization, about coastal communities making a good living from the ocean in a way that's a climate solution and bolsters a lot of these places that suffered, because of, say, the collapse of cod fishing. To eat wild animals at that scale will never be sustainable.[196] And waterfront access and infrastructure for the coastal economy have been lost in many places. The vision of New England fishing, with all these boats going out in the morning and coming in with a catch of the day for the local restaurants, is a niche exception these days, almost mythological. But regenerative ocean farming could be a way to restore coastal economies to thriving again. I feel like you can see clearly in your mind's eye what a reborn <u>working waterfront</u> might look like.

Bren: I can picture it because it's been taken from me. I've seen each piece disappear over the years, turned into hotels and maritime museums. It's a stunning direct relationship—the fewer boats, the more museums.

If we get it right, working waterfronts would be grounded in farmer-owned infrastructure. That's hatcheries, that's processing, that's offloading places. And then there would be a community college, a strong maritime school, infrastructure for engine repair, electric power stations for the boats, and processing hubs. Hopefully we're going to have millions of people doing this work. Every farmer out there growing seaweed and shellfish needs someone to repair their engine, someone to manufacture the moorings, to weave the rope, and on and on and on. These are the secondary jobs that are created.

I come from Newfoundland. I used to work in Alaska. Those waterfronts were vibrant places. They were busy. It was chaotic. That's what I want, that sort of chaos and bustle and the energy that comes with that, and people trained and working in different sectors. I go to fishing places and ports now, and it's so goddamn calm. It's like Teslas and people hold-

---

196 94% of global fish stocks are maximally exploited or overfished.

ing hands walking on the dock. Get rid of that. Let's drive those folks off the dock. They can go somewhere else.

One of the biggest challenges we're facing is all that waterfront property is locked up in private hands, and the places that aren't are so remote that it changes the economics totally in terms of bringing crops to market.

Ayana: Are there examples you look to as inspiration for this transition?

Bren: The offshore wind farm industry is incredible. They have training programs. They're in the process of formatting working waterfronts. They have the permitting figured out, they have the investment, the policy. New London, Connecticut, is a completely reimagined place because of all the wind turbines that are going to come out of this port.

And, instead, what we in regenerative ocean farming have done is a hatchery here, a farm here, a processor there, people fighting to get permits and then going to raise money. It's such a patchwork when it really needs to be systematic and strategic, where everybody's planning at once and then the funding comes in and they all move together. None of us can do it on our own. This is industrial planning and development at the end of the day.

Ayana: All this can only come to pass with successful development of the market, of the demand. We're going to eat all these oysters, obviously, but people in the U.S. don't eat much kelp. What are we gonna do with all this seaweed?

Bren: I don't want to be too pollyannaish about this. It's a challenge. GreenWave's got a market development program right now, and we're building these markets for a zero-waste policy—using everything we harvest and finding multiple uses of the seaweeds. The strategy includes specialty products, kelp mustards and stuff like that. That's all wonderful, but very boutique and the prices are too high. Then you have the kelp burgers, things like that, riding this wave of plant-based diets.

What I'm interested in now is what's in the kelp, because there are all these nutrients that the world needs that are in the ocean, and kelp gathers them up. For example, kelp is a stunning biostimulant for soil health and plant health.

Recently, I was at a regenerative agriculture conference of like 200 people, and I was the only ocean person there. And all the land farmers were saying, "We have a nutrient crisis. We can't get the nutrients we need.

* We have to use fertilizers. It's a real problem." And I was like, "Shit, we got 'em all in the ocean. You poured 'em into the ocean, I'll collect them with kelp. I'll sell it back to you."

Ayana: Yep, that over-applied fertilizer washes downstream and causes algae blooms and dead zones in the ocean.[197]

Bren: Creating a nutrient loop—harvesting all that phosphorus, nitrogen, carbon, and other micronutrients, and getting them into land-based agriculture—is an incredible market opportunity.

The power of seaweed to me is that we don't have to just eat it. We can take this regenerative crop and weave it into land-based agriculture, into animal feed and fertilizer. Seaweed is used for beekeeping. It can be used in pharmaceuticals and cosmetics. We can weave it through all these industries.

Ayana: And then there's the critical financing piece, which seems like it's not yet sorted.

Bren: Right now there are two kinds of funding, private capital (like venture capital) and government money for scientific research. What's not there is the support for farmers, for the folks who are doing all the work in the water. So, GreenWave thinks there needs to be a source of funding that helps farmers scale, and pays them for the environmental work they do.

We created the Kelp Climate Fund as a philanthropic subsidy to farmers. They get up to $25,000 a year based on how much kelp they plant. And they use that to weather the storms—if their farm gets wiped out, if a processor screws 'em and doesn't honor a forward contract, interruptions in the market. They've got something to fall back on, and also to scale, to buy more gear the next year. And in return, the farmers collect data and share it with GreenWave.

We're trying to build a culture of data within the farming community—something that I did not believe in ten years ago. When regulators and scientists used to come to my farm, I was supposed to keep all these rec-

---

197 Not to mention that industrial fertilizer production is energy-intensive and based on using fossil fuels, "natural" fossil gas in particular. Far from ideal.

ords, and right before they'd show up I would've had four different colored pens and my paperwork, and I'd fill out all the "data," just like random numbers.

Ayana: Oh, Bren, your air quotes around "data" are killing me.

Bren: But, it wasn't just me being an asshole. This is about science not being farmer-forward. It was science for science's sake.

Ayana: What would farmer-forward science look like?

Bren: We built a My Kelp app, with farmers, where they take pictures of the kelp every month, at three places on the farm, and then they take some weight measurements. As farmers, we know how to read kelp. You show me a piece of kelp, I can tell you by the coloration if it's starving. What's happening in the nutrients, the stipe length, what kind of biofouling, the bubbles on it, whether it needs more sunlight. So we're collecting as many pictures of kelp around the country as possible. We've sort of become the Facebook of kelp. We believe kelp is the best environmental sensor ever invented.

Ayana: You're reading the tea leaves, but they're kelp fronds.

Bren: Haha, exactly. And we can teach computers to see what I see when I see a piece of kelp. And that means we'll be able to measure impact. We'll know how much kelp is in the water, how much is coming out, and then be able to extrapolate local impact and national impact for climate benefit, as well as predict yields—all based on photos and not million-dollar sensors.

And I know this is going to sound silly, but we kind of need less science. Yes, science is absolutely core for us to stay ahead of the climate curve, for us to be better farmers and get good data. But every state is doing the same studies, and then there are national studies at the same time. Tens if not hundreds of millions of dollars are going to academic research institutions and not to farmers. And it takes a long time to do these studies.

We need fast science. We need to agree that kelp does good shit and get kelp in the water and then let the scientists haggle over how good that shit is.

Ayana: Doing that fast science should also provide data helpful for making the case to investors.

Bren: Sure, but the truth is that no agricultural system in the world makes money. It's a total myth. Name one agricultural country that doesn't subsidize their ag industry.

Ayana: That's the big, open secret. The subsidies are enormous and we need to start being thoughtful about *what* we subsidize.

* **Bren:** Everybody asks, "Are the farms profitable?" And they are, but that's not the right question. The question is, as we look at this whole ecosystem: What parts of the regenerative ocean farming system need to be nonprofit? Where are the real investment opportunities for private equity? Which are high-risk and need insurance and subsidies?

One of my concerns about where things are going is everyone's trying to shift the risk onto farmers. I was talking to an investor and he was like, "I'm into bioplastics and patenting seed." And I'm thinking, who's going to get the kelp in the water? Then, I realized, oh, this is the Cargill model. Big ag owns the grain silos, the tractors, the seed, and the markets. And the farmers farm. And society, the government, doles out the vast majority of subsidies to corporate agriculture. What an insane system that is.

Ayana: That's a bad model.

! **Bren:** And if we let the Cargills and Monsantos shape this, it's going to take that energy, that blue-collar innovation, that emergence of a new climate workforce, and it's gonna suck the air out of it. That's not even a political critique, I think it's a bad strategy to centralize.

Ayana: Yeah, that makes it more fragile.

Bren: That's a great word for it.

Everybody asked me why I founded a nonprofit. The last thing I ever wanted to do was be involved in a nonprofit. But part of resiliency is not creating a fragile model. To get thousands, millions of people doing this, we believe there needs to be this pre-competition stage, all of us working together, sharing knowledge to create . . . what's the opposite of fragility?

Ayana: Robustness, redundancy, interconnectedness . . .

* **Bren:** *That* is going to be key to viability and survival. Not building walls around solutions, but distributing them. That's a core principle, I think, to climate resiliency.[198]

---

198 As Bren put it in a GreenWave newsletter, it's "collaborate or die."

Ayana: What is the nerdiest, least sexy, most esoteric thing we need to do to have this vision of the future become reality?

Bren: Revenue-based financing.

Ayana: Ha! Okay, tell us more.

Bren: It's basically low-interest loans that get paid back when the farmer, or any business, begins making money. And that could take a hundred years or one year. It's a sharing of the risk between the folks who have the money and the folks who are doing the work. And the key is it's zero-collateral. So if the loan fails, if you can't pay it back, they don't go take people's boats and buildings. It's a particular kind of collaborative financial instrument that is in tune with farmers and the goal of making sure we all succeed.

Ayana: That seems reasonable and promising. There's also policy change needed to support this new industry. We literally need to write the rules. My understanding is there are some places where regenerative ocean farming isn't banned, per se, but there's no pathway for getting a permit, which leaves folks in this limbo. How would you describe the ideal policy framework?

Bren: We ask farmers to do too much. Farmers should be able to just be farmers. But they've gotta fix their engine, grow the kelp, run a social media feed, sell the kelp, and apply for these permits with thirty-eight pages of instructions. It's crazy. Policy can play a huge role here. It would look much more like we've seen in the evolution of wind farms—large batches of pre-permitted areas. We need organized permitting.

Crop insurance like they have on land is also absolutely key. And working waterfronts are absolutely key. As we retreat from our coast due to sea level rise, let's make sure coastal infrastructure remains able to service the folks working on the water.

And then there's other pieces. We are beginning to see the privatization of seed. Folks are starting to patent kelp seed already. There are applications in at the patent office right now, and these are our *friends* doing this. I can't fight that as a farmer. We need regulation. That's also why GreenWave is rushing to create farmer-owned seed banks.

We have to avoid privatizing the commons. We can't shift all the risk to the farmers so everyone else is making money but farmers are at the whims of storms and crop failures. And we can't allow vertical integration. Incentives and leasing structures need to ensure that it's not out-of-state owners, it's not like three guys at the top owning a thousand-acre farm and we all just work for them. Horizontal is key.

Ayana: We don't want a big monopoly in the ocean. There's more resilience, too, perhaps, in having hundreds of small farms instead of a few huge ones.

Bren: Exactly. When you think about monoculture and privatization, and how it's only a few folks at the top benefiting, in other industries that's a really depressing story. It's not a depressing story in our sector, because it hasn't happened yet. I feel so bad for everybody in other industries. They have to unravel all this bad shit before they can build. Starting from scratch is such a luxury.

Ayana: I know you're a big advocate of worker cooperatives, co-ops. How do you see that fitting in here?

★ **Bren:** We need to ground this in co-ops. I've been in a co-op for fifteen years. I grew up next to a fisherman's co-op, and they're essential. They're not just cute; co-ops are key in terms of power for farmers for a few reasons. One is it's legal to price fix in a co-op. You can actually say, "Okay, we're not going to sell for below this price." You can't do that if we're all individual farmers or individual businesses. We need to be like the oil cartels. We need to set that price.

Two, being able to group purchase. Sounds like a small thing, but it's huge. It's expensive for me to buy a hundred boxes for my shellfish. For us to buy a thousand boxes, it's cheap. So group purchasing is key. And then shared infrastructure, pooling capital together so that you own some of that value chain.

And then three is standards. The amazing thing about co-ops is there's a lot of peer pressure to make sure quality stays high. Because if one person's producing crappy shellfish or crappy seaweed, we all suffer because we're under a similar brand. GreenWave has partnered with Organic Valley, a billion-dollar dairy co-op in the U.S., to help us think through and work with ocean farmers to set this up the right way, and learn from their experience. It's exciting. We have a chance to build, plan, and start all these processes early.

Ayana: This is why I wanted to interview you. When we think about climate solutions, we're going to have to develop a bunch of new sectors and industries, which gives us a chance to get it right from the jump. We can learn lessons from the past and build something new that actually makes sense for this current era. As we do this, are there other key things we should make sure to include?

Bren: One is this idea of creating a pre-competitive space. In an early industry, when you're trying to build the foundation and you're trying to build the right way, the last thing you want everybody doing is trying things alone, making the same mistakes or trying to build walls around what they learn. Because we're only going to succeed in this together. Maybe someday we'll all be competing, but in this stage one, this is about collaboration, and farmers are embracing that.

The other thing is let's not miss the opportunities to weave economic ✱ and social justice into this industry. There are a whole bunch of people who did well in the industrial economy who have been completely left behind. That includes the White working class. They were sold a bill of goods about globalization that completely decimated their communities. That includes fishermen, fishing communities. And there are folks who were never, ever included and were systematically excluded from the benefits of the Industrial Revolution, and that's folks of color.

There's an opportunity here to create new politics, because when you ✱ get people working together, talking together, get people knowing each other's stories, you find common politics. There are Trump voters growing incredible amounts of seaweed who are doing more for the environment than Greenpeace staffers. That's a fascinating thing. And we've got White working-class fishermen going down and working with the Indigenous group Shinnecock Kelp Farmers Cooperative to set anchors. This is promising—we don't have to replicate this ridiculous political situation we have online.

Ayana: There's something so special about working together on a tangible solution that can help break down a lot of these other walls and divisions.

Bren: There is. Solidarity is created. We have to make sure everybody benefits. You know me, I'm not an environmentalist, I'm not a foodie.

But when I went to the Climate March with my boat, I looked around at people solarizing the hollers in Appalachia, and urban farmers in Detroit, and I was like, "Oh my God, these are my people; I belong here." They welcomed me in my half-feral self, while I didn't fit in culturally in the ocean conservation space or the food space. And that says something about the promise of the new politics of climate solutions.

Ayana: We need everyone.

Bren: We really do. And I'm not sure elites understand that they need us. I was in a funder meeting the other day with a board of folks and literally, one of the things that was said was, "There's three ways to analyze a solution. One is people, one is economic benefit, and one is environmental benefits." And then this person said, "I don't care about people at all. Millions of people are going to die. They die all the time. That's irrelevant."

Ayana: Wow.

Bren: Second, they said, "Economic development, yeah, that's important but all that really matters is improving environmental conditions and climate health." And I was like, "You just don't understand that you need us to get the politics. Your kids are gonna starve, burn, and die if that's your analysis of what's needed." It was stunning.

Ayana: In their bunker in New Zealand, who's going to bring them snacks? What are they thinking?

Bren: Exactly.

Ayana: You seem so animated and enlivened by getting to do real, tangible work in the world, problem-solving, and making solutions happen. Is there a particular feeling you get when you think about ocean farming and the possibilities? I know you're not a feelings-forward kind of dude, but maybe indulge us here.

Bren: This is actually complicated, and a hard question. And the reason is we can't lie to ourselves. I gave a presentation to a bunch of high school and college kids in Alaska, and I asked, "Okay, what's going to stop regenerative ocean farming from happening?" And they said, "This is gonna be taken over by the corporations. This is gonna just look like everything else. They're gonna trick us and none of us or our communities are gonna benefit." That was their analysis.

Ayana: Kids these days. Reasonably jaded.

Bren: They could well be right. So how do we tell a story to ourselves that acknowledges the harm we've done to the planet and to people, and is realistic about how hard it's going to be to unravel that, and how there's a glimmer of hope that we need to lean into and thrive around even if we don't succeed? I don't want to be lying to people that we're going to fix climate change—the world's going to look so different, people are going to suffer. We need bundles of solutions. Ocean farmers are gonna be part of it. All I want is to be on my boat and have that pride of working with ♥ the water and working with people in my community, shoulder to shoulder, to try to build a more beautiful and better place. But what I say to myself is, "It's okay if it doesn't work out. It was still a day worth living." Is this making sense?

Ayana: It makes perfect sense. And it's not a binary of complete success or total failure. I mean, yeah, it's gonna be hard and, yeah, there's a bunch of jerks who just wanna make a bunch of money and screw everybody over and don't actually care about people and think they can somehow techno-utopia our way into being saved or whatever. But if we can get 80% of the way there with solutions, or even 70% or, hell, even 60%, I'll take it. Because so many lives hang in the balance. And that's helpful to me to see it more as a spectrum of possible futures, with those increments really mattering.

Bren: Absolutely. I'm trying to foster my tolerance for ambiguity, the unknown, living with the hope of a better future. We gotta enjoy this journey. Because part of not knowing where it's gonna go is that every day we need solidarity and we need love for each other and we need to laugh ♥ and have joy. And that's not normally the way I speak. And I wouldn't have spoken that way years ago.

Ayana: No, you're so salty. I've *never* heard you talk about feelings before.

Bren: Climate has changed me a little bit, the way I feel about the world and want to interact with folks. Like the toxicity we see in our culture, folks attacking each other and trying to tear each other down. That's not how I wanna die. Let's live and die without regret.

Ayana: Yeah, you gotta try. "It's worth a shot" is kind of where I land on these things.

Bren: And we fucking destroyed the ocean 'cause we're really good at trying and pulling it off.

Ayana: Well, there's the cause for hope! Look how good we did at over-fishing when we put our minds to it.

Bren: Yeah, Jesus, I killed a lot of fish. But seriously, we're really good at what we do as humans, so let's repurpose it. And that's why we want an

✱ army of people tapping that blue-collar innovation, and getting the politics right, and making sure we empower communities. Because the answers are going to come out of those communities.

Ayana: What are the top three things you wish everyone knew about climate solutions?

✱ **Bren:** One, believing in climate change is not about believing climate change is real. It's about tackling it on a granular level. What do you need to change about your life? We have all these folks saying they're not in climate denial, but they are because everybody's flying in the same way. Everybody's investing in the same way. Everybody's eating in the same way.

♥   Speaking from the culture I grew up in, we're going to have to say goodbye to some things we hold close to our heart. For me, that's chasing fish, being on the high seas. That search and the hunt is hard for me to say goodbye to, but it is what I have to give up to make a better world for my kid.

    I suspect everybody on the planet is gonna have their own version of something like that. It's cultural things we're gonna have to give up, and those are the most painful. We need to look in and admit to ourselves that we are climate deniers and ask ourselves, "What do I need to change?"

Ayana: We need to live as though we understand this crisis is real.

✱ **Bren:** Two, there is <u>working-class environmentalism</u>. It exists, it's happening. It's not about the left wing and progressives and the cultural left. It's real and has been there for generations, and we need to honor it and bring it into the center.

    The third one is we don't know what our future is going to look like. We don't know what's gonna work and what won't. And we have to have a higher tolerance for risk of failure, failing and just continuing on. This happened in the New Deal. One of my favorite pieces of history is that Roosevelt once did one of his fireside-y things and said to everybody, "We're going to try a ton of stuff and a lot of it's going to fail, but the people have voted me in to try a lot of things." That's the moment we're in. God bless you for failing right now.

Ayana: You have such a strong and clear vision. Are there any main influences who have shaped it?

Bren: "Vision" is a tough word because I'm such a granular person. But I remember a rancher I was working with, and he wanted me to cut this steel cable. I was young and I went over and I tried for a while and then went back to him like, "Yeah, I can't cut it with this." He walked over, sawed for twenty-five minutes without talking. Went through the cable, dropped it, walked away. He was saying to me, "Your definition of trying is just ridiculous."

To Be of Use

Marge Piercy

The people I love the best
jump into work head first
without dallying in the shallows
and swim off with sure strokes almost out of sight.
They seem to become natives of that element,
the black sleek heads of seals
bouncing like half-submerged balls.

I love people who harness themselves, an ox to a heavy cart,
who pull like water buffalo, with massive patience,
who strain in the mud and the muck to move things forward,
who do what has to be done, again and again.

I want to be with people who submerge
in the task, who go into the fields to harvest
and work in a row and pass the bags along,
who are not parlor generals and field deserters
but move in a common rhythm
when the food must come in or the fire be put out.

The work of the world is common as mud.
Botched, it smears the hands, crumbles to dust.
But the thing worth doing well done
has a shape that satisfies, clean and evident.
Greek amphoras for wine or oil,
Hopi vases that held corn, are put in museums
but you know they were made to be used.
The pitcher cries for water to carry
and a person for work that is real.

Climate Oath

With Oana Stănescu

"Do no harm." In conversations within this book, the need for a new Hippocratic Oath in the face of the climate crisis was mentioned repeatedly. The original ancient Greek version was for doctors. Paola raised such an oath as critical for architects and designers. Régine mentioned the need for something equivalent for investors. Rhiana brought up doing no harm with respect to the pursuit of justice via policy, Bryan as it applies to justice via design, and Colette regarding renewable energy. And Jade spoke of the need for each of us to make bold commitments to changing our lifestyles. She used the word "devotions"—what a beautiful way to conceive of reshaping our roles. What do we want to be devoted to? What people, community, place, ways of living?

Based on that quorum of mentions, friend/architect/philosopher Oana Stănescu and I put our heads together to create such an oath. It is not solely focused on how we might best comport ourselves professionally. Although that is a powerful place to start, what this moment calls for is beyond any professional practice, and broadly harkens back to the precautionary principle.

This oath aspires to open up conversations, and to make space for celebration, commitment, and accountability. We need new rituals for our climate-changed world: What food and music, what locations and relationships, what aspects of ceremony might you want to incorporate into your devotions, into making or deepening your vow to be part of climate solutions? All this change swirling around us is a chance to consider: What do we want to take with us? What should we leave behind?

In developing what follows, we realized something critical must come before avoiding harm: committing to the collective. Asking "What can *we* do?" instead of "What can *I* do?" The climate crisis is not a one-to-one doctor-to-patient, doctor-as-singular-hero scenario. This is a human-to-humankind-to-all-of-life scenario. So, the imperative to "do no harm" is second on our list. This is an oath meant to be taken with others.

The original Hippocratic Oath starts with swearing to the healing gods; we instead choose elements of life on Earth we hold particularly dear. Substitute in those that reverberate deeply with you, those you would be mortified to let down and elated to make proud. We offer this model for you to make your own, and pass along:

> On the majesty of turquoise seas, and fireflies, and aspen trees,
> On the honor of our parents, our ancestors, and humans-to-come,
> On the wonders of laughter and sunshine,
> I make these devotions to climate solutions for my community
>   and for our magnificent planet:

First, move from "I" to "we."
> We will expand our sense of interdependence.
> We will rein in our sense of individualism.
> We will ask, "What should we do, together?"
> Survival is collective, our fates are intertwined.

Second, do no harm.
> We will restore and heal, not pollute and deplete.
> We will regenerate ecosystems and our own resolve.
> We will live lightly, as part of the Earth.
> Accountability, generosity, and sweetness.

Third, less is more.
> We will expand our creativity and contract our consumerism.
> We will conserve, and distinguish between needing and wanting.
> We will be gentle with our own imperfections and others'.
> There is such a thing as enough. Basta.[199]

Possibility exists.
> This is a world of our making.
> We can remake it, remix it, restore it, rebalance it.
> The path of least resistance is only one of many paths.
>
> I will be part of getting it right.
> We will be part of getting it right.

---

199 My father's favorite Italian/Spanish term.

The Joyous Work

From possibility to transformation, perhaps you can now not only see what the future could hold, but also see yourself held in it. We do need dreamers. Lean into your imagination. Don't look only at the list of problems—look also, right next to them, for the possibilities. Dream not only about the world we want to live in, but also how we can build it, together. Because, of course, we also need *doers*—billions of them. Be both. Dream, then make your dreams come true.

Throughout these essays and conversations, themes emerged about what's important for the journey ahead. Out with: maximizing short-term profits, lollygagging and delay, bankrolling climate pollution, the primacy of shareholders and private property, playing small, and individualism. In with: ecological regeneration, accountability, doing no harm, re-localizing, justice, and collectivism. Nature first. Community foremost. And with maximal swiftness.

Here I'll offer what I believe to be the sexiest word in the English language: <u>implementation</u>. Because, as you've seen, we already have most of the solutions we need. We, all of us, need to get to work, and to pick up the pace. Even though you probably didn't personally do much to cause the climate crisis, we need your help cleaning up this gargantuan mess. We need to build the biggest, strongest <u>team</u> possible to take on this gnarliest challenge humanity has ever faced. And we must do it with gusto.

Another word I love, that is perfect for this moment, is "<u>leaderful</u>." Social movements can be fragile when they're helmed by just a few prominent leaders. We need many leaders to accelerate the transformation from an extractive economy to a regenerative one. Maybe you are one of those leaders. And if you don't want to be out front, that's more than okay. It's about being useful and advancing solutions, not about being a hero or an influencer. Well . . . maybe be an influencer—not in the Instagram model way, but in leading by example. There is a soft power in how you spend your time and money, how you look out for each other, how you roll up your sleeves and make change.

What joyous work can you do to help address the climate crisis? How can you be of use? I am thrilled to say this is what I get asked about most these days (now surpassing "How fucked are we?"), as more and more eyes are wide open. So, let's revisit the climate action Venn diagram. Each person interviewed in this book is living at the heart of theirs. How soon, and how fully, will you find your way to yours?[200] Actually draw it out:

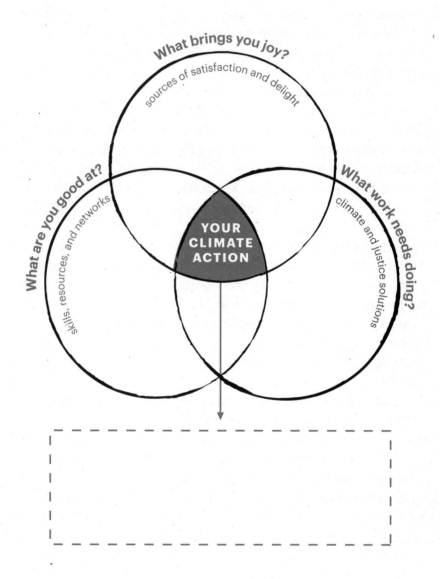

---

200 Visit climatevenn.info and @climatevenn for inspiration.

If we each find that sweet spot, that will create way more progress on climate solutions than if each of us gets sucked into obsessing about our individual carbon footprints. Starting now, reside at the intersection of ✱ these three circles for as many minutes of your life as you can. Choose something bigger than yourself, something that can have positive ripples. And don't just diagram your way to what you're already doing. Think about this expansively—start from scratch to consider where you can have the most impact.

Note: This is about *action*. I've noticed that many people draw their way to some version of "communicate about climate," but informing and being informed is not enough.[201] We need farmers and teachers and mutual aid organizers and green building retrofitters and wind turbine installers and nurses and accountants and engineers and designers and local politicians and project managers and train conductors and activists. Climate change has changed everything. We must remake the world.

This doesn't necessarily mean quit your job! Or go start a nonprofit.[202] Quite possibly you are most powerful within your existing roles, where you already have specialized knowledge and robust networks, where you have the most power to make significant change, through updating processes, protocols, and decision-making frameworks, through changing the ways of doing business. We need employees pushing their employers to change. What would it look like to lean into your talents? Can you help your company, school, church, or town charge ahead with climate solutions? Because what we need is change in every sector and in every community.

The thing at the center of your Venn diagram can change over time. (It certainly has for me.) And you can have multiple diagrams at once. (I take a portfolio approach to my work.) But keep going toward gaps you can fill. My wish for you is that you find your climate purpose—and encourage others to find theirs. And here's my advice, guidance I try to follow myself, for you to consider along the way:

---

201 Yes, I fully appreciate the irony, given that this is an entire book communicating about climate.

202 Truly, the last option should be to start your own nonprofit. A lack of nonprofits is not what's holding us back.

1. **Keep showing up.** This is not a spectator sport. It's not just billionaires and politicians who will decide our future—it's small business owners and students and citizens; it's whoever steps up and whoever you bring along with you. Take breaks but keep going. We shape the future.[203]

2. **Bring your superpowers.** Be gentle with yourself on the "What are you good at?" question. Put your insecurities aside and simply consider what you have to offer—in your personal life, professional life, and civic life. If we each harness our superpowers, *that* will actually enable the radical changes we need.

3. **Join something.** Contribute your skills to an existing effort—make it possible. Build the website, raise the funds, recruit the talent, plan the events. As Bill McKibben puts it, "Faced with the kind of crises that we face, the most important thing that an individual can do is to not always be an individual." Move from "I" to "we."

4. **Find your people.** Someone recently asked me, "Who are your people?" And without a moment of hesitation, directly from my soul came these words:

   My people are conjurers. They don't stop at dreaming, they make something where there was nothing, something needed. They make magic in the real world.

   It gives me goosebumps to think how lucky I am to get to say that sincerely. Support your people and love them, and hold on. Lean into possibility together.

5. **Be a problem solver.** What this moment in history requires is a relentless focus on solutions. Whether your purview is finance, energy, urban planning, manufacturing, construction, law, food, administration, or transportation, there is nothing more attractive than a problem solver.

---

203 Special shoutout to all the realists and pessimists leaning into possibility and trying to build a better world.

6. **Choose your battles.** That's how my dad put it. My mom prefers the cornier, "Do your best and don't worry about the rest." And it's the best advice my parents have given me. Keep things in perspective by keeping at least one eye firmly on the future of life on Earth. Choose something to fight *for*, and, please, no friendly fire.

7. **Nourish joy.** There are so many things that need to be done—don't pick something that makes you miserable! It's imperative to avoid burnout, so choose what enlivens and energizes you. Take climate change seriously, but don't take yourself too seriously. The work can and should be gratifying and punctuated with joy.

8. **Love nature.** And remember that you are a part of it. I can't say it better than Rachel Carson: "Those who contemplate the beauty of the Earth find reserves of strength that will endure as long as life lasts."

Keep showing up. Join something.
Find your people. Bring your superpowers.
Be a problem solver. Choose your battles.
Nourish joy. Love nature.

Here's what I know: There are innumerable possible futures. I know that we each get some say in which future we'll collectively have, and a chance to help build it. I know that every tenth of a degree of warming we prevent, every centimeter of sea level rise we avoid, every increasingly unnatural disaster we avert, every species we save, every bit of nature we protect and restore, matters. I know that our efforts add up, and that our fates are intertwined. Averting climate catastrophe, this is the work of our lifetimes. So, go where there is need and where your heart can find a home. And when it all feels too much, return to this simple guiding question: What if we act as if we love the future?

Be <u>tenacious</u> on behalf of life on Earth.

Away from the Brink

If we get it right, the world is a lot more green, more full of life. Damaged ecosystems are on the mend. Cycles of water, carbon, and nitrogen are rebalanced. Beavers are admired for helping landscapes absorb water and diminish wildfires. Insects, a precious foundation of our food web, are rebuilding their numbers and doing their many jobs. Soils are accumulating instead of eroding; full of roots and bugs and microbes, they can resist droughts and deluges. We are fostering intact forests, plains, and watersheds—working at landscape scale, regardless of geopolitical borders. Half of nature, of each ecosystem, is protected from plunder in perpetuity. The success of the Land Back movement means Indigenous peoples are once again stewarding their ancestral lands, often in collaboration with local and national governments. The spewing of pollution and poisons, the proliferation of extinctions have ended. Regenerative organic farms are thriving; subsidies for toxic megafarms are done. We have replaced tenuous monocultures with resilient polycultures. Composting and mulching and cover crops are the new/old convention. Lawns are a thing of the past and golf courses are rewilded, replaced with Climate Victory Gardens and diverse native flora. Pollinators buzz. Life is luscious.

If we get it right, the combustion phase of humanity is over. We no longer burn things to make energy (except some hydrogen); we have electrified. Gen Z brought down the fossil fuel industry, ruthlessly calling out oil and gas corporations, mocking the destructive absurdity of their ways, and refusing, en masse, to work for them. The air is cleaner. The water is cleaner. Asthma and cancer are more rare again. Communities of color are no longer overburdened with pollution. There are no more oil spills. Turning forests into "biomass" pellets for energy is an absurd "remember when. . . ." Renewable energy projects can no longer be blocked by wealthy property owners. Wind farms and solar arrays and transmission infrastructure are aesthetically pleasing additions to our landscape. Even though energy is renewable, we conserve it—to minimize both the amount of metals mined and the amount of nature that is disrupted with panels and turbines. Batteries and solar panels are recycled and newly mined materials are justly sourced. We have figured out fusion.

If we get it right, we have re-localized and we eat well. We no longer eat Washington State apples in New York State and New York apples in Washington. Our food is fresh and flavorful, ample and accessible to all. We don't fight the seasons, but rather enjoy what the time and place offer. Our regionalized supply chains are more resilient to global shocks. We look back in horror at the age of shipping bottles of water from Fiji and France to stores all over the world. (But fear not, we are hydrated!) We've grown beyond "reduce, reuse, recycle" and now think first of re-fuse, repair, and repurpose. Every neighborhood has a repair shop. We value materials, from food scraps to metals to fabrics. Biodegradable is the norm, and composting is ubiquitous. Landfills aren't needed. Trash is out and circularity is in. Homemaking (preparing food, mending things, caretaking, and so on) is esteemed. Many people have moved nearby to their loved ones, relocating to more climate-safe places, re-gathering diasporas. Living is intergenerational, people with disabilities are cared for, and elders are cherished.

If we get it right, our homes are comfortable. Green roofs support biodiversity and keep us cool in heat waves. Heat pumps and solar panels and induction stoves are the norm. We shake our heads remembering how we once spewed fossil gas into our kitchens, ruining indoor air quality to cook. Buildings are all well-insulated, and are not overheated in winter and overcooled in summer. Construction materials are more locally and regionally sourced. We have found replacements for high-energy concrete and steel. Instead of demolishing, we retrofit or deconstruct—materials are salvaged for reuse, like taking LEGOs apart for the next project. We've got good insulation dialed. Energy bills are much lower; energy conservation is much higher. The future is not drafty.

If we get it right, there is no traffic in cities, because there are so few cars in cities. Thanks to extensive, elegant, and free public transit systems, and with much of what we need within walking or biking distance, personal vehicles are rarely needed. Many parking lots have been removed, replaced by community gardens, gathering places, and bicycle parking. Glorious systems of bike lanes have many people pedaling. Electric hydrofoil ferries are common. Maybe trolleys are a thing again. From semi-trucks to vintage sports cars, we figured out vehicle conversion, conserving their metal husks as we move beyond combustion engines. Wildlife corridors bridge roadways and railways. High-speed trains are the preferred mode for long journeys. Travelers arrive with a sense of place, having traversed ecosystems at eye level, at a pace that allows for digesting the geography. Air travel is both pollution-free and less common—our ample vacation days (and airships!) make it easy to abandon planes. Locomotion happens with fewer decibels. Stealthily silent electric motorcycles are the midlife-crisis transport of choice. There is a sailboat renaissance.

If we get it right, coastlines are greener too. Mangroves and wetlands and seagrasses and dune grasses have been replanted. So have oyster reefs—billions of oysters are thriving in harbors and bays, filtering seawater and lessening wave impacts. These salty ecosystems are soaking up carbon and protecting us from storms. Where this green infrastructure alone is not enough, we supplement with gray infrastructure (seawalls and such), but sparingly because we have helped coastal communities move in and up, out of harm's way. We show respect for the rising seas and have ceased trying to hold back the entire ocean, ceased the foolhardy rebuilding after each storm. Flood insurance premiums accurately reflect risks. No new buildings are constructed in flood zones. Coastlines are for working waterfronts and recreation. When coastal parks flood, there's a delayed soccer game, not destroyed homes. Hurricanes are less deadly, even though our changed climate means they are more intense. And on a clear day, if you squint at the horizon, miles in the distance you just might glimpse offshore wind turbines. Woosh.

If we get it right, the tyranny of the minority is over; the climate-concerned majority rules. Our democracies are robust and truly representative. Every vote is equal and counted and easily cast. Politicians have been wrested from the stranglehold of the fossil fuel industry and big ag. They caught up with popular opinion that we must do more, and faster, on climate. Subsidies for fossil fuels are long gone. The filibuster has been abolished. Climate-science deniers are unelectable, perceived as laughably unfit for office. Because some places are no longer habitable (too hot, too stormy, too wet, too dry), policies reflect the humanitarian necessity for people to migrate. Governments ensure that people have all the basics covered (food, housing, education, healthcare, childcare, and eldercare), so that we can roll with the climate punches. In tandem with investing in community services, incarceration has plummeted, freeing time, labor, budgets, and human potential. Ditto for hyper-militarization, with all its high-cost energy-intensiveness. There are no more petro-states enabling petro-dictators—decentralizing energy production decentralized political power. The gulf in pay between executives and workers has narrowed—all jobs pay a thriving wage. As both climate and technological change remade our world, we instituted a universal basic income.

If we get it right, there are fewer desk jobs. We are out in the world, remaking society together and (re)figuring out how to live within nature instead of on top of it. Through climate corps programs, governments around the world employ hundreds of millions for efforts like managing fire risks, deploying clean energy, and restoring coastal wetlands. Access to healthcare is universal, not tied to a job, freeing people to relocate away from areas prone to extreme weather and toward necessary and meaningful work. Green jobs are the status quo. The trades—like carpentry, plumbing, electrical, welding—are revered. Our biceps are buff. Vocational training is easily accessible, enabling people to build skills for the shifting job market and economy. A single income can support a family, allowing time for the work of adapting to all the changes that are already here and those that are rushing toward us. Everyone has a role to play, contributing to a society powered by wind and sun and waves, and an ever-accelerating resurgence of nature. Electricians do very well on dating apps.

If we get it right, the pace of life is more humane. Time that had been spent dealing with health- and flood-insurance paperwork, advocating for renewable energy, being stuck in traffic, and otherwise butting up against outdated and broken systems, is now used to grow food, prepare for extreme weather events, and care for each other. All streets, not just those in wealthier areas, are lined with trees (including fruit and nut trees), providing shade and beauty and photosynthesis (and snacks). Rain gardens and bioswales line streets, ready to absorb and divert storm waters. We linger outside, in parks and on sidewalks, with friends and neighbors. We have time to make meals at home or consume them at a cafe. "To-go" is uncommon, instead we meet eyes as we chew. We know plastic recycling is mostly bullshit and have abandoned disposables—instead we (gasp!) wash the dishes. No longer frenzied with meaningless to-dos, we find ease amidst the generational work of making our planet livable. As we spend more time outside, our appreciation for nature grows with immersion, inspiring ever more creative adaptations to our changed climate. Biophilia and biomimicry flourish in a virtuous cycle with the thriving of biodiversity. We are unrushed, chill even.

If we get it right, culture has caught up with our climate reality. Hollywood and celebrities are all-in—climate is the context, the cachet is in solutions, and implementation is sexy. The obsession with endless economic growth is considered wildly out of touch. Corporations are beholden to the limits of nature, not to maximizing quarterly profits for shareholders. Our healthy retirement accounts invest in the just transition, not fossil fuel extraction. The big money is in zeroing out and drawing down carbon pollution. Mainstream media brims with actionable climate information. Social media is greener, too—an enticing scroll of ecological and electrification projects, luring more and more people into this important work. We share snazzy retrofitting hacks and extreme weather prep tips. We help our neighbors. Influencers get canceled for hawking shit we don't need. Consumerism is so uncool. A victorious campaign for clothing durability requirements ended fast fashion. Fabrics are no longer made from fossil fuels. Artificial intelligence is reined in, and its computing power helps to optimize energy efficiency, maintain electricity grids, and generally advance climate solutions—and to free us from administrative tasks. Design is planet-centric, not human-centered.

If we get it right, even with all these changes, the world and our homes are still familiar. It's the invisible systems that undergird our lives—our energy, food, and transit systems; our methods of engineering, manufacturing, and building; and the policies that organize it all—that have been overhauled. Re-greening the world and re-rooting in communities feels good. There is more collaboration, more knowing and being known, more sweetness. We lean into trial and error, into replication more than scaling. The world is a mashup of traditional and high-tech, old ways made new. The extractive, fossil-fuel economy is out; the renewable, re-generative economy is in. Humanity has backed off, made more room for other species, learned to share this magnificent planet. Our surround-ings are verdant. Spring is not silent; it's cacophonous. We are putting the pieces back together, adapting to the climate-changed world with eyes and hearts open wide. We embrace possibility, continually moving away from the brink and toward answers to the grand question: What if we get it right?

We need joy as we need air. We need love as we need water. We need each other as we need the earth we share.

—Maya Angelou

Behind the Scenes

It takes a village. Here is the core team around me and behind the book, the magnificent humans making edits in the documents with me.

Jenisha Shrestha, my wondrously multi-talented chief of staff, was the research assistant for this book, and contributed a sharp editing eye. She put her master's degree in environmental science and policy to very good use going down gnarly rabbit holes to draft the lists of problems and possibilities. Jenisha stood with me in the muck of creating and keeps an unfathomable number of balls in the air.

Chris Jackson is editor of this book and editor-in-chief of One World. After attending a few of the Science & Society events I host at Pioneer Works in Brooklyn, he said: *This* is the book—the thing you do that is special is bring to us people we should be listening to, put them in conversation, and help us see what they see and grapple with what that means for us and the world. Without Chris this book would not exist.

Rachel Waldholz is an audio climate journalist who helped me prepare for the interviews within this book. We were colleagues on the *How to Save a Planet* podcast, and she is now climate editor for local news at NPR. Rachel interviewed me to pull the "Away from the Brink" chapter out of me when I was stuck answering my own big question. She's also a whiz at naming things—late-night-text-message-chapter-title brainstorms to the rescue.

Simon Sullivan is the VP and director of art and design at Random House. He was the interior designer for *All We Can Save*, and it was a pleasure to work directly with him to shape the layout and style of this book. His patience with my attention to details is noteworthy.

Greg Mollica is the art director for Random House, One World, and Roc Lit 101. Collaborating with him on the cover design, from my sketch with markers to the final that you see, was a meandering journey for which I cannot imagine a better companion.

Rachel Ward is the audio producer who ensured that we got good sound of the interviews ("Ayana, are you recording?"), helped come up with interview questions, and nudged me to ask key follow-ups. Thank goodness for her work making sure everything was faithfully transcribed.

James Gaines is a science journalist and fact-checker living in Seattle, Washington. He grew up in a cabin in the woods and, as the child of two librarians, loves a good footnote. He has helped ensure the correctness of these fact-filled pages, and any lingering errors you may find should be attributed to my incessant rewriting after his checking was completed.

Louise Maher-Johnson is my mom, and my first and always editor. She has been dreaming about and living into this title question for decades. She was a high school English teacher for thirty-seven years in NYC inner-city schools before moving upstate to devote herself to creating a regenerative farm homestead, a model for getting it right in our climate-changed world. She planted many of the seeds that sprout in these pages, through the books she has passed to me, our kitchen-table conversations, and the example of her life. She grew me. I love her more than anyone, and she loves me slightly more than her chickens.

CONTRIBUTOR BIOS

Abigail Dillen, JD, is the president of Earthjustice, which represents more than five hundred clients free of charge, harnessing the power of law to force climate solutions while protecting healthy communities and ecosystems. Abbie has litigated precedent-setting cases that have held polluters accountable and cleared the way for clean energy nationally. • earthjustice.org

Adam McKay is an Academy Award–winning writer, director, and producer of films including *Don't Look Up*, *The Big Short*, *Vice*, and *Anchorman*. He is an executive producer of TV shows, including *Succession*, was head writer at *Saturday Night Live*, founded Yellow Dot Studios, and is on the board of Climate Emergency Fund, where he works to get much-needed funds and resources into the hands of climate activists committed to civil disobedience. • yellowdotstudios.com

Ayisha Siddiqa is a Pakistani human rights and climate defender, co-founder of Fossil Free University, and climate advisor to the UN Secretary-General. Ayisha's work focuses on uplifting the rights of marginalized communities while holding polluting companies accountable. She has led campaigns for the creation of the Loss and Damage Fund and for a conflict of interest policy on Big Oil at the UNFCCC, and is working on the Rights of Future Generations.

Bill McKibben is an author, environmentalist, and activist. In 1989, he published *The End of Nature*, the first book for a common audience about global warming. Bill is founder of the nonprofit Third Act, which organizes people over the age of sixty for action on climate and justice, and a co-founder and senior advisor emeritus at 350.org, a climate campaign that works in 188 countries. • billmckibben.com • thirdact.org

Bren Smith is a former commercial fisherman turned ocean farmer who pioneered the development of regenerative ocean farming. Born and raised in Newfoundland, he left high school at the age of fourteen to work on fishing boats from the Grand Banks to the Bering Sea. He is the owner of Thimble Island Ocean Farm and executive director of the nonprofit GreenWave, which trains new ocean farmers. • greenwave.org

Brian Donahue and his family co-own and operate Bascom Hollow Farm in Gill, Massachusetts, which raises pastured beef and pork, pumpkins, and timber. He is professor emeritus at Brandeis University, where he was chair of the Environmental Studies program. Brian co-founded and directed Land's Sake, a community farm in Weston, Massachusetts, and was director of education at The Land Institute in Salina, Kansas.

Bryan C. Lee Jr. is an architect and advocate based in New Orleans who established the contemporary design justice movement. He is founder and principal of Colloqate Design, a nonprofit architecture and urban planning practice focused on expanding the field of design justice. He is a design critic at the Harvard Graduate School of Design and 2025–2026 president of the National Organization of Minority Architects. • colloqate.org

Colette Pichon Battle, JD, is an attorney, climate justice organizer, and a generational native of Bayou Liberty, Louisiana. She serves as the Vision and Initiatives partner for the nonprofit Taproot Earth, which she co-founded. Colette spearheads efforts around equitable climate-disaster recovery, migration, community economic development, and energy democracy. • taproot.earth

Erica Deeman is a visual artist and educator who considers the liminal and transitory spaces in which Black identities are formed. Deeman is concerned with the multiple ways selfhood manifests through queer, transnational, and hybrid modes; and how we find a sense of belonging and "home" through migratory patterns, memory, personal biography, and ancestral legacy. • ericadeeman.com

Franklin Leonard is a film and television producer, cultural commentator, and entrepreneur. He is the founder and CEO of the Black List, a company that celebrates and supports great screenwriting and the writers who do it via film production, its annual survey of best unproduced screenplays, an online marketplace, live staged script readings, screenwriter labs, and film-culture publications. • franklinleonard.com • blcklst.com

Jacqueline Woodson is an American writer of books for adults, children, and adolescents, best known for her National Book Award–winning memoir *Brown Girl Dreaming* and her Newbery Honor–winning titles *After Tupac and D Foster*, *Feathers*, and *Show Way*. She has served as the Young People's Poet Laureate and the National Ambassador for Young People's Literature. She has been awarded the Hans Christian Andersen Medal and named a MacArthur Fellow. • jacquelinewoodson.com

Jade Begay, of Tesuque Pueblo and the Diné Nation, works at the intersections of climate and environmental justice policy and Indigenous rights. Jade has worked with Indigenous-led organizations and Tribes from the Amazon to the Arctic to advance Indigenous-led solutions and self-determination through advocacy campaigns, research, storytelling, and narrative strategies. She is currently developing a research project on the impacts of climate change on Native people in the United States. • jadebegay.com

Jean Flemma is an expert in ocean and climate policy and the politics of achieving policy change. She is a co-founder of Urban Ocean Lab and an advisor for nonprofits and foundations. Jean spent two decades advising members of the U.S. Congress on the conservation and sustainable management of ocean and other natural resources. • urbanoceanlab.org

Jigar Shah is a clean-energy entrepreneur serving as director of the Loan Programs Office at the U.S. Department of Energy. Previously, he was co-founder and president at Generate Capital. He also founded SunEdison, a company that pioneered "pay as you save" solar financing, and served as the founding CEO of the Carbon War Room, a global non-profit that helps entrepreneurs address climate change.

Judith D. Schwartz is a Vermont-based author and journalist who tells stories to explore and illuminate scientific concepts and cultural nuance.

She takes a clear-eyed look at global environmental, economic, and social challenges, and finds insights and solutions in natural systems. Whatever she's doing, she will stop to listen to the song of the hermit thrush. • judithdschwartz.com

K. Corley Kenna is VP of communications and policy at Patagonia, where she is responsible for the development and execution of Patagonia's global communication strategy and the advancement of key policy issues related to the company's business and advocacy priorities. She has over twenty years of experience working in government, politics, and the private sector. • patagonia.com/actionworks

Kate Marvel, PhD, is a climate scientist and writer living in New York City. Her first book, *Human Nature,* will be published by Ecco Press in 2025. Kate has a doctorate in astrophysics, so she knows Earth is the best place in the entire universe. • marvelclimate.com

Kate Orff, RLA, is a landscape architect focused on climate adaptation and biodiversity in the built environment. She is the founder and principal of the landscape architecture practice SCAPE, which pioneered a joint approach to community engagement and ecological infrastructure. Kate is also a professor at Columbia University with a joint appointment in the School of Architecture and the Climate School. • scapestudio.com

Kelly Sims Gallagher, PhD, is dean *ad interim* and professor of energy and environmental policy at The Fletcher School, Tufts University. She directs the Climate Policy Lab and the Center for International Environment and Resource Policy at Fletcher. She works on low-carbon, climate-resilient growth strategies in major emerging economies around the world. • climatepolicylab.org

Kendra Pierre-Louis is an award-winning climate reporter currently on staff at *Bloomberg.* Previously, she reported for the Gimlet/Spotify podcast *How to Save a Planet, The New York Times,* and *Popular Science* (*Pop-Sci*), where she wrote about science, the environment, and, occasionally, mayonnaise. • kendrawrites.com

Leah Penniman is a Black Kreyol farmer, mother, soil nerd, author, and food justice activist who founded Soul Fire Farm with the mission to end racism in the food system and reclaim our ancestral connection to land. She facilitates food sovereignty programs including farmer trainings, food distribution, and organizing toward equity in the food system. • soulfirefarm.org

Marge Piercy is the author of nineteen volumes of poetry and seventeen novels, including the *New York Times* bestseller *Gone to Soldiers*. Her newest book of poetry is *On the Way Out, Turn Off the Light*. Born in center-city Detroit, educated at the University of Michigan and Northwestern University, the recipient of four honorary doctorates, she is active in antiwar, feminist, and environmental causes. • margepiercy.com

Mustafa Suleyman is a founder, venture investor, and technology entrepreneur. He is the co-founder and CEO of Inflection AI, a new startup that creates Pi, a personal AI. He previously worked at Google as VP of AI Products and AI Policy. Before that, he co-founded DeepMind, which was bought by Google in 2014. • mustafa-suleyman.ai

Oana Stănescu is a Romanian architect, curator, writer, and educator based in New York and Berlin. Her studio's work is an undisciplined collection of projects across scales, programs, and collaborators, always in the pursuit of the affect of physical space. She is co-founder of the Blueprints of Justice studio at MIT and curator of the 2024 Architecture Biennial Beta Timișoara. • oanas.net

Olalekan Jeyifous examines contemporary socio-political, cultural, and environmental realities through the tradition of architectural utopianism, from a sci-fi-inspired and Afro-surrealist perspective. Olalekan regularly combines digital illustration and 3D computer models with photographs, hand drawings, and collage to produce detailed illustrations, photomontages, VR experiences, adaptable and mobile architectural sculptures, and large-scale public installations that examine architecture and its relationship to place and community through an eco-political lens. • jeyifo.us

Paola Antonelli is senior curator in the department of architecture and design of the Museum of Modern Art, where she also serves as the founding director of research and development and as a member of the Web3 taskforce. From Milan, Italy, Paola is a trained architect who describes herself as a "*pasionaria* of design." • instagram.com/design .emergency

Régine Clément is a Brooklyn-based entrepreneur who advances climate and sustainability solutions. She is co-founder and CEO of CREO Syndicate, a think-and-do tank that catalyzes private capital into the decarbonization transition. Under Régine's leadership, CREO Syndicate has become a global knowledge center for climate investing, and, in partnership with a community of family offices, family foundations, and family-controlled businesses, is facilitating investment in the climate marketplace, de-risking solutions, and decarbonizing their operations. • creosyndicate.org

Rhiana Gunn-Wright is a policy expert and Chicago native who helped to develop the Green New Deal. She is a senior fellow and resident author with the Roosevelt Society, and was previously director of climate policy at the Roosevelt Institute, and a 2013 Rhodes Scholar. Rhiana aims to create, in collaboration, a body of work that examines the role of economic policy and large-scale economic transformation in catalyzing just and rapid responses to the climate crisis. • rooseveltinstitute.org

Samantha Montano, PhD, is an assistant professor of emergency management at Massachusetts Maritime Academy. She primarily studies nonprofits, volunteerism, and informal aid efforts in disaster. She is the co-founder of Disaster Researchers for Justice. • disaster-ology.com

Steve Connell is an acclaimed spoken-word artist who believes story is an act of survival. He is a two-time National Poetry Champion whose work explores democracy and technology, climate and racial justice, family and grief, purpose and possibility, and, of course, love. In addition to registering more than a million voters while touring with the Declaration of Independence, he is Ayana's favorite cousin. • steveconnellcreates.com

Wendell Berry is an American novelist, poet, essayist, environmental activist, cultural critic, teacher, and farmer. He is the author of more than fifty books of poetry, essays, and fiction. For fifty-nine years, he has been farming and writing with his wife at Lane's Landing Farm in Kentucky. As a prominent defender of agrarian values and supporter of the local economy, he sees industrial farming, strip-mining, and war as opposite to the health of the world.

Xiye Bastida is a climate justice activist from an Otomi-Toltec Indigenous community located in Central Mexico. She is co-founder and executive director of the international youth-led organization Re-Earth Initiative, and a recent graduate of the University of Pennsylvania. Since 2018, she has become a leading speaker, organizer, and author who is driven to make the climate movement more inclusive and diverse. • xiyebeara.com

ACKNOWLEDGMENTS

To all those whose wisdom and wit and research fills these pages, thank you for your vision and generosity, for illuminating our themes and the way, and for bringing us in and along. (Note: citations are at getitright.earth.) To Jenisha Shrestha, Chris Jackson, and the whole ace book team (see page 436), heaps of gratitude.

Thank you to those who took care of my heart ♥ so I could create this: Kate Gage, Nathan Golon, Heidi Hackemer, Frank Heidinger, Julia Ioffe, Boris Khentov, Franklin Leonard, Nadia Nascimento, the Seavers, Oana Stănescu, Sarah Stillman, Amber Tamblyn, and Akaya Windwood. Shoutout to the group texts.

Thank you to those who have shaped my thinking over my lifetime and in various ways shaped this book, including: Zack Anchors, Emily Atkin, Melissa Bell, Mia Birdsong, Alex Blumberg, adrienne maree brown, Mona Chalabi, the Chouinards, Chrissy and Gotha, Marlene Clary, Chris Earney, Gabriel Florenz, Miriam Goldstein, Marce Gutiérrez-Graudiņš, dream hampton, Gabriela Hearst, Zach Henige, Jeremy Jackson, Eric King, Naomi Klein, Nancy Knowlton, Inger Kristiansen, Forest Lewis, Gina McCarthy, Anne Miltenburg, Noah Murphy-Reinhertz, Andrea Olshan, Varshini Prakash, Cameron Russell, Marquise Stillwell, Cyndi Stivers, Maggie Thomas, Carri Twigg, Elizabeth Warren, the team at Urban Ocean Lab, and my students at Bowdoin. Thank you to Katharine Wilkinson, whose collaboration expands my thinking and reminds me how much the heart matters in it all.

Thank you to the One World crew—shoutouts to Carla, Lulu, Sun, and Angie. To my team at CAA, especially Anthony Mattero and Kip Ludwig, my literary and speaking agents, and their assistants. To Sarah Lewis, for introducing me to editor Chris Jackson. To the folks at Pioneer Works, where the concept for this book emerged. Thank you to the Monarch, Joe and Clara Tsai, and Overbrook Foundations and Beth Braun for investing in this vision. To the Rockefeller Foundation, Natasha Ziff, Maine, and Mom for the quiet writing nooks ensconced in nature.

Thank you to all who write and speak and *act* on climate and justice, who have sparked or remixed my thoughts, who have asked me incisive questions, and whose ideas continue to reverberate within me. Thank you, reader, for not giving up on this magnificent planet and all who share it. Onward.

And turn to the very end—I made you a mixtape.

With love,

Ayana

CREDITS

Grateful acknowledgement to the following for permissions to reprint previously published material:

Wendell Berry: "A Vision" by Wendell Berry. Reprinted by permission of the poet.

Steve Connell: "This Living Earth" by Steve Connell. Reprinted by permission of the poet.

Alfred A. Knopf, an imprint of the Knopf Doubleday Publishing Group, a division of Penguin Random House LLC: "To be of use" from *Circles on the Water* by Marge Piercy, copyright © 1982 by Middlemarsh, Inc. Used by permission of Alfred A. Knopf, an imprint of the Knopf Doubleday Publishing Group, a division of Penguin Random House LLC.

Ayisha Siddiqa: "On Another Panel About Climate, They Ask Me to Sell the Future and All I've Got is a Love Poem" by Ayisha Siddiqa. Reprinted by permission of the poet.

Jacqueline Woodson: "Dear Future Ones" by Jacqueline Woodson. Reprinted by permission of the poet.

———

This book also draws upon these previously published works from the author:

"Ayana Elizabeth Johnson—What If We Get This Right?," *On Being* with Krista Tippett, June 9, 2022.

"Be Tenacious on Behalf of Life on Earth," *Time*, June 2, 2023.

"After COP27, a Plea—and a Pathway—for Urgent Corporate Action," *Atmos*, December 1, 2022.

"Why Our Secret Weapon Against the Climate Crisis Could Be Humour"(co-authored with Adam McKay), *The Guardian*, January 13, 2022.

"There Is Nothing Naïve About Moral Clarity," *Medium*, September 27, 2019.

"The Big Blue Gap in the Green New Deal" (co-authored with Chad Nelsen and Bren Smith), *Grist*, July 15, 2019.

"An Origin Story of the Blue New Deal," *How to Save a Planet* podcast, Gimlet Media, June 27, 2021.

"What I Know About the Ocean," in *Black Futures*, edited by Kimberly Drew and Jenna Wortham (New York: One World, 2020).

"Our Hero: Trying to Address the Climate Crisis Without the Ocean Will Not Work," Patagonia, May 31, 2023.

"In a Time of Hurricanes, We Must Talk About Environmental Conservation," *Scientific American*, October 2, 2017.

"I'm a Black Climate Expert. Racism Derails Our Efforts to Save the Planet," *The Washington Post*, June 3, 2020.

"We Can't Solve the Climate Crisis Unless Black Lives Matter," *Time*, July 9, 2020.

"How to Find Joy in Climate Action," TED Talk, April 2022.

Index

catastrophes, 350
See also disasters
CDFIs (community development financial
institutions), 384
CEJST (Climate Economic Justice Screening
Tool), 385
certified benefit corporations, 167
Chevron, 263*n*
Chicago, what transformation could look
like, 295–96
childcare, 286–87
children, decisions about having, 200
China:
air pollution in, 268–69, 279
climate policy collaboration with,
278–80
economic development in, 268, 269,
271, 276–78
emissions responsibility, 260, 263, 271,
276
circularity:
economic principles, 89, 161
embodied energy and architectural
design, 114
Citelli, Anna, 112
citizenship. *See* civic engagement; political
participation
Citizens United, 153–54
civic engagement, 157–58, 166, 195–96,
221–22, 230
opportunities for, 157–58, 223, 227,
228
support and training for, 220, 222–23,
384–85
See also political participation
Civilian Climate Corps, 311, 317, 322
clams, 401
See also ocean farming
Clean Air Act, 325, 327, 329–30, 339
Clean Development Mechanism, 275
clean energy development, 27, 28, 139,
151, 172
clean energy employment, 139, 196,
261, 345
Department of Energy funding programs
and priorities, 172, 180–96
economics of, 139, 151, 169*n*, 183–84,
268, 332
environmental trade-offs, 28, 339–40
financing needs, 145–47, 172, 183, 340
foregrounding just sourcing of, 370, 378

in the Global South, 145–47
green industrialization in developing
countries, 264, 268–71, 273, 276–78
hydrogen and hydrogen fuels, 184–86,
187, 336
Inflation Reduction Act funding, 154,
165, 172, 183, 261, 336
infrastructure needs and optimization,
27, 29, 126–27, 190, 339
localized/decentralized production and its
impacts, 152, 154–55, 299
local regulatory obstacles, 195–96, 332
new technologies and their
commercialization, 184–87, 195
politics of, 154, 298, 332
public/community control of, 299,
377–78
public policy to support, 332, 339, 340
socioeconomic impacts of, 277–78,
286–87, 289
the sufficiency of current technologies,
187–88, 195
what transformation looks like, 428–29
See also decarbonization; nuclear power;
solar energy; wind energy
Clean Water Act, 325, 327, 328–29, 339
Clément, Régine, 169–70, 421
interview, 170–78
climate action. *See* action and advocacy
climate action Venn diagram, 5–6, 424–25
climate anxiety and grief, 236, 255, 344
on hope, 397–99
climate apocalypse skills, 382
climate assessment reports, 23, 24, 25,
141
climate communication/conversations,
4–5, 203–4
vs. action, 425
climate coverage in the conventional
media, 200, 201, 205–6, 219
how the Green New Deal shifted the
narrative, 289, 290–91, 292
the importance of narrative, 290–91,
322
in the movies, Leonard and McKay on,
204–17
op-eds as a channel for, 306
Pierre-Louis on climate journalism,
219–36
problems and possibilities, 200, 201
See also media

decarbonization (*cont'd*):
   Shah on industrial decarbonization, 187, 189–90, 192
   what transformation looks like, 428–29
   See also carbon capture and storage; clean energy development; emissions reductions; transformation
decentralization. *See* localization/ decentralization
deconstruction, 89, 114
Deeman, Erica, 64, *65*
deep decarbonization, 162
deepfakes, 129–30
DeepMind, 120, 124, 132–33
Deepwater Horizon oil spill, 349
deforestation, 12, 22–23, 34, 51
democracy, 220
Department of Energy. *See* DOE
deserts, 20, 44–45
design. *See* architecture and design
Design Emergency, 105, 106–7
   See also Antonelli, Paola
design justice, 100
determination, 399
development and development financing, Gallagher on, 264, 268–76
devotions, 388
diet, 35, 56, 303, 409
   meat and dairy problems and possibilities, 13, 34, 35, 56
Dillen, Abigail (interview), 324–41, 376*n*
Disaster Researchers for Justice, 351
disasters, 4
   community-based response and recovery, 95, 355, 356–57, 371
   disaster justice, 348, 351–52, 375
   displacement due to, 251, 344, 364
   frequency/prevalence of, 344, 347, 365
   local emergency management funding, 352–53, 354, 361–62
   Montano on disasters and response/ recovery, 347–62
   in the movies, 211
   mutual aid, 95, 389
   in the news, 219, 220, 225–26
   Pakistani floods of 2022, 251–52, 269–70
   problems and possibilities, 344, 345
   related debt and financial hardship, 148, 252, 270, 344, 357, 358
   related policy needs, 283, 359–62, 375

   systemic factors in, 350–51
   See also FEMA; flooding; *other types of disasters*
disinformation, in the news media, 231–32
diversity:
   in climate science, 27–28
   diversifying and revitalizing rural communities, 56, 61–63
   See also biodiversity
divestment (from fossil fuel investments), 138, 139, 142, 263*n*
   See also finance
DOE (Department of Energy) Loan Programs Office, 180
   Shah on funding programs and priorities, 181–96
domestic violence, 251–52
Donahue, Brian (interview), 49–63
donate, 5
Don't Look Up (film), 204, 208–9
droughts, 19, 20, 34, 40
durability, 161

Earthjustice, 318*n*, 324, 325
   Dillen on environmental law, litigation, and the courts, 325–41
ecological forestry, 53–54, 58
ecological restoration, 38
economic justice. *See* justice
economy:
   circularity, 89, 161
   consumerism and its impacts, 115–16
   economic benefits of conservation/ restoration, 35
   the economic case for clean energy, 332
   economic growth, 175, 179
   economic impacts of climate change, 138, 260, 272*n*
   economic impacts of decarbonization, 289
   economic importance of ecosystem services, 34
   the economics of farming and land access, 59–61, 70–71
   Gallagher on green economic development, 264, 268–76
   rural economic declines, 49, 55
   what economic transformation looks like, 431
   See also agriculture; capitalism; clean energy development; corporations and corporate responsibility; finance; labor

About the Author

Dr. Ayana Elizabeth Johnson is a marine biologist, policy expert, writer, and teacher working to help create the best possible climate future. She co-founded and leads Urban Ocean Lab, a think tank for the future of coastal cities, and is the Roux Distinguished Scholar at Bowdoin College. Ayana co-edited the bestselling climate anthology *All We Can Save*, co-created and co-hosted the Spotify/Gimlet podcast *How to Save a Planet*, and co-authored the Blue New Deal, a roadmap for including the ocean in climate policy. She earned a BA in environmental science and public policy from Harvard University, and a PhD in marine biology from Scripps Institution of Oceanography. She serves on the board of directors for Patagonia and GreenWave and on the advisory board of Environmental Voter Project. Above all: Ayana is in love with climate solutions.

––––––––

What If We Get It Right?: Visions of Climate Futures (author)

Best American Science and Nature Writing of 2022 (editor)

All We Can Save: Truth, Courage, and Solutions for the Climate Crisis
(co-editor with Katharine K. Wilkinson)

ANTI-APOCALYPSE MIXTAPE

Anthems for victory, love songs to Earth,
tunes for tenacity, and sexy implementation vibes.

♥ ♥ ♥

7 – Prince • Tell Me Something Good – Rufus & Chaka Khan
Our Day Will Come – Fontella Bass • Keep On Pushing – The Impressions
It Ain't Over 'Til It's Over – Lenny Kravitz • The Air That I Breathe – The Hollies
Stir It Up – Bob Marley & The Wailers • Here Comes the Sun – Nina Simone
Sunny – Bobby Hebb • Weather Report – The Tennors
Ain't No Mountain High Enough – Marvin Gaye, Tammi Terrell
Through the Fire – Chaka Khan • Back to Life – Soul II Soul, Caron Wheeler
New World Water – Mos Def • Another Sad Love Song – Khalid
Town & Country – Bibio • Keep a Shelter – Justin Jones
Pulaski at Night – Andrew Bird • Better Than – Lake Street Dive
Light On – Maggie Rogers • Right Down the Line – Lucius
Believer – Emily King • All I Need Is a Miracle – Mike + The Mechanics
Foundations – Kolidescopes • Another Try – Haim
All for Us – Labrinth, Zendaya • How I Feel – Wax Tailor
Didn't I – Darondo • Come and Be a Winner – Sharon Jones & The Dap-Kings
Love and Peace – Quincy Jones • Lovely Day – Bill Withers
Replay – Tems • Biking – Frank Ocean, Jay-Z, and Tyler, the Creator
Work – Charlotte Day Wilson • In Tall Buildings – John Hartford
Love Is Tall – Oshima Brothers • How Deep Is Your Love – Bee Gees
Working My Way Back to You – George Nooks & Trinity
9 to 5 – Dolly Parton • Bitch Better Have My Money – Rihanna
Higher Love – Kygo, Whitney Houston • Hold Up – Beyoncé
You're All I Need to Get By – Aretha Franklin
Dedicated to the One I Love – The Mamas & The Papas
Sunset Lover – Petit Biscuit • A Change Is Gonna Come – Sam Cooke
This Magic Moment – The Drifters • Joy to the World – Three Dog Night
Move On Up – Curtis Mayfield • We Can Work It Out – Stevie Wonder
Hard Road to Travel – Jimmy Cliff • Let Me Let You Go – Mega
Don't Forget to Breathe – Stormzy, Yebba
You're Still the One – Caleb Hawley

♥ ♥ ♥

Listen to the playlist at
www.get it right.earth